# Re-entry and Pla[illegible]

# Physics and Technology

## II / Advanced Concepts, Experiments, Guidance-Control and Technology



*Edited and authored by W. H. T. Loh*

NORTH AMERICAN ROCKWELL CORPORATION

DOWNEY, CALIFORNIA

*Springer Science+Business Media, LLC 1968*

ISBN 978-3-642-50084-8 ISBN 978-3-642-50082-4 (eBook)
DOI 10.1007/978-3-642-50082-4

Originally published bySpringer-Verlag New York Inc. in 1968.

Softcover reprint of the hardcover 1st edition 1968

Library of Congress Card Number 68-18103

Title No. 3893

**Volume 3**

**APPLIED PHYSICS AND ENGINEERING**

**An International Series**

# Re-entry and Planetary Entry Physics and Technology

*To*

PRESIDENT JOHN F. KENNEDY

*for his decision to land men on the moon*

*in this decade*

*and*

NASA ADMINISTRATOR JAMES E. WEBB

*for his achievements*

*to carry out our national goal in space*

# *Preface*

During the last decade, a rapid growth of knowledge in the field of re-entry and planetary entry has resulted in many significant advances useful to the student, engineer and scientist. The purpose of offering this course is to make available to them these recent significant advances in physics and technology.

Accordingly, this course is organized into five parts: **Part 1, Entry Dynamics, Thermodynamics, Physics and Radiation; Part 2, Entry Ablation and Heat Transfer; Part 3, Entry Experimentation; Part 4, Entry Concepts and Technology;** and **Part 5, Advanced Entry Programs.** It is written in such a way so that it may easily be adopted by other universities as a textbook for a two semesters senior or graduate course on the subject. In addition to the undersigned who served as the course instructor and wrote Chapters, 1, 2, 3 and 4, guest lecturers included: Prof. FRANKLIN K. MOORE who wrote Chapter 5 *"Entry Radiative Transfer,"* Prof. SHIH-I PAI who wrote Chapter 6 *"Entry Radiation-Magnetogasdynamics,"* Dr. CARL GAZLEY, Jr. who wrote Chapter 7 *"Entry Deacceleration and Mass Change of an Ablating Body,"* Dr. SINCLAIRE M. SCALA who wrote Chapter 8 *"Entry Heat Transfer and Material Response,"* Mr. DAVID G. STONE who wrote Chapter 9 *"Entry Flight Research and Experimentation,"* Dr. JOSEPH G. LOGAN who wrote Chapter 10 *"Lifting Re-entry Concepts,"* Dr. JOHN F. MCCARTHY, Jr. who wrote Chapter 11 *"Earth Entry from Lunar and Planetary Missions,"* Dr. KRAFFT A. EHRICKE who wrote Chapter 12 *"Braking Entry of Mars and Venus,"* Dr. ROBERT W. BUSSARD who wrote Chapter 13 on *"Entry Propulsion and Power Technology,"* Dr. ROBERT C. DUNCAN who wrote Chapter 14 *"Entry Guidance and Control Technology,"* Dr. WILLIAM T. THOMSON who wrote Chapter 15 *"Dynamics Problem of Planetary Approach,"* and Dr. JOSEPH F. SHEA who wrote Chapter 16 *"Entry Programs Management."* The material in Setion 8 of Chapter 3 on *"High Temperature Properties of Gases"* was written by Dr. KENNETH G. SEWELL of LTV Research Center and the material in Section 2 of Chapter 4 on *"Orbital Maneuver"* was written by Dr. GEZA S. GEDEON of TRW Systems.

Los Angeles, California
University of California

W. H. T. LOH

## List of Contributors

ROBERT W. BUSSARD
Corporate Chief Scientist
Electro-Optical Systems Inc.
A Division of Xerox
Pasadena, California

ROBERT C. DUNCAN
Chief, Guidance and Control Division
National Aeronautics and Space Administration
Cambridge, Massachusetts

KRAFT A. EHRICKE
Assistant Division Director
Autonetics Division
North American Rockwell Corporation
Anaheim, California

JOSEPH G. LOGAN, JR.
Director
Aerodynamics and Propulsion Research Laboratory
Aerospace Corporation
Pasadena, California
Currently:
Special Assistant to
Director of Research and Development
Douglas Aircraft Company
Santa Monica, California

JOHN MCCARTHY, JR.
Vice President
Research, Engineering and Test
North American Rockwell Corporation
Downey, California

JOSEPH F. SHEA
Apollo Spacecraft Program Office
Manned Space Center
National Aeronautics and Space Administration
Washington, D.C.

DAVID G. STONE
Manager, Flight Reentry Programs Office
National Aeronautics and Space Administration
Langley Research Center
Hampton, Virginia

WILLIAM T. THOMSON
Chairman and Professor
Department of Mechanical Engineering
University of California
Santa Barbara, California

# Contents

*Contents of Re-Entry and Planetary Entry*
I / DYNAMICS, PHYSICS, RADIATION, HEAT TRANSFER AND ABLATION

## Introduction



## Part One. Entry Dynamics, Thermodynamics, Physics and Radiation



## Part Two. Entry Ablation and Heat Transfer

# Part Three

## *Entry Experimentation*

# 9

# *Entry Flight Research and Experimentation*

DAVID G. STONE, Manager
*Flight Reentry Programs Office*
*Langley Research Center*
*National Aeronautics and Space Administration*

## [9-1] INTRODUCTION

For several years NASA has devoted a major effort to the study of aerodynamic heat transfer and methods for thermal protection. An important phase of this effort is through the use of rocket-propelled, free-flight models. Heat-transfer investigations are needed at high velocities with the high enthalpy conditions of flight in the real environment which rocket models provide. Rocket model investigations are therefore needed to confirm or evaluate heat-transfer formulae obtained from theoretical analysis or from ground facility experimental investigations. They are also needed to obtain data from which empirical relations can be developed for problem areas where theory is not well established. Although rocket models are poorly suited for parametric studies, they are valuable to proof-check the heat-transfer predictions for specific shapes and specific environmental flight conditions.

Many factors must be considered in planning and carrying out aerodynamic heating investigations in free flight using multistage rocket propelled vehicles. The purpose of this paper is to discuss some of the techniques associated with configuration, trajectory planning, instrumentation, calorimetry, nature of the heating problem, and to present examples and experience with the rocket-vehicle techniques from low supersonic speeds to earth-escape speeds.

## [9-2] MODEL CONFIGURATION

One of the first requirements is that the rocket vehicle have a high degree

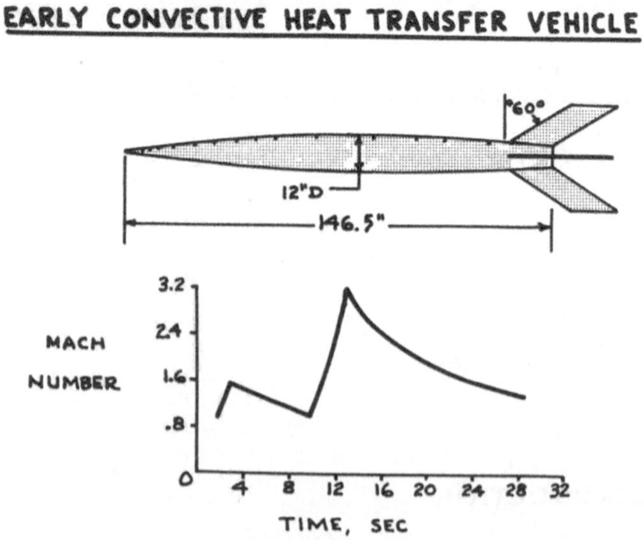


**Fig. 9-1.**

**Fig. 9-2.** Nose shape test.

of aerodynamic stability. With a multistage vehicle, the final stage plus its heat-transfer model should, in particular, be very stable, so that any disturbance due to final stage ignition will damp out rapidly, and the model will be at zero angle of attack during the high-speed, data collecting part of the flight. An example of an early convective heat-transfer vehicle is shown in Fig. 9-1. This vehicle was known as the RM-10 and was used for convective heating studies at supersonic speeds as a function of both laminar and turbulent flow along the body. Note the large fins to give the high degree of aerodynamic stability and that the body was instrumented the full length to obtain skin temperature and hence heat-transfer measurements. This body shape was tested in both wind tunnels and free flight at supersonic speeds and good correlation of data was obtained. This will be discussed further in Section 9-5.1.

Typical early rocket models and vehicles are illustrated by the following figures. Fig. 9-2 shows a two-stage vehicle carrying a model to investigate the effect of heating on various nose shapes. Fig. 9-3 shows a single-stage vehicle used to investigate heating on wing leading edges. The test specimens in this case are the stabilizing fins. In other vehicles of this type, elaborate

**Fig. 9-3.** Wing leading-edge test.

**Fig. 9-4.** Simulated wing test.

wings in the stabilizing location were constructed to investigate heating effects through various types of wing construction configurations. Fig. 9-4 shows a model designed to investigate the heating on a highly swept delta wing at angle of attack. A three-sided pyramidal shape was used to obtain zero resultant lift on the configuration. The faces and leading edge were instrumented as the principal test surface. Fig. 9-5 shows a configuration designed to investigate the heating on the wings and bodies of two aircraft-type configurations at angle of attack. The models were mounted on a rocket at an angle to the vehicle center line with the model lift vectors opposing each other. The two wing planforms were not identical but the areas were proportioned to produce equal lift. Therefore, this two-stage configuration flew at zero trim angle without oscillations and the desired heat-transfer data were obtained.

**Fig. 9-5.** Test of wings at angle of attack.

## [9-3] TRAJECTORIES

With a multistage rocket vehicle a wide choice of test conditions can be obtained by changing the launch angle and the times of ignition of the stages, as indicated in Fig. 9-6. The maximum Mach number can be made to remain approximately the same, but different trajectories produce widely different test conditions of Reynolds number and heat rate. For instance, high Reynolds numbers and rapidly increasing heat rates can be obtained by using the first stages to provide altitude and by firing the later stages after peak altitude along a re-entry-type trajectory. This trajectory technique is now used for simulating re-entry conditions for earth environment and for simulating entry into planetary atmospheres. Lower heating rates which change much less rapidly can be obtained at lower Reynolds numbers by firing the later stages in a shallow climb just before peak altitude is reached. The type of trajectory chosen will, of course, depend on the objectives of the investigation. In any case, the trajectory used for design of the model calorimeter must be an accurate prediction of the actual flight. Therefore, the trajectory

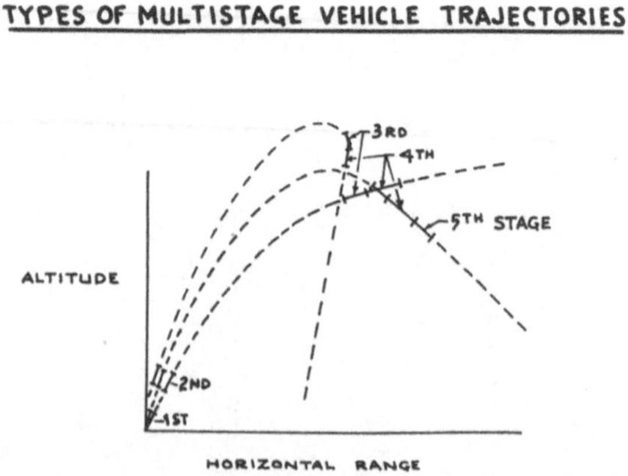


**Fig. 9-6.**

computations must be based on accurate predictions of the weights, motor performances, and vehicle aerodynamics. Also, the effects of winds on the trajectory must be computed so that the launch conditions can be adjusted for the winds existing at the time of the test. Generally speaking, a heat-transfer test should be successful if the maximum Mach number and its altitude match the predicted values within about 10 percent and 10000 feet, respectively. However, to stay within these limits requires careful preflight computations of the trajectory, and ignition of the later stages must occur very close to the prescribed times.

## [9-4] INSTRUMENTATION

Because high velocity is generally of primary importance in a heat-transfer investigation, space and weight limitations require that only very essential instrumentation be carried. The flight conditions; that is, the velocity and ambient air conditions, are obtained primarily from ground instrumentation such as tracking radar and sounding rockets, rather than from instrumentation in the model. However, the model instrumentation should include orthogonally mounted longitudinal, normal, and transverse accelerometers. The magnitude of an oscillating angle of attack could be determined from these instruments alone only if they were located at the model center of gravity and the model roll rate was essentially constant. However, the measurements will indicate the duration and frequency of oscillations in angle of attack and will provide an estimation of the magnitude of a trim angle of attack.

The basic heat-transfer data are, of course, the skin temperatures, where the skin is functioning as a calorimeter in the classic sense. Thermocouples are unquestionably the best available means of measuring skin temperatures. In order to obtain temperature measurements at several locations in the calorimeter, the voltages from several thermocouples are usually commutated on one telemeter channel. Along with the thermocouple voltages, a series of known voltages equivalent to temperatures spaced over the expected temperature range can also be commutated on the channel. This provides an in-flight calibration for the telemetered temperature data.

The kind of thermocouple wire used will depend on the temperature range to be covered and on the millivolts output required by the telemeter for reasonable sensitivity. Iron-constantan is generally used for temperatures up to about 900 °F and chromel-alumel is used if the temperatures are expected to approach 2000 °F. Tungsten-iridium has been used in tests to measure temperatures of 3200 °F on graphite. The smaller the thermocouple wire size, the less heat it will conduct away from the skin. However, with wire smaller than .010 inch in diameter, mechanical problems develop with the wire during installation and model handling; such that, wire breakage becomes a practical design consideration. Wire size of .005 inch in diameter was used successfully in the NASA Project Fire thermocouples even though mechanical breakage was still experienced as a major problem. Redundancy and numerous measurements were designed-in to circumvent the breakage problem. Some methods of installing thermocouples on the calorimeter skin are shown in Fig. 9-7. Each wire of the thermocouple is attached separately so that the temperature measured is that of the skin surface itself, rather than that of a bead in contact with the surface.

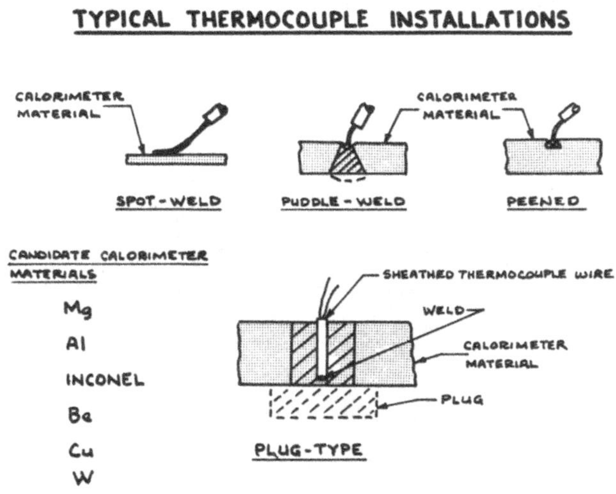


Fig. 9-7.

The best thermocouple installation in relative thick skins, from the standpoint of accurate determination of heating rate, is the technique of the plug type. The important features in the plug type are that the plug material is the *same* material as the skin, the wires are very small in a very small hole and accurately determined depths, and a smooth flush finish is obtained on the heated side of the skin.

In ablation investigations, accelerometer measurements to aid in determining the flight conditions are usually the only data telemetered in addition to the ablation measurements. Various types of sensors for measuring the ablation of the skin have been devised. One type, known as a break-wire system, utilizes a fragile wire embedded at a known depth in the material and its destruction, when the skin ablates to that depth, breaks an electric circuit. This type of sensor is usable in principle in a variety of ablating materials. In another type devised for use in charring ablators, the charred layer which is electrically conducting, completes the circuit between two wires embedded a known depth in the skin when the charred layer reaches that depth. Neither of these types has been completely satisfactory because the break wire sometimes fails to break immediately upon exposure, and, in the other type, the char layer contact with the wires sometimes is not good enough to complete the electrical circuit.

An example of a spacecraft telemetry system is shown in Fig. 9-8 in a simplified functional block diagram form. This FM/FM telemetry system was of the type used in the Project Fire of the NASA (refer to Section 9-9). This system had the three types of telemetry that have been used on past and present systems; and therefore, is a good example of the different types of FM/FM systems. Shown on the left of Fig. 9-8 is a block representing the numer-

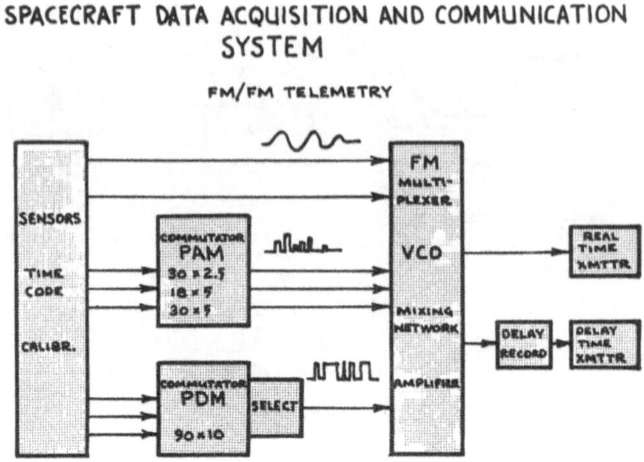


**Fig. 9-8.**

ous sensors, an on-board timing function, and calibrate signals as described previously. The scheme indicated across the top could be called an "analog method" in that the output is directly and continuously (no commutation) proportioned with time to the measured quantity as indicated by the wave train sketched in. The second scheme is the *P*ulse *A*mplitude *M*odulation method in that the output amplitude is a function of the commutated measured quantity as indicated by the wave train sketched in. The $30 \times 2.5$ translates into 30 samples every 2.5 revolutions per second of the commutator; therefore, the product gives the number of samples per second, i.e., 75. The third scheme is the Pulse *D*uration *M*odulation method in that the output duration is a function of the commutated measured quantity as indicated by the wave train sketched in. This system has a greater information capacity as indicated, by a $90 \times 10$ commutator, i.e., 900 samples per second. Also, as indicated, a selector may be used to select one commutator at a given time in the flight to transmit data that is required only when desired. The voltages are then fed into a multiplexer where the voltages modulate the subcarrier oscillators, the information is then amplified and sent to the transmission system. Since re-entry flights go through a radio blackout period which usually occurs at the most vital measurement time, a delay-time storage capacity is provided by an endless-belt type tape recorder or solid-state memory to play the data out through a separate transmitter after radio blackout. The amount of data that the system can communicate as described requires automatic computing equipment to reduce the computation time not only from the standpoint of quantity, but also because of the complex decoding of the measurements.

## [9-5] CALORIMETRY

For each total heat measurement investigation the design of the calorimeter is a particular problem or application. For example, consideration of a thermally thin-skin or a thermally thick-skin calorimeter must be made. This decision rests on the heat rate versus survivability, material thermal properties, practical fabrication techniques, and data required from the test.

### [9-5.1] Thin-Skin Calorimeters

The main advantages of a thin-skin calorimeter are that the heat conduction along the skin is negligible and there is no temperature gradient through the skin. Therefore, the measurements are not dependent on the previous history of heating which makes possible the determination of the location of the boundary layer transition from laminar to turbulent along the body by the sudden change of skin temperature. This type of thin-skin calorimetry was utilized at high supersonic speeds on the body-shape vehicles illustrated by the RM-10 in Fig. 9-1. The convective heat transfer using thin-skin calorimetry is determined by graphically reading the slopes of the measured skin-temperature time curve as is illustrated by the heat balance equation of Fig. 9-9. Listed below each of the terms of the equation are the

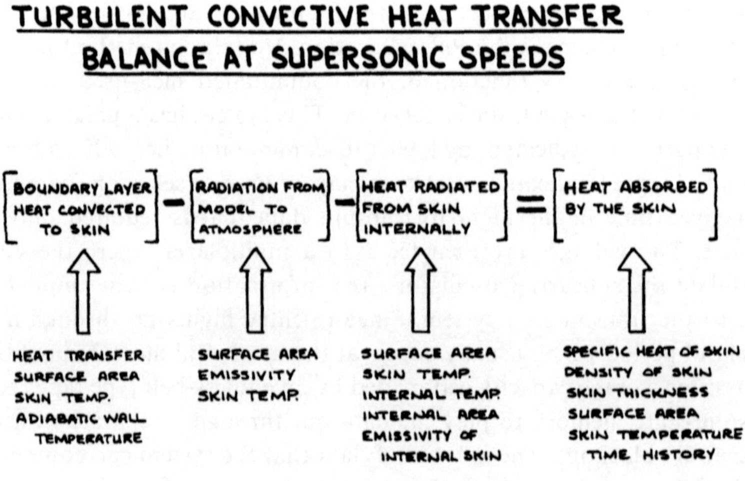


Fig. 9-9.

items that must be either measured during the flight or determined before the test. Accurate values of the emissivity and specific heat of the calorimeter material must be known for the range of the environment. For example, the specific heat must be known over the temperature range and it must be free of anomalies, and emissivity must be stabilized over the temperature range. These requirements for thin-skin calorimetry have been met using Inconel where the entire surface of the vehicle was the calorimeter. Other materials have been used extensively in heat-transfer investigations; such as aluminum, magnesium, copper, beryllium, and tungsten. Materials having high heat capacity per unit skin area are good for high heating rates and usually have high thermal conductivity which minimizes temperature gradients through the skin. However, the application of thin-skin calorimetry is usually utilized to determine local hot spots and possibly boundary-layer transition; therefore, it is preferable to employ a low-conductivity material, such as Inconel, so that the variations in heating along the surface are not obscured by conduction along the skin. In some tests, such as the heating on wings, to minimize the conduction along the surface the skin thickness was varied approximately in proportion to the anticipated heat-transfer distribution.

Convective heat-transfer data along bodies are usually correlated in terms of Nusselt, Prandtl, and Reynolds number. For the sides of bodies the heat-transfer parameter $(Nu)\,(Pr)^{-\frac{1}{3}}$ is primarily a function of Reynolds number rather than body station; that is, results obtained at different body stations were the same when the Reynolds numbers were equal. These parameters are usually based on conditions of the air in the undisturbed free stream ahead of the model. A correlation using these parameters is illustrated in Fig. 9-10. By solving the correlation equation,

$$(Nu)\,(Pr)^{-\frac{1}{3}} = f\,(RN), \tag{9–1}$$

for heat transfer in definition of the individual terms of the parameters and ignoring the exponent, the convective heat transfer is mainly a function of density, velocity, and the specific heat of air.

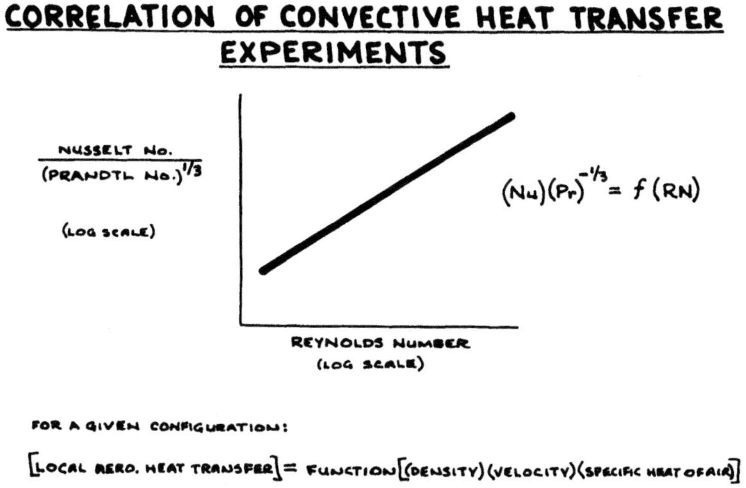


**Fig. 9-10.**

## [9-5.2] Thick-Skin Calorimeters

At hypersonic velocities the heating rate and heat load become so large that thick-skin calorimeters are required to give a survival time adequate for measurements of temperature rise before melt or change of state of the calorimeter material. This then requires the use of the thermocouple plug principle so that temperature rise through the skin can be determined.

Assuming the use of a thick-skin calorimeter material that will give one-dimensional transient heat conduction, then the governing partial differential equation is

$$\frac{\partial}{\partial x}\left(k\frac{\partial T}{\partial x}\right) = \rho c \frac{\partial T}{\partial t} \tag{9–2}$$

where the local temperature $T(x, t)$ is a function of both position $x$ and time $t$. The properties of the calorimeter material are represented by the specific mass density $\rho$, the specific-heat capacity $c$, and the thermal conductivity $k$.

According to Fourier's law of heat conduction, the one-dimensional heat flux $\dot{q}(x, t)$ at any depth $x$ and time $t$ is given by

$$\dot{q}(x, t) = -k\frac{\partial T}{\partial x} \tag{9–3}$$

where the minus sign indicates that a positive heat flux takes place in the direction of decreasing temperature. If Eq. (9–3) is substituted into Eq. (9–2), the governing differential equation involving the heat flux becomes

$$-\frac{\partial \dot{q}}{\partial x} = \rho c \frac{\partial T}{\partial t} \tag{9–4}$$

Considering the physics of the heat balance in presence of both radiative and convective heat inputs, heat stored, heat radiated away, and heat transferred out the back, Eq. (9–4) can be developed into the heat balance relationship shown in Fig. 9-11. Listed below each of the terms are the items

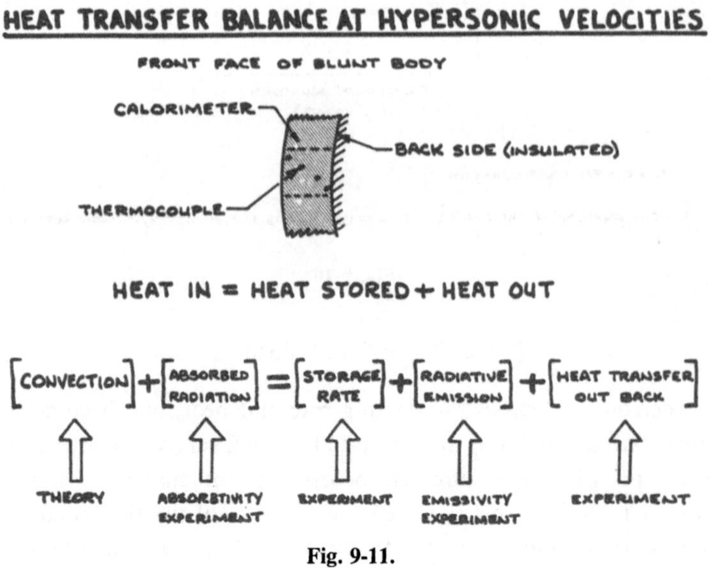


Fig. 9-11.

that must be either measured during the flight or determined before the test. Accurate values of the absorptivity, emissivity, and specific heat of the calorimeter material must be known over the spectrum of the heating environment to be encountered.

### [9-5.3] Isolated Calorimeters

Heat-transfer measurements by calorimeter plugs in non-metallic skins, such as ablation material, has shown that the resulting discontinuities in temperature along the skin surface make measurements of heat transfer most difficult. For the NASA Project Fire re-entry heating program an accurate isolated calorimeter was developed which may be used in non-metallic materials with excellent results as long as there is no contamination of the surface of the calorimeter; for example, deposition of ablation products. This calorimeter consists of a 0.13-inch-thick-gold disc which has a high emittance surface coating of nickel-chromium oxide. The calorimeter functions as a direct heat sink. Gold was chosen because its high thermal conductivity permits the surface temperature of the thin disc to be measured directly by thermocouples mounted on the rear face. Heat losses due to conduction to surrounding structure are minimized by supporting the gold disc on a ring of paper and containing the entire assembly in a very low thermal conductivity housing of Lava Stone (aluminum silicate). To keep the operating temperature of the gold sensor down, the exposed surface is coated with high emittance oxidized nichrome. This permits radiation of a determinable quantity of heat back to space.

## [9-6] EXAMPLES OF ROCKET VEHICLE HEATING INVESTIGATIONS

Several years ago much interest was manifested in the importance of blunt noses as a means of reducing heat transfer to high velocity missiles. Also, there are many indications that extremely blunt shapes foster longer runs of laminar flow than do sharp shapes. Therefore, the NASA has investigated the effects of nose shape on heating. The rocket vehicle part of this investigation covered a speed range of Mach number 2 to 17.8 and included a parametric change in the nose radius. Typical nose detail of the payloads is shown in Fig. 9-12. The calorimeters varied from Inconel, copper, to Inconel skins on either copper or stainless steel. In general, the calorimeters had a surface finish of 20 microinches. Typical of the results obtained from this investigation is that of the effect of nose radius on stagnation heating. This system-

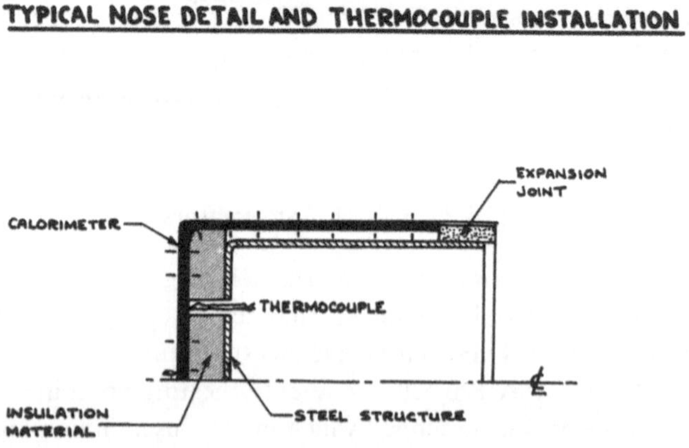


Fig. 9-12.

atic series of tests showed that the stagnation-point heating rates for the flat face are about 50 percent of those for the hemisphere.

A current example of heating investigations are the re-entry tests made with the use of the NASA Scout vehicle. The five-stage Scout is flown on trajectory so that the last three stages are fired down in a re-entry path. Typical of the Scout payloads flown to about 28000 feet per second are shown in Fig. 9-13. Total heat-transfer measurements have been made with a highly instrumented Inconel nose-cap calorimeter which melted off and

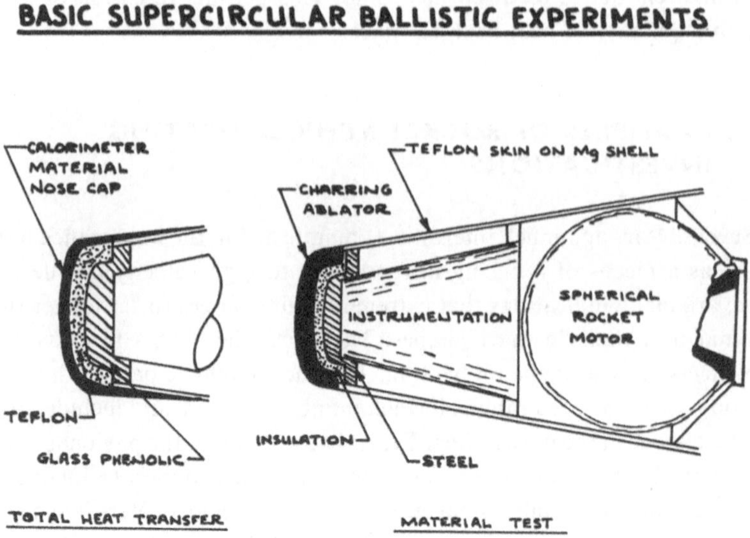


Fig. 9-13.

then the payload was protected by the teflon ablation nose so that the payload would survive and then, the data gathered by delayed telemetry after re-entry. Also, the vehicle has been used for the proof testing of charring ablators after the material has had an extensive ground research investigation. For ablation tests, ablation sensors and instantaneous thermocouple readings in the material are utilized, as well as temperature measurements in the steel strongback.

## [9-7] NATURE OF HYPERVELOCITY HEATING AND PROBLEMS OF MEASUREMENT

The present state of the art of measurements of heat transfer at velocities near escape velocity does not allow experimental test facilities, such as shock tubes and ballistic ranges, to duplicate the combined temperature and density environment associated with flight conditions. In light of this background, large-scale flight experiments will provide a needed definition of the heating in the velocity region greater than satellite speeds.

Fig. 9-14 shows the factors having a primary effect on the convective heating rate, the Mach number, and the Reynolds number. The Mach number is the simulation parameter for the inviscid flow field about the body and the Reynolds number is the parameter that largely governs the type of boundary layer in which the convective interchange of heat takes place. It is

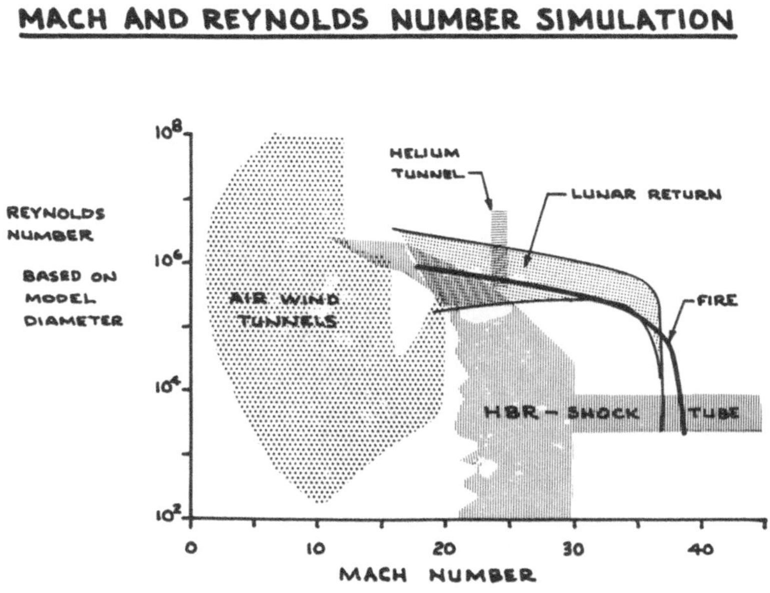


Fig. 9-14.

to be noted that thermodynamic and chemical simulation do not necessarily follow from the terms given in Fig. 9-14. The lunar return corridor is shown together with the capabilities of several facilities which come closest to simulating the re-entry environment. It is not the purpose of this figure to compare flight test capabilities of other facilities so much as to show that the flight tests not only overlap the full-scale conditions but also the facilities as well and, therefore, provide a set of well-controlled flight test data to use in validating results.

At hyperbolic velocities, above 30000 feet per second (9.14 km/sec), the total heating is the sum of the radiative heating from direct radiation from the hot gas-cap and the convective heating conducted to the body from the heated gas. Above 34000 feet per second the nature of the heating problem begins to change drastically due to the onset of this radiation of hot air between the shock front and the spacecraft. The heating increases roughly as velocity to the twelfth or possibly to the twenty-fourth power, as opposed to convective heating, which increases approximately as the square of the velocity. This variation in heating is shown in Fig. 9-15 for the Apollo shape.

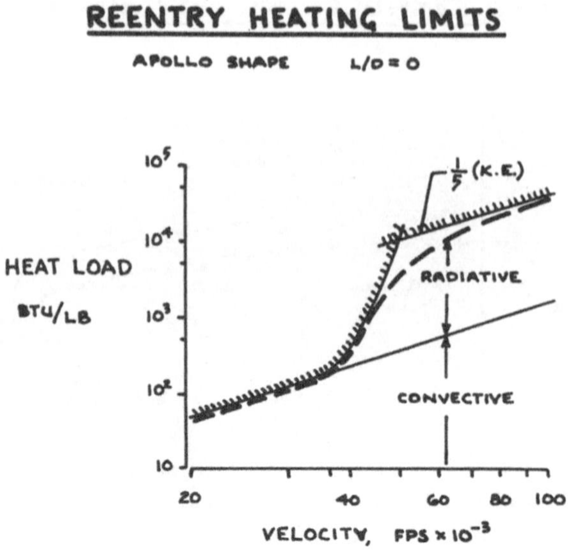


Fig. 9-15.

The radiative heat load cannot continue to increase indefinitely at the rate indicated because the heat energy rate would soon exceed the rate at which energy can be transmitted to the plasma by motion of the vehicle. It can be postulated, on the basis of theory and a few applicable experiments, that ultimately the total heating load, which is predominantly radiative, will be of the order of $\frac{1}{5}$ of the initial kinetic energy.

Basically, the problems of predicting the radiant heat input are as follows: (1) Equilibrium radiant heating is directly proportional to the shock stand-off distance, and hence the nose radius. (2) The factors affecting the absorptivity characteristics of the radiant gas, and the interplay of the gas dynamic properties with respect to the shock stand-off distance.

The models used for experimental tests in ground facilities are usually very small which means small shock stand-off distances with attendant difficulties in measuring radiant heating and more importantly, scaling the results. Furthermore, simulation of the hyperbolic velocities in ground facilities usually means the physical characteristics of the gas constituents are not exactly like those experienced during the actual re-entry or are not in the same thermal-fluid scale as in re-entry.

The mechanism of convective heating has always been better understood than radiant heating. However, large-scale tests are required to confirm that at hyperbolic velocities and with opportunity for significant coupling of the radiative heating to the convective, the convective heating continues to be inversely proportional to the square root of the nose radius, and that it increases in direct proportion to approximately the square of the velocity as predicted by theory.

The variation of theoretical radiative and convective stagnation heating with nose radius is shown in Fig. 9-16 for a blunt body with constant cross-sectional area. The different available radiant heating theories will change the level of radiative variation up or down depending upon the ingredients that are emphasized in the make-up of the mathematical model of the gas dynamics. From this variation it can readily be appreciated that large body size is particularly significant in measuring radiative heating.

## [9-8] RADIO ATTENUATION

Electron concentrations which are, by many times, larger than those in the ionosphere and which bring plasma frequencies into the microwave regime, are caused when the air is heated by friction and by the shock wave formed around a body travel at hypersonic speeds. This causes the "re-entry communication blackout" for which no applied engineering solution has yet been found. Raising the frequency beyond the plasma frequency which seems the obvious solution has its limitations; first, because it becomes difficult to build sufficiently powerful transmitters when the wavelength approaches the millimeter regime; and second, because some constituents of the air (water vapor, carbon dioxide) become absorbers of wave energy.

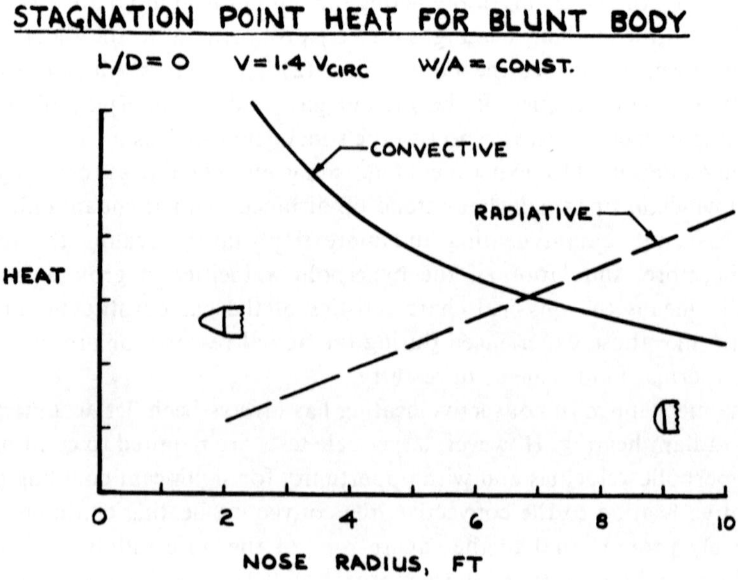


Fig. 9-16.

Generally, when the plasma frequency is less than the radio transmission frequency the signal will penetrate the sheath. At high altitudes and high velocities the boundary is sharp between transmission and cutoff where at lower altitudes considerable absorption takes place and the signal is attenuated before complete cutoff. Plotted in Fig. 9-17 is the electron concentration versus plasma frequency for the high altitude case, about 100000 feet and above. Spotted on the plasma frequency scale are the VHF telemetry and C- and X-band radar frequencies. For re-entries at hyperbolic velocities an electron concentration of $10^9$ per $cm^3$, or a plasma frequency greater than VHF, extends considerably back in the wake so that the VHF signals will even be cut off rearward. Similarly, raising the communication frequency to X-band (10000 mc/sec) would overcome plasma electron concentrations of $10^{12}$ per $cm^3$ which means some rearward and maybe some sidewise propagation links could be established.

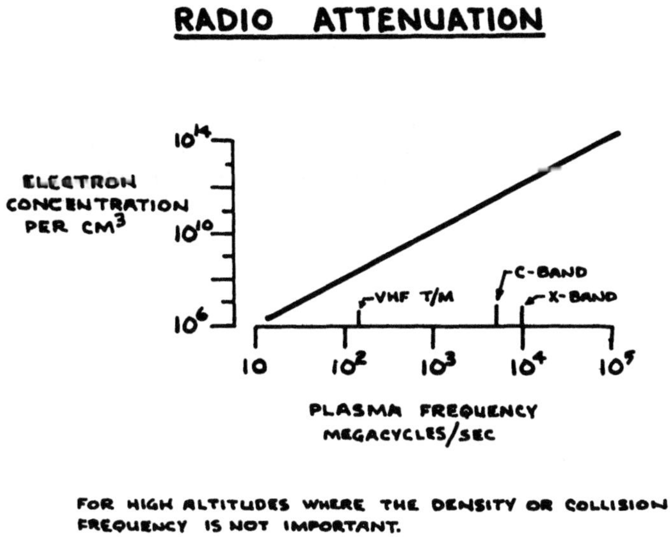


Fig. 9-17.

## [9-9] DESCRIPTION OF PROJECT FIRE

Project Fire was undertaken by the National Aeronautics and Space Administration to investigate the heating environment of vehicles entering the earth's atmosphere at velocities slightly higher than return velocities from a lunar mission. The primary purpose of this flight re-entry experiment was to determine the hot-gas radiance and the total-heat-transfer rates on a blunt-nosed body of fairly large scale. The resulting data have provided anchor points for comparison with results obtained from ground facilities and theoretical prediction methods.

The Fire space vehicle, which incorporated an Atlas D launch vehicle, is shown in Fig. 9-18. The Atlas was used to place the spacecraft on a coast ellipse which delivered it to a predetermined point in space at which the Antares II solid fuel rocket motor was ignited to provide the final velocity increment prior to re-entry. The spacecraft system is also shown in Fig. 9-18. It consisted of an aerodynamic shroud which protected the re-entry package during exit and was jettisoned at Atlas booster-engine cutoff, a separable structure containing guidance components, an Antares II rocket motor, and the Fire re-entry package. The spacecraft system, less adapter and shroud, was separated from the Atlas following vernier-engine cutoff. A cold-gas control system activated by inertial guidance components reoriented the spacecraft to the proper re-entry attitude, which was maintained during the coast period of approximately 20 minutes duration. At a point some 4000 nautical miles downrange and at an altitude of about one million feet, the guidance was terminated; the spacecraft system was spun to approximately three revolutions per second to provide stability during Antares motor separation from the guidance shell and through ignition and burn of the motor.

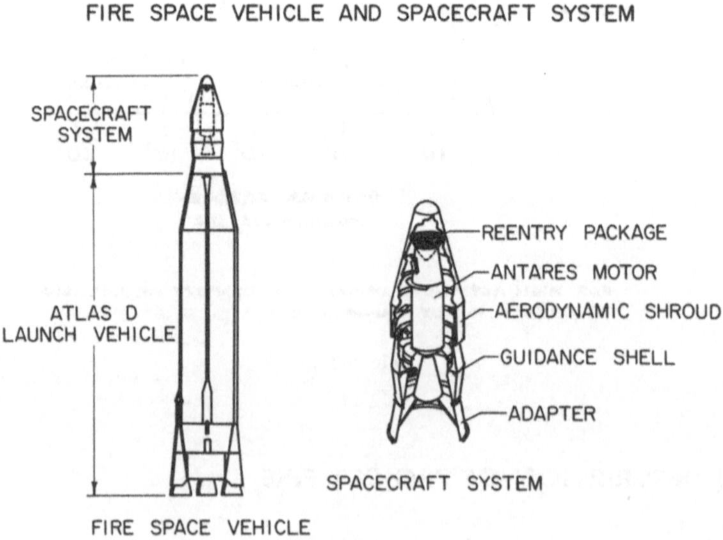


**Fig. 9-18.**

The Antares motor provided a velocity increment of about 17000 feet per second. At an altitude of approximately 450000 feet the re-entry package was separated from the spent Antares motor, which was then decelerated, despun, and tumbled to insure its physical separation from the re-entry package. The data-gathering portion of the re-entry took place between 400000 feet and 100000 feet altitude. The velocity at the start of the re-entry, which is defined at an altitude of 400000 feet (121.9 km), was indicated to be 37970 feet per second (11.57 km/sec) for Fire I and 37239 feet per second (11.35 km/sec) for Fire II. Each re-entry was indicated to be a few tenths of a degree shallower than the $-15°$ nominal re-entry angle, which is well within the tolerance set before the flight.

The atmospheric properties up to 320000 feet altitude were obtained from soundings by means of rocket vehicles launched from Ascension Island following each re-entry. These atmospheric properties were generally close to the 1962 U.S. Standard Atmosphere.

The Fire re-entry package is shown in Fig. 9-19. At the start of the re-entry

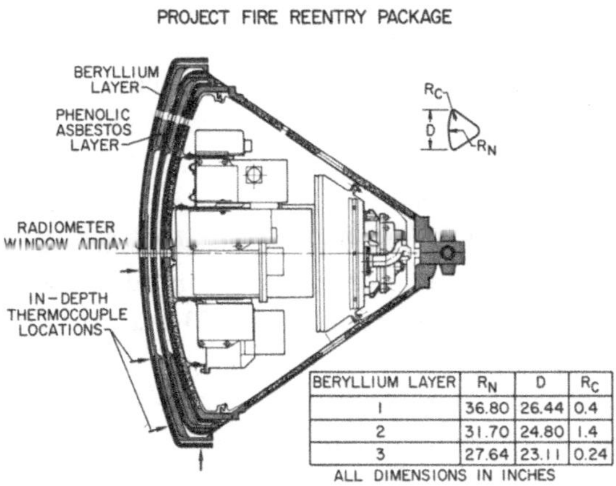


| BERYLLIUM LAYER | $R_N$ | D | $R_C$ |
|---|---|---|---|
| 1 | 36.80 | 26.44 | 0.4 |
| 2 | 31.70 | 24.80 | 1.4 |
| 3 | 27.64 | 23.11 | 0.24 |

**Fig. 9-19.**

it weighed approximately 190 pounds and was 26.44 inches in diameter. The heat shield was a novel arrangement of six layers, alternately consisting of beryllium calorimeters and high-density laminated phenolic-asbestos abalation layers. Each layer was extensively instrumented with thermocouples distributed over the face and in depth in order to obtain temperature time histories. Fused-quartz optical windows for the onboard radiometers were also provided in each layer. The beryllium calorimeters were polished to a surface finish of eight micro inches (root-mean-square value). The multiple heat shields permitted making local measurements of the calorimeter heating, which represents the sum of the convective heating and the absorbed radiation, during three different periods of the re-entry. The re-entry began at an altitude of 400000 feet with the first calorimeter exposed. After reaching melting temperature some 20 seconds later, heat protection for the re-entry package was afforded by the first ablation layer. This layer was mechanically ejected just prior to occurrence of the expected maximum heating rate, thus exposing a clean, cool calorimeter and also fresh optical windows for the forebody radiometers. The second beryllium calorimeter melted after several seconds, after which the second ablation layer protected the last calorimeter until it was exposed on the decaying side of the heat pulse. Signals triggering the mechanical ejection of the ablation layers were generated by preflight settings of an onboard timer which was started at a predetermined deceleration level. Use of the beryllium calorimeters permitted making the heating measurements in a clean environment, uncontaminated by products of ablation. A total of 222 thermocouples were installed in the six heat-shield layers at depths from 0.010 to 0.200 inch. The decrease in the re-entry package weight which accompanied the respective melting-off and jettisoning of the beryllium calorimeters and the ablation layers closely matched the decreasing diameter such that the ballistic coefficient, $W/C_DA$, remained substantially constant during the re-entry at about 32 to 33 pounds per square foot.

Radiometers were installed to measure the radiation from the heated shock layer at the nominal stagnation point, and at points located 16° off-center on the heat shield and on the afterbody. The fused-quartz optical window array effectively transmitted all the radiation between wavelengths of 0.23 micron and 4 microns. A spectral radiometer which scanned radiation between 0.29 micron and 0.62 micron with a resolution of 50 angstroms was also located at the stagnation point.

Other measurements which were made during both Fire re-entries consisted of afterbody temperature time histories and static pressures, and accelerations and angular rates about the three principal body axes.

The data system consisted of 11 standard IRIG channels which were broadcast by two VHF telemetry systems in real time and after a 45-second delay provided by an onboard tape recorder. Each radiometer was assigned a continuous channel; however, the calorimeter temperature time histories were commutated. Following emergence from blackout, the record and erase functions of the tape recorder were disabled, thus permitting continuous replays of the data obtained during the blackout period until splash occurred. The re-entry package was not recovered.

An example of the calibration knowledge which must be known prior to a re-entry heating experiment is that required for Project Fire. Shown in Fig. 9-20 is the manner in which the beryllium calorimeters and the total

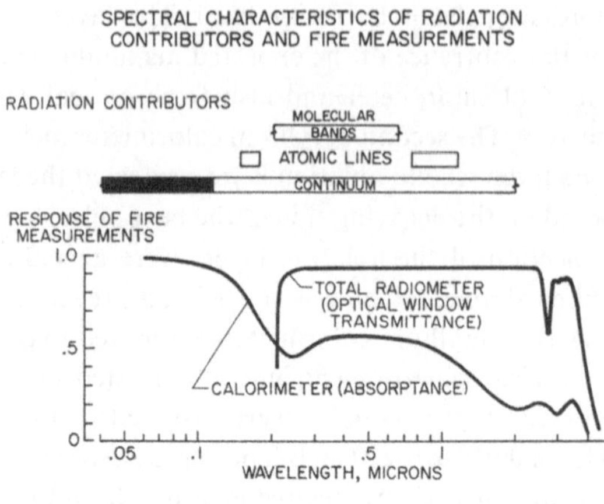


Fig. 9-20.

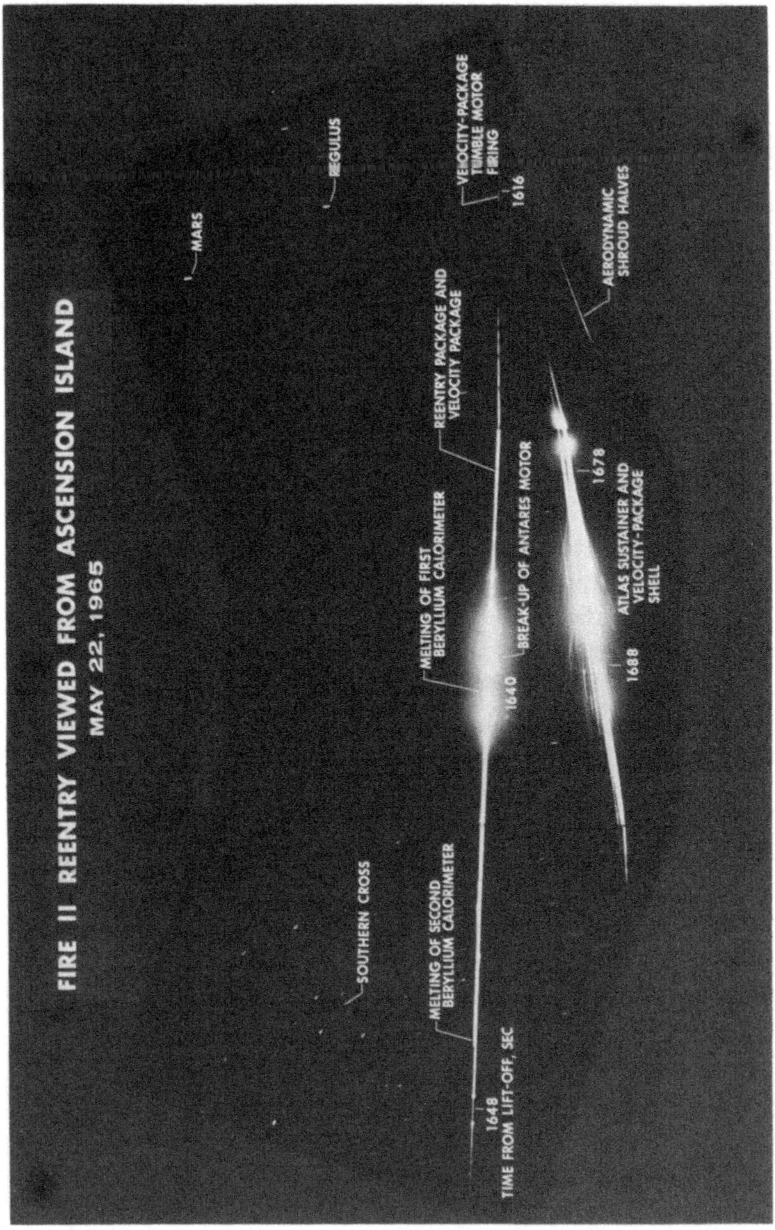
FIRE II REENTRY VIEWED FROM ASCENSION ISLAND
MAY 22, 1965
MARS
REGULUS
VELOCITY-PACKAGE TUMBLE MOTOR FIRING
1616
REENTRY PACKAGE AND VELOCITY PACKAGE
AERODYNAMIC SHROUD HALVES
MELTING OF FIRST BERYLLIUM CALORIMETER
BREAK-UP OF ANTARES MOTOR
1678
ATLAS SUSTAINER AND VELOCITY-PACKAGE SHELL
1640
1688
SOUTHERN CROSS
MELTING OF SECOND BERYLLIUM CALORIMETER
1648
TIME FROM LIFT-OFF, SEC

**Fig. 9-21.**

radiometers spectrally respond to the radiation from the shock layer. At the top of Fig. 9-20 are shown several bars representing the spectral regions in which the radiative heating from the different sources expected is primarily distributed. On the lower half of Fig. 9-21 are curves showing the spectral response of the principal Fire sensors. The total radiometer is limited by transmittance of the fused-quartz windows. The beryllium calorimeters are able to sense radiative heating at all wavelengths and their absorptance is dependent upon the wavelength.

The Fire re-entries were photographically recorded from Ascension Island. An example of a streak photograph, Fig. 9-21, is the composite photograph of the meteor-like Fire II re-entry recorded on May 22, 1965 by three ballistic cameras. The camera shutters were opened prior to the re-entry and the paths of the different parts of the Fire space vehicle were recorded as they penetrate the earth's atmosphere. The upper streak was made by the re-entry package and the spent Antares motor, which travel a similar path as they reenter at a speed in excess of 37000 feet per second. At the beginning of the re-entry (far right of photograph), the firing of small rocket motors used to position the Antares away from the instrumented re-entry package has been recorded as a short burst of light. Radiation from the heated air surrounding the re-entering bodies begins at an altitude of about 300000 feet. Flares associated with the breakup of the Antares motor and with the removal of heat shields from the re-entry package are apparent later on in the re-entry. The lower streaks were made by the Atlas sustainer and the velocity-package airframe shell as they re-enter at a steeper angle some 50 seconds later and at a speed of 21000 feet per second. Slightly to the right of this brilliant display may be seen faint streaks made by the shroud halves that protected the spacecraft during the launch phase of the flight.

The successful NASA Project Fire re-entries at speeds of 37970 and 37239 feet per second have provided valuable data anchor points defining the re-entry heating in this severe environment.

# Part Four

## *Entry Concepts and Technology*

# 10

# *Lifting Entry Concepts*

J. G. LOGAN, Jr., Director*
*Aerodynamics and Propulsion Research Laboratory*
*Aerospace Corporation*

## [10-1] MISSION CONSIDERATIONS

As Becker[1] has recently pointed out, the ballistic entry vehicles of the ICBM program have almost achieved full maturity in the short span of about ten years. This has been accomplished through a combination of clearly defined requirements and a national research and development effort. With similar simple and clear requirements, the first manned orbital vehicles were developed by straightforward extensions of ICBM technology. This growth pattern now appears to be changing as the requirements for future systems become less clearly defined and, at the same time, become much more diverse and complex.

During the next decade, there will be increasing interest in earth orbital operations and the development of hypersonic vehicles capable of long-range flight within the earth's atmosphere. Assuming eventual operational missions and use of multi-manned vehicles it is anticipated that a basic requirement will be survivable entry and soft landing. The advantages of employing lifting re-entry or maneuvering concepts are obvious:[2,3]

a. The recovery point need not lie in the orbit plane.
b. A greater precision of impact can be achieved.
 (1) Mobile recovery forces not required;
 (2) Land recovery possible with precision;
 (3) Night recovery becomes possible;
 (4) Security can be maintained;
 (5) Rapid call-down becomes possible.

As a consequence of the greater flexibility which can be achieved through the use of large maneuver capability, current DOD and NASA programs are exploring the potential of lifting concepts. Although much of this activity is

* Currently, Special Assistant to Director, Research and Development, Missiles and Space Systems Division, Douglas Aircraft Company, McDonnell Douglas Corporation.

classified, it can be stated that NASA research centers around the developing lifting body entry vehicles such as the Ames M-2 and the Langley HL-10. The Air Force Research and Technology Division is developing a high $L/D$ configuration at the Flight Dynamics Laboratory[3] and the Air Force Space Systems Division is developing a lifting body shape and also exploring high $L/D$ concepts.

The technical knowledge required for the design of re-entry vehicles from earth orbit is currently available and the mechanics of design reasonably well understood. It is obvious that many configuration concepts can be evolved and that a single vehicle concept cannot be identified without first considering the specific mission or missions and establishing the performance capability required.

Sufficient unclassified literature is currently available so that it is possible to describe certain obvious potential missions and establish mission requirements.[4] The missions which require use of re-entry techniques may appropriately be divided into manned and unmanned categories.

As space operations become increasingly routine, rescue and recovery capability becomes increasingly important. Rescue from space will be essential for all manned operations and recovery over land from space, whether manned or unmanned, is an obvious necessity for an efficient practical earth-orbital logistic system.

## [10-2] TRAJECTORY CONTROL

For a re-entry system to be useful, a controlled trajectory must be flown. Uncertainties in system parameters and errors in system operation can lead to unacceptably large dispersion errors in the landing point. These errors can arise from errors in the process of deboost from orbit as well as uncertainties in re-entry vehicle aerodynamic characteristics. Some of the more important aspects of re-entry from earth orbit are briefly reviewed.

The return from an earth orbit can be characterized by three phases. The first phase extends from the point of deflection from orbit to re-entry into the earth's atmosphere. The second phase consists of the portion of the re-entry trajectory from re-entry into the sensible atmosphere (50 to 70 n mi altitude). The third phase includes the terminal control portion of flight and ends with touchdown at or near the desired landing site. These phases need not be distinct.

This discussion is only intended to provide a brief review of this area and indicate some of the more important re-entry system design considerations. Chapters 2 and 4 and Refs. 5-27 provide more detailed information.

## [10-2.1] Descent From Orbit

Investigations of the characteristics of transfer trajectories have been carried out in Refs. 5, 6 and 7. The major results of these studies are summarized for descent from circular orbits. In order to accomplish the return from a satellite orbit, the vehicle flight path must be deflected toward the earth by aerodynamic or propulsive forces. The use of propulsive deboost for orbital deflection is generally preferred over the use of aerodynamic forces because it provides more positive control over the re-entry trajectory and because it is applicable to both low and high altitude orbits.

Using the geometry in Fig. 1, and assuming motion about a spherical, non-rotating earth, altitude changes that are small compared to the radius of the earth, and velocity changes that are small compared to satellite velocity, simple expressions may be found which describe the situation at re-entry in terms of the orbital and deboost parameters. The deboost velocity increment is assumed to be impulsive since the burning times for practical deboost rockets are negligible compared to the orbital period. The details of the derivations are given in Ref. 7 and reviewed by Friedenthal, Ref. 5.

The flight path angle at re-entry, $\gamma$, referred to the local horizontal, is (Ref. 5)

$$\gamma = \sqrt{\delta^2 \sin^2 \eta - \zeta^2 - 4\delta\zeta \cos \eta} \tag{10–1}$$

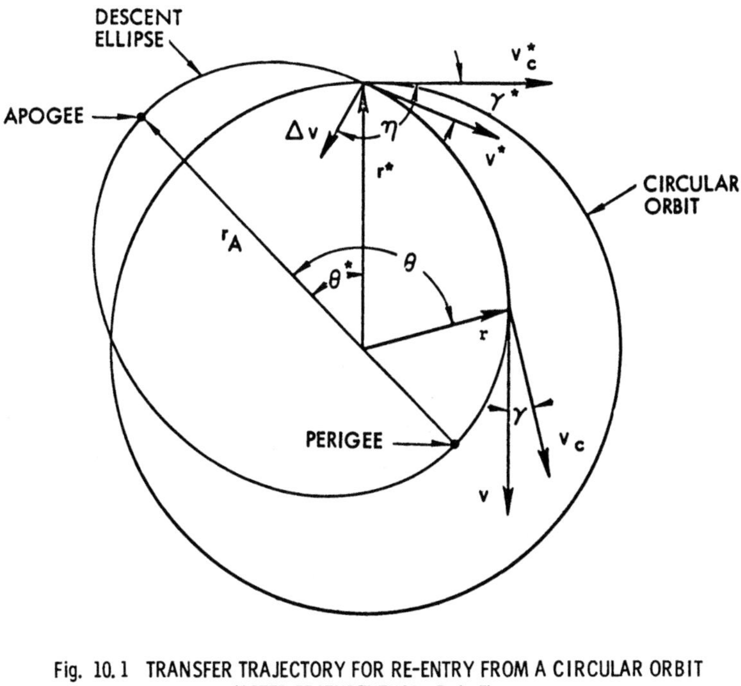


Fig. 10.1 TRANSFER TRAJECTORY FOR RE-ENTRY FROM A CIRCULAR ORBIT (AFTER FRIENDENTHAL, Ref. 5)

where $\delta$ is the deboost velocity increment magnitude normalized in terms of the local circular velocity at the deboost point ($=\Delta v/v_c^*$). $\eta$ is the deboost velocity increment orientation in the orbital plane, measured positive downward from the local horizontal and referred to the direction of motion. $\zeta$ is the fractional change in the radial distance from the earth's center between the deboost point and the re-entry point $\zeta=(1-r/r^*)$.

The distance, $S$, traveled from the point of deboost is

$$S = r(\theta - \theta^*), \quad \eta < 180° \tag{10–2}$$

$$S = r(\theta - \theta^*), \quad \eta > 180° \tag{10–3}$$

where the range angle variables $\theta$ and $\theta^*$ are found from the expressions

$$\cos\theta = -\frac{2\cos\eta + (\zeta/\delta)}{\sqrt{1 + 3\cos^2\eta}}\left[1 + \delta\left(2\cos\eta + \sqrt{1 + 3\cos^2\eta}\right)\right] \tag{10–4}$$

and

$$\cos\theta^* = -\frac{2\cos\eta}{\sqrt{1 + 3\cos^2\eta}}\left[1 + \delta\left(2\cos\eta + \sqrt{1 + 3\cos^2\eta}\right)\right] \tag{10–5}$$

The velocity magnitude at re-entry is given by

$$\frac{v^2}{(v_c^*)^2} = 1 + 2\zeta + 2\delta\cos\eta \tag{10–6}$$

The time of flight from the point of deboost is

$$t = \frac{r^*}{v_c^*}(1 + 3\delta\cos\eta)\left[\cos^{-1}\left(\frac{-\frac{\zeta}{\delta} - 2\cos\eta}{\sqrt{1 + 3\cos^2\eta}}\right) \mp \cos^{-1}\left(\frac{-2\cos\eta}{\sqrt{1 + 3\cos^2\eta}}\right)\right] + \sqrt{\delta^2\sin^2\eta - \zeta^2 - 4\zeta\delta\cos\eta} \mp \delta\sin\eta \tag{10–7}$$

where the negative sign applies for $\eta < 180°$ and the positive sign applies for $\eta > 180°$. Deboost velocity increment orientation angles less than 180 degrees result in transfer trajectories which descend directly from the initial orbit. Deboost angles greater than 180 degrees provide trajectories which first increase the orbital altitude and then descend.

The expressions presented above for the descent from a circular orbit may be modified to apply to elliptical orbits. The actual deboost velocity increment, $\Delta v$, at an angle $\eta$ can be replaced by a fictitious increment $\Delta v'$ at an angle $\eta'$.

Referring to Fig. 2, the circular velocity at the deboost point is $v_c^*$, the velocity in the elliptical orbit at the deboost point is $v_s^*$ and the final velocity after the application of the deboost increment $\Delta v$ at an angle $\eta$ with respect to the local horizontal is $v^*$. The same final velocity could have been obtained by deboost from a circular orbit with a velocity increment $\Delta v'$ at an angle $\eta'$.

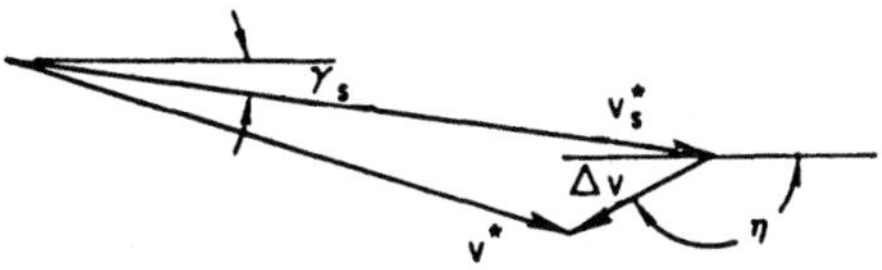


(a)

VELOCITY INCREMENT FOR ELLIPTICAL ORBIT

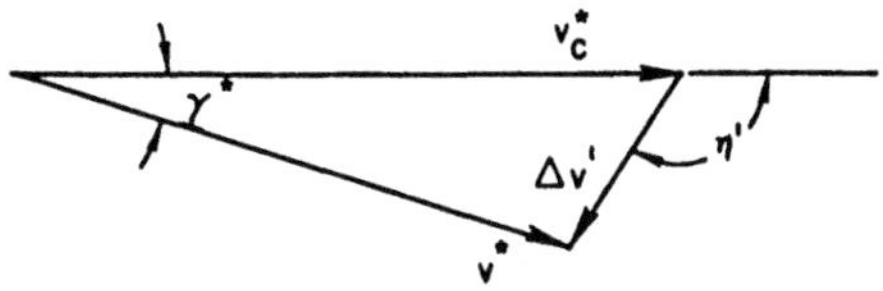


(b)

FICTITIOUS INCREMENT FOR CIRCULAR ORBIT RESULTING IN SAME FINAL VELOCITY AS GIVEN IN (a) ABOVE

Fig. 10.2 DEBOOST FROM ELLIPTICAL ORBIT (AFTER FRIEDENTHAL, Ref 5)

If the conditions at the apogee of the original elliptical orbit are known, then the velocity $v_c^*$ and flight path angle $\gamma^*$ may be found from

$$V^2 = V_A^2 - 2\left(1 - \frac{1}{\varrho}\right) \tag{10–8}$$

and

$$\tan\gamma = \sqrt{\frac{V_A^2 - 2}{V_A^2}(\varrho)^2 + \frac{2\varrho}{V_A^2} - 1} \tag{10–9}$$

where

$$V_A = (v_A/v_{cA})$$

$$\varrho = r/r_A$$

The distance from apogee and time of flight are obtained from

$$\cos\theta = \frac{V_A^2 - \varrho}{\varrho(V_A^2 - 1)} \tag{10–10}$$

and

$$t = \frac{r_A}{v_{cA}} \frac{1}{(2 - V_A^2)^{\frac{3}{2}}} \cos^{-1}\left[\frac{(2 - V_A^2)\varrho - 1}{1 - V_A^2}\right] + \frac{V_A \tan\gamma}{2 - V_A^2} \qquad (10\text{–}11)$$

From the geometry of Fig. 2, the following relations may be found which describe the deboost from an elliptical orbit in terms of the fictitious "equivalent" deboost from a circular orbit.

$$\left(\frac{v^*}{v_c^*}\right)^2 = \left(\frac{v_s^*}{v_c^*}\right)^2 + \delta^2 + 2\,\frac{v_s^*}{v_c^*}\,\delta\cos(\eta - \gamma_s) \qquad (10\text{–}12)$$

$$\gamma^* = \gamma_s + \cos^{-1}\left[\frac{\left(\frac{v_s^*}{v_c^*}\right)^2 + \left(\frac{v^*}{v_c^*}\right)^2 - \delta^2}{2\,\frac{v_s^* v^*}{(v_c^*)^2}}\right] \qquad (10\text{–}13)$$

$$\delta' = \left[\left(\frac{v^*}{v_c^*}\right)^2 + 1 - 2\,\frac{v^*}{v_c^*}\cos\gamma^*\right]^{\frac{1}{2}} \qquad (10\text{–}14)$$

$$\eta' = \cos^{-1}\left[\frac{\frac{v^*}{v_c^*} - 1 1(\delta')^2}{2\delta}\right] \qquad (10\text{–}15)$$

where

$$\delta' = \Delta v'/v_c^*$$

The velocity along the elliptical orbit at the deboost altitude can be found from Eq. (10–8), and the corresponding flight path angle is found from Eq. (10–9). For a given velocity increment and orientation, Eqs. (10–12) and (10–13) may be solved for the final velocity and flight path angle at the deboost point which, in turn, are used to solve for the fictitious increment and orientation, $\Delta v'$ and $\eta'$. These values are used in Eqs. (10–1) through (10–7) to find the trajectory conditions at the re-entry altitude. These expressions are only applicable to those cases for which the descent trajectory is nearly circular. The deboost impulse may also be minimized for a fixed range or fixed re-entry angle by proper selection of the deboost orientation, Ref. 8.

Numerical results from Ref. 5 are useful for illustration. Calculations based on a 200 nautical mile circular orbit are shown in Fig. 3. This figure shows the minimum deboost velocity increment and the required orientation as functions of flight path angle at re-entry. The figure also shows that the minimum required deboost increment increases markedly for larger re-entry

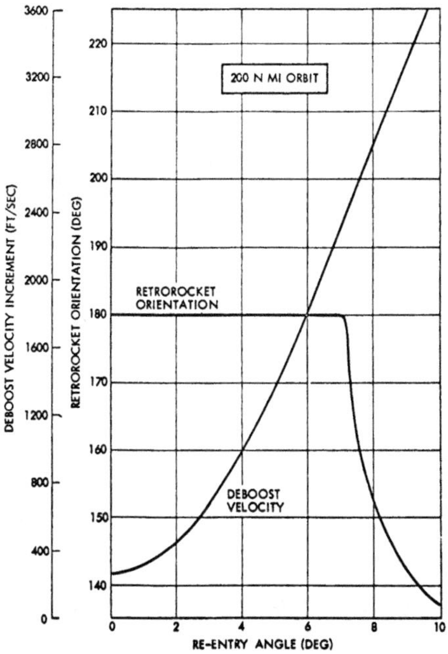


Fig. 10.3 DEBOOST VELOCITY INCREMENT AND ORIENTATION BOTH AS A FUNCTION OF RE-ENTRY ANGLE (AFTER FRIEDENTHAL, Ref. 5)

angles. For the 200 nautical mile orbit, the minimum deboost increment should be applied opposite the direction of motion for re-entry angles less than 7 degrees. For the same deboost situations, the range angle and velocity at re-entry are shown in Fig. 4. The error coefficients for errors in range due to deboost increment and orientation errors are given in Fig. 5. From this figure, the error coefficient for range error due to deboost increment errors is seen to decrease substantially as the re-entry angle increases. The error coefficient for range errors due to deboost orientation errors is nearly constant over a wide range of re-entry angles. The magnitude of this error coefficients points out the need for accurate attitude control during deboost. A crossplot and extension of the data of Fig. 3 is given in Fig. 6 and shows the

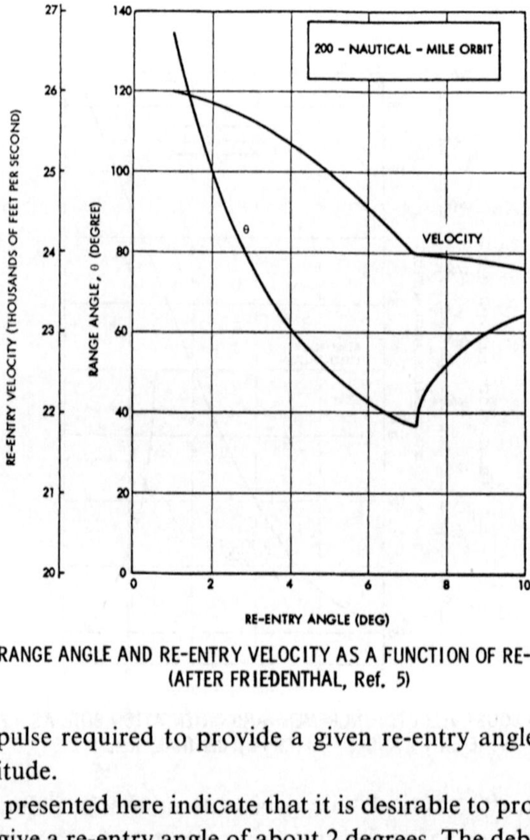


Fig. 10.4 RANGE ANGLE AND RE-ENTRY VELOCITY AS A FUNCTION OF RE-ENTRY ANGLE (AFTER FRIEDENTHAL, Ref. 5)

minimum impulse required to provide a given re-entry angle as a function of orbital attitude.

The results presented here indicate that it is desirable to provide a deboost increment to give a re-entry angle of about 2 degrees. The deboost increment required for re-entry angles of this size can be obtained by rocket motors of reasonable size and the sensitivities to deboost errors are small enough so that the alignment and impulse requirements are not unrealistic as far as current technology is concerned.

### [10-2.2] Transition to Equilibrium Glide

After entry into the sensible atmosphere, at altitudes from 50 to 70 nautical miles, the aerodynamic characteristics of the re-entry vehicle will determine the method of trajectory control. Control over the re-entry trajectory can be obtained by the modulation of the aerodynamic forces. The modulation should be performed so that the heat input and input rates do not complicate the heat shield design, the accelerations placed on the vehicle are acceptable for a man or a delicate payload, and the landing point can be accurately predicted.

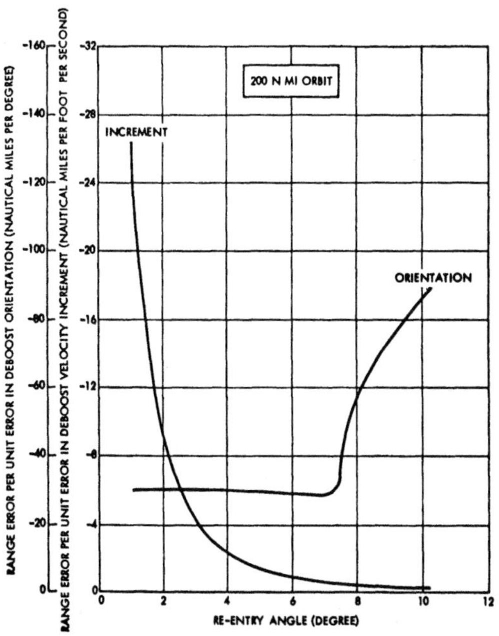


Fig. 10.5 RANGE ERROR DUE TO ERRORS IN DEBOOST ORIENTATION AND VELOCITY INCREMENT BOTH AS A FUNCTION OF RE-ENTRY ANGLE (AFTER FRIEDENTHAL, Ref. 5)

Aerodynamic lift may be provided by the use of a center of gravity offset such as will be used for the Apollo capsule, through the use of flared cone configurations, using plain flaps, or using lifting bodies such as the Eggers *M*-1 shape.[1,2,3] Relatively simple configurations can provide adequate lift capability and not complicate the vehicle structure and heat shield designs.

The use of aerodynamic lift does introduce additional complication in that the trajectory for flight within the atmosphere is determined by the lift to drag ratio and the program of modulation. For this reason, there has been considerable effort devoted to the shaping of re-entry trajectories which can be used in guiding the re-entry vehicle to the landing point.

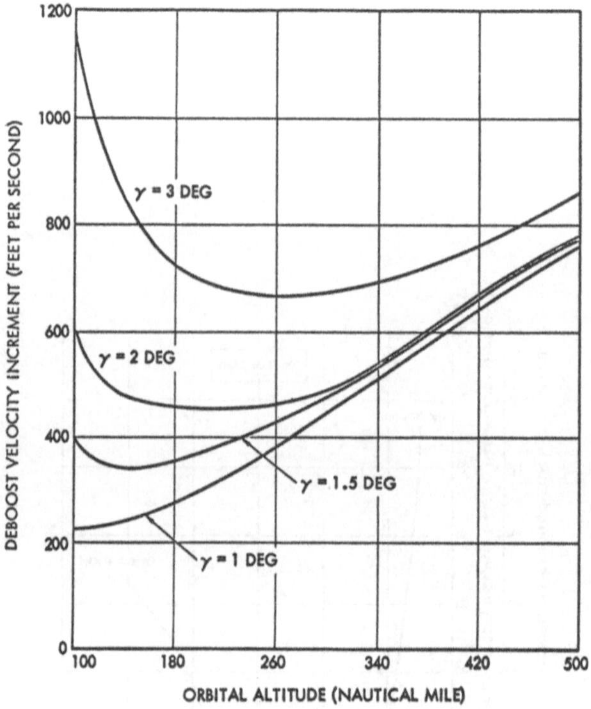


Fig. 10.6 DEBOOST VELOCITY INCREMENT AS A FUNCTION OF ORBITAL ALTITUDE (AFTER FRIEDENTHAL Ref. 5)

The motion of a vehicle entering the atmosphere with constant lift and drag coefficient, Fig. 7, is oscillatory, (Chapter 2). Typical oscillatory (skip) trajectories are shown in Fig. 8. Modulation of the vehicle lift can result in trajectories which do not exhibit this oscillatory character. An important class of non-oscillatory trajectories suitable for the control of the re-entry trajectory are termed equilibrium glide trajectories.

### [10-2.3] Equilibrium Glide (Refs. 25-31)

The equilibrium glide trajectory concept discussed in Chapter 2 provides a basis for the prediction of the range capabilities of lifting re-entry vehicles. This trajectory is characterized by a small flight path angle and a small rate

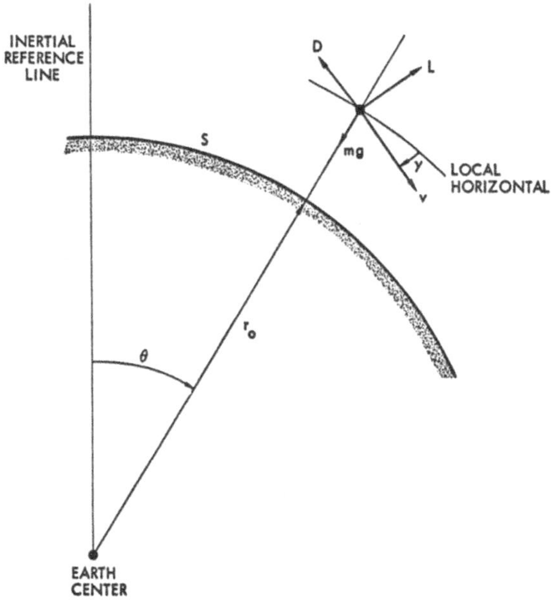


Fig. 10.7 COORDINATE GEOMETRY FOR PLANAR MOTION

of change of the flight path angle. For a given vehicle with fixed $W/C_DA$ and $W/C_LA$, there is a unique combination of atmospheric density, flight path angle, and velocity for which an equilibrium glide can be established. Assuming small angles, and neglecting the rate of change of the flight path angle, the conditions for an equilibrium glide may be approximated by the following equations

$$\frac{dv}{dt} = -\frac{g\varrho_0\sigma}{2W/C_DA}v^2 + g\gamma \qquad (10\text{–}16)$$

$$v\frac{d\gamma}{dt} = -\frac{g_0\varrho_0 v^2}{2W/C_LA} + g\left(1 - \frac{v^2}{rg}\right) \cong 0 \qquad (10\text{–}17)$$

The gravitational acceleration, $g$, and the radius, $r$, are assumed constant at values which represent their averages over the trajectory. From Eq. (10–17),

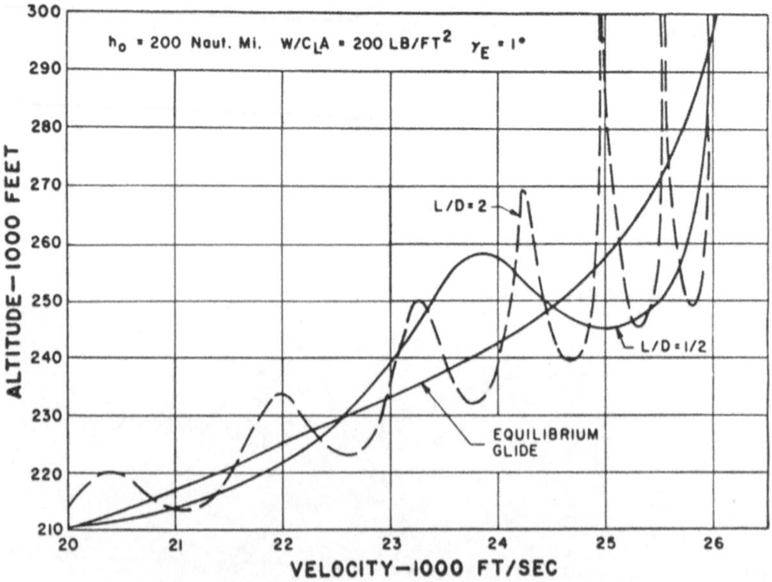


Fig. 10.8 COMPARISON OF SKIP AND GLIDE PATHS (AFTER GALMAN, Ref. 6)

the density corresponding to the equilibrium glide altitude is

$$\varrho_0 \sigma = \frac{2 \dfrac{g}{g_0} (1 - \bar{v}^2) \dfrac{W}{C_L A}}{v^2} \tag{10–18}$$

$$\bar{v} = \frac{v}{v_c} = \frac{v}{\sqrt{rg}}$$

If the independent variable is changed in terms of the flight distance, $S$, Eqs. (10–16) and (10–17) may be combined to yield

$$\frac{v \, \mathrm{d}v}{g \, \mathrm{d}S} = \gamma - (1 - \bar{v}^2) \frac{D}{L} \tag{10–19}$$

Differentiation of Eq. (10–18), (with $\mathrm{d}\sigma/\mathrm{d}S = -\gamma \, \mathrm{d}\sigma/\mathrm{d}y$) yields

$$\gamma \varrho_0 \frac{\mathrm{d}\sigma}{\mathrm{d}y} = \frac{2}{\bar{v}^4 v_c^2} \left( \frac{W}{C_L A} \right) \frac{\mathrm{d}(\bar{v}^2)}{\mathrm{d}S} \tag{10–20}$$

Combining Eqs. (10–19) and (10–20) yields

$$\gamma = \frac{\dfrac{1 - \bar{v}^2}{L/D}}{1 - \dfrac{v^4}{4g \left( \dfrac{W}{C_L A} \right)} \cdot \dfrac{\mathrm{d}\varrho}{\mathrm{d}y}} \tag{10–21}$$

which can be reduced by assuming an exponential density lapse to

$$\gamma = \frac{\left|\dfrac{1 - \bar{v}^2}{L/D}\right|}{1 + \dfrac{v^4 \beta \varrho_0 \sigma}{4g\left(\dfrac{W}{C_L A}\right)}} \tag{10–22}$$

An alternate technique can be used for the estimation of the equilibrium glide conditions, Ref. 5. The relationships between density, velocity and flight path angle are found from the standard assumption of zero acceleration normal velocity vector and an additional constraint on its first derivative, i.e.,

$$v\dot{\gamma} \simeq 0 \tag{10–23}$$

$$\frac{\mathrm{d}}{\mathrm{d}t}(v\dot{\gamma}) \simeq 0 \tag{10–24}$$

These relationships hold strictly for an equilibrium glide at low speed in a constant density atmosphere but yield results in good agreement with the approximation derived previously. For the flight path angles of interest the component of gravity along the velocity vector may be neglected compared to the drag acceleration.

$$\frac{\mathrm{d}v}{\mathrm{d}t} \simeq \frac{-g_0 \varrho_0 \sigma v^2}{2W/C_D A} \tag{10–25}$$

Using Eqs. (10–17) and (10–24) the equilibrium glide velocity is

$$v \simeq \sqrt{\frac{gr}{\dfrac{\varrho_0 \sigma g_0 r}{2(W/C_L A)} + 1}} \tag{10–26}$$

Using Eqs. (10–17), (10–23), and (10–25), the flight path angle may be written as

$$\gamma \simeq \frac{\varrho_0 \sigma g_0}{\beta W/C_D A} + \frac{2}{\beta r}\left(\frac{D}{L}\right) \tag{10–27}$$

These expressions can be used to find the equilibrium glide path velocity and angle in terms of the density ratio, $\sigma$, or equivalently, the altitude. The expression for the velocity shows that the equilibrium glide velocity is the circular satellite velocity modified by the lift capability of the vehicle and the flight altitude.

Figs. 9 and 10 show equilibrium glide altitude and flight path angle as functions of velocity. As the velocity decreases, the altitude of equilibrium glide decreases and the flight path angle becomes steeper. At the start of the equilibrium glide, when the velocity is still nearly circular satellite speed, the flight path angle is on the order of .25 degrees for the example shown.

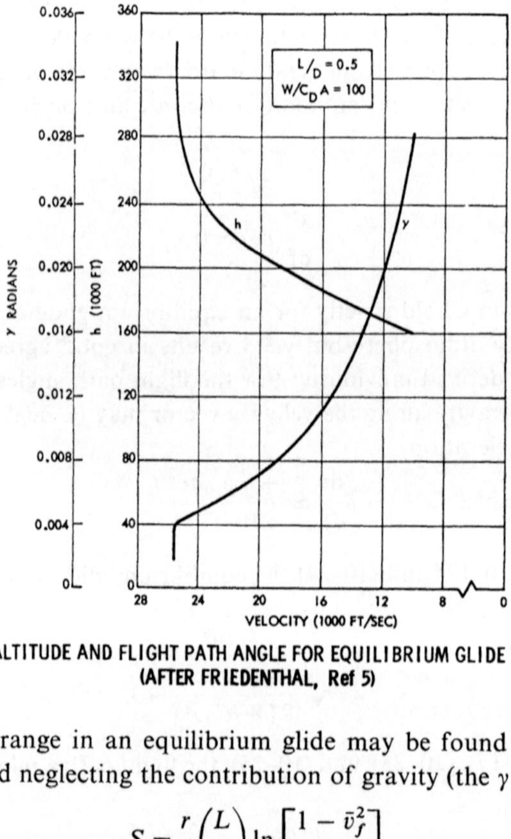


Fig. 10.9 ALTITUDE AND FLIGHT PATH ANGLE FOR EQUILIBRIUM GLIDE TRAJECTORY (AFTER FRIEDENTHAL, Ref 5)

The vehicle range in an equilibrium glide may be found by integrating Eq. (10–19) and neglecting the contribution of gravity (the $\gamma$ term).

$$S = \frac{r}{2}\left(\frac{L}{D}\right) \ln\left[\frac{1 - \bar{v}_f^2}{1 - \bar{v}_i^2}\right] \tag{10–28}$$

The time of flight in an equilibrium glide may be approximated by neglecting the axial component of gravity and the rate of change of flight path angle and combining Eqs. (10–17) and (10–25)

$$t = -\sqrt{\frac{r}{g}}\left(\frac{L}{D}\right)\int_{\bar{v}_i}^{\bar{v}_f} \frac{d\bar{v}}{1 - \bar{v}^2} = \frac{1}{2}\sqrt{\frac{r}{g}}\left(\frac{L}{D}\right) \ln\left[\frac{(1 + \bar{v}_i)(1 - \bar{v}_f)}{(1 + \bar{v}_f)(1 - \bar{v}_i)}\right] \tag{10–29}$$

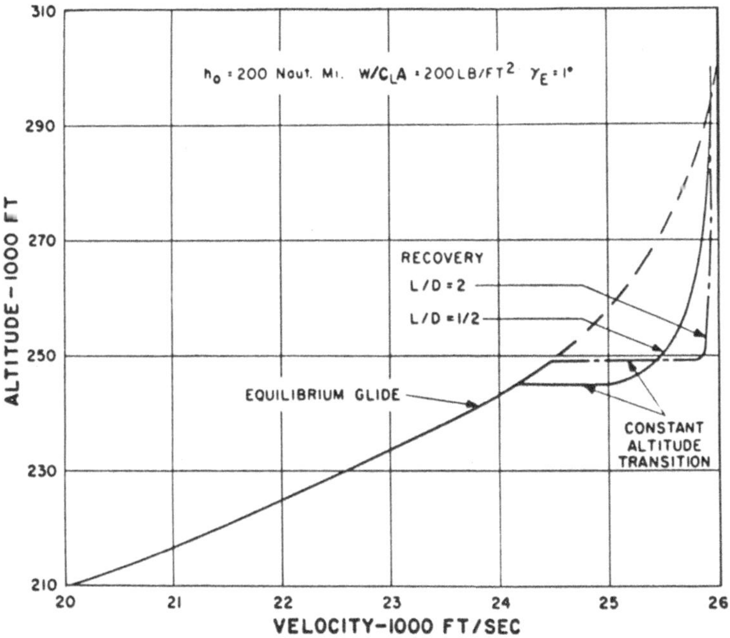


Fig. 10.10 RECOVERY AND TRANSITION TO EQUILIBRIUM GLIDE (AFTER GALMAN, Ref. 6)

The practical application of equilibrium glide trajectories to the guidance of re-entry vehicles requires the determination of a lift modulation program which will establish the equilibrium glide, Ref. 24.

## [10-2.4] Maneuvering (Refs. 23-28)

Perhaps the most important potential advantage possessed by lifting hypersonic vehicles is the ability to maneuver within the atmosphere. The range of glide vehicles is primarily dependent on $L/D$. The range which can be achieved by a lifting vehicle can be estimated by considering that the total lift of the vehicle is generated in one plane and that lateral forces are developed by banking through an angle $\phi$, as shown in Fig. 11. The lateral range is denoted by $N$ and the lateral deflection of the flight path is through the angle $\psi$. The vertical component of the lift force is therefore $L \cos \phi$ and the turning force is $L \sin \phi$. The results for planar motion may be applied to turning trajectories if $L$ is replaced by $L \cos \phi$.

The in-plane force balance is therefore written

$$F_V = L \cos \phi = mg\,(1 - \bar{v}^2) \tag{10–30}$$

and the lateral force which turns the trajectory is

$$F_N = L \sin \phi = mg\,(1 - \bar{v}^2) \tan \phi \tag{10–31}$$

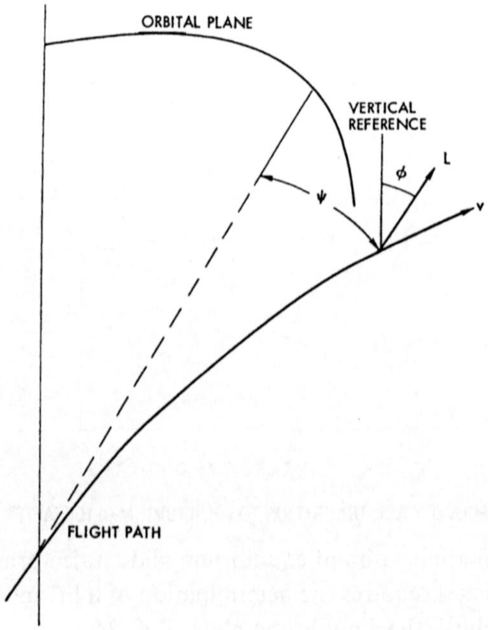


Fig. 10.11 COORDINATE GEOMETRY FOR TURNING TRAJECTORY

The turning rate is, therefore,

$$\dot{\psi} = \frac{F_N}{mv} \tag{10–32}$$

The turn angle is found by integrating Eq. (10–32) using Eq. (10–31)

$$\psi = \int_{t_i}^{t_f} \frac{g(1-\bar{v}^2)}{v} \tan\phi \, dt \tag{10–33}$$

For small flight path angles, $\gamma$,

$$dt \simeq -\frac{dv}{D/m}$$

and

$$\frac{L\cos\phi}{mg} = \left(\frac{L}{D}\cos\phi\right)\frac{D}{mg} = (1-\bar{v}^2)$$

for small $\gamma$ and $\dot{\gamma}$.

Therefore, for constant $L/D$ and bank angle,

$$\psi = -\frac{L}{D}\sin\phi\int_{v_i}^{\bar{v}_f}\frac{dv}{v} = \frac{L}{D}\sin\phi\ln\left(\frac{\bar{v}_i}{\bar{v}_f}\right) \tag{10–34}$$

The lateral range is found from

$$N = \int v\sin\psi\,dt \tag{10–35}$$

The lateral acceleration is

$$\frac{d}{dt}(v\sin\psi) = v\dot{\psi}\cos\psi + \dot{v}\sin\psi \tag{10–36}$$

Since

$$\frac{d\psi}{\tan\psi} \gg \frac{dv}{v}$$

the lateral acceleration equation may be written

$$\frac{d}{dt}(v\sin\psi) \cong v\dot{\psi}\cos\psi \tag{10–37}$$

Using Eqs. (10–30) and (10–31),

$$m\frac{d}{dt}(v\sin\psi) = L\sin\phi\cos\psi \tag{10–38}$$

Dividing Eq. (10–38) by Eq. (10–25)

$$\frac{d(v\sin\psi)}{dv} = -\frac{L}{D}\sin\phi\cos\psi \tag{10–39}$$

The lateral range can therefore be found from

$$N = r\left(\frac{L}{D}\right)^2\int_{\bar{v}_i}^{\bar{v}_f}\frac{1}{1-\bar{v}^2}\int_{\bar{v}_i}^{\bar{v}_f}\sin\phi\cos\psi\,d\bar{v}\,d\bar{v} \tag{10–40}$$

for constant $L/D$ vehicles. For small turn angles ($\psi < 10°$), and constant bank angles, the lateral range is approximated by

$$N = \frac{r}{2}\left(\frac{L}{D}\right)^2\sin\phi\left\{\bar{v}_i\left[\ln\frac{(1-\bar{v}_f)(1+\bar{v}_i)}{(1-\bar{v}_f)(1-\bar{v}_i)} - \ln\frac{(1-\bar{v}_f^2)}{(1-\bar{v}_i^2)}\right]\right\} \tag{10–41}$$

Fig. 12 shows, for $\bar{v}_i = 0.8$, a typical example of maneuverability in terms of $L/D$ and bank angle. This figure must be interpreted as representing an idealized case, since obviously ranges approaching zero cannot be obtained without incurring large heat flux and deceleration, although no specific limitations are shown on the figure. It is apparent, however, that relatively small changes in $L/D$ can provide range changes probably adequate to compensate for trajectory perturbations. For cross range maneuvering, a similar statement can be made, in that relatively low $L/D$'s can provide substantial correction potential.

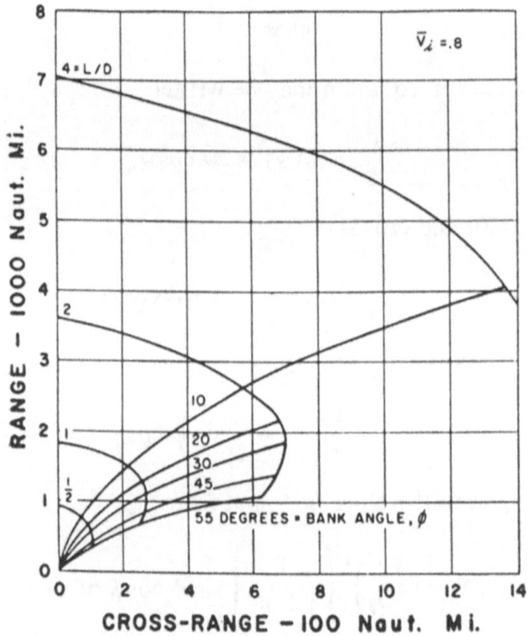


Fig. 10.12 LANDING POINT SELECTIVITY WITH CONSTANT BANK ANGLE (AFTER GALMAN, Ref. 6)

Another point illustrated in Fig. 12 is that the nominal mission range must be less than the maximum range capability of the vehicle, since cross range maneuvering is accompanied by a corresponding range loss. In addition, some range increase potential is obviously necessary to correct for dispersion in the range plane itself.

Just how much maneuverability or landing point selectivity is required of a particular vehicle is dependent to a large extent on non-aerodynamic factors, such as the accuracy of orbital and atmospheric navigation, guidance, and control systems. Before a particular configuration selection is finally confirmed, a dispersion analysis of the primary mission should be made, reflecting the various error sources.

### [10-2.5] Approach and Landing

Lifting vehicles can be considered to fall into three general categories. The "semi-ballistic" vehicle which depends on body lift, maintaining angle of attack with aerodynamic control surfaces, offset c.g., or reaction devices. A typical semi-ballistic shape might be a flat-topped half-cone of relatively high bluntness. Trimmed subsonic $L/D$'s of 0.5 to 2.0 might be attainable. The second category can be described as thick wings or blended wing bodies. The third category can be termed "winged" vehicles, since they tend to look like highly swept airplane configurations and employ lifting surfaces in addition to the body or "fuselage." Subsonic $L/D$'s from 2 to 6 seem attainable with winged vehicles.

The semi-ballistic vehicle, because of its very low $L/D$, is not normally considered to have runway capability, but must be landed by parachute just as if it were pure ballistic.

After the glide vehicle has decelerated to about 5000 ft/sec, it no longer is subject to appreciable centrifugal force and hence begins to descend quite rapidly if no auxiliary power is available, the glide path angle being a function of maximum $L/D$ as shown by Fig. 13. $L/D$ is a strong function of Mach number below 5000 ft/sec, so that an integration method is necessary to obtain range. If this variation is neglected the range can be obtained as a

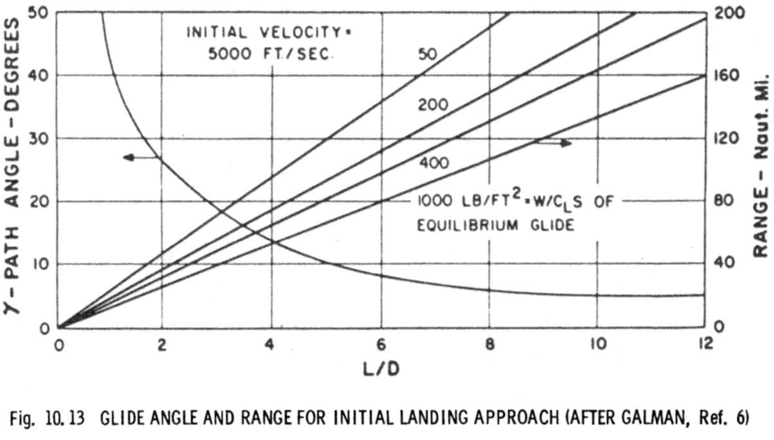


Fig. 10.13 GLIDE ANGLE AND RANGE FOR INITIAL LANDING APPROACH (AFTER GALMAN, Ref. 6)

function of altitude and hence $W/C_L A$ of equilibrium glide. It is apparent from Fig. 13 that low $L/D$ vehicles have an "approach" range on the order of only 100 N.M. even though their initial approach altitude is about 100000 ft. In considering landing speeds, the parameter $W/C_L A$ again becomes important, since

$$V_{st} = [(2/\varrho)(W/C_L A)]^{\frac{1}{2}}$$

As shown by Fig. 14, an increment of 15 percent above stall speed is normally assumed for touchdown speed. From information available in the literature, final approach speed at the beginning of flare may be as high as 1.6 $V_{st}$ for high speed configurations. For low aspect ratio delta configurations, ground effect is significant in reducing vertical impact velocities to below 5 ft/sec.

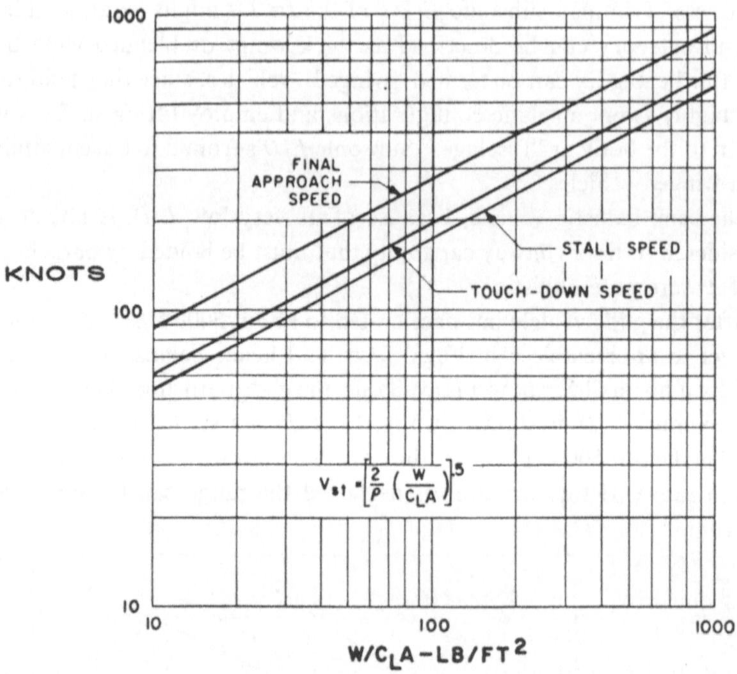


Fig. 10.14 GLIDE VEHICLE LANDING SPEEDS (AFTER GALMAN, Ref. 6)

If a wing loading of 30 lb/ft$^2$ and a maximum effective $C_L$ of 0.5 are representative of a typical landing configuration, Fig. 14 shows the estimated touchdown speed to be 150 knots, which does not seem unreasonable.

One aspect of the landing problem is considered in Fig. 15 in which a summary of landing approach criteria (just prior to flare) derived from pilot evaluation is presented in terms of wing loading and subsonic $L/D$. These

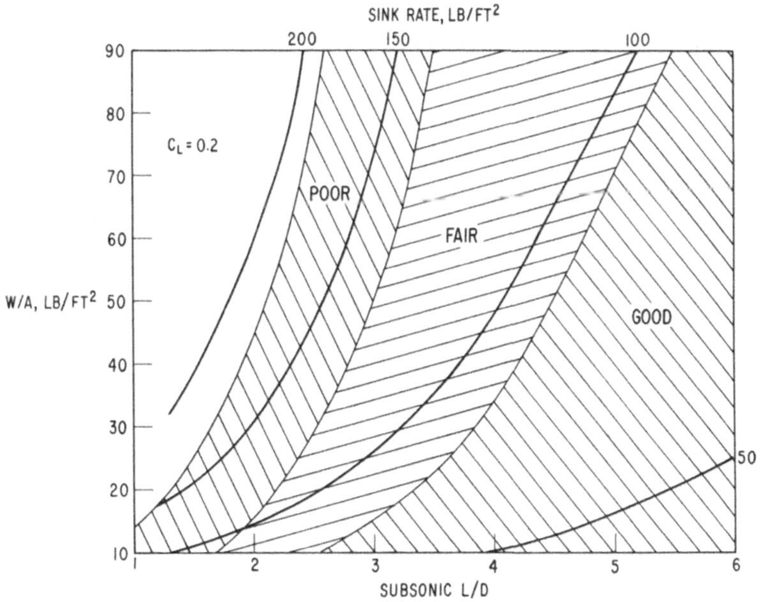


Fig. 10.15 LANDING APPROACH CRITERIA FROM PILOTS' EVALUATIONS (AFTER LOVE, Ref. 32)

results are for a $C_L$ of 0.2 which is representative of the lower values of $C_L$ encountered at this point in the landing approach. The ratio of $C_L$ at touch-down to $C_L$ at subsonic $(L/D)$ max is also important in determining the ease at which landing can be accomplished. Similar evaluations of this criterion show a preference for a subsonic $L/D$ of about 4 or greater.

## [10-3] CONFIGURATION CONSIDERATIONS

The past decade has seen research, design and development of several classes of re-entry vehicles for various applications. For typical examples see Refs. 1-4, 32-36. Unmanned military configurations have progressed from the blunt heat-sink type through the sphere-cone-cylinder ablation protected designs to the more recent slightly blunted cones. This progression has been related to improved heat protection methods and more stringent military objectives. Manned vehicles have undergone a corresponding trend from the Mercury shape to Gemini and Apollo, with the introduction of lifting re-entry in order to widen the re-entry corridor, reduce re-entry loads and provide good maneuverability. All of the vehicle designs have had to face the various compromises between good volumetric efficiency, good re-entry qualities and for manned applications, suitable landing characteristics.

A manned system should provide precision and flexibility in its landing characteristics and be capable of routine launch and return without a large recovery task force. These criteria must be satisfied without sacrifice of payload volume or weight or without reduction of the reliability of re-entry protection. In general, these requirements conflict since efficient entry vehicles (blunt lifting bodies) possess poor landing characteristics, whereas fixed geometry winged configurations which possess good landing characteristics have low volumetric efficiency and may have serious re-entry design problems.

For return from earth orbit, a wide range of entry velocities is of interest. Fig. 16 (Ref. 34), presents the effect of orbit altitude (circular orbit) or apogee

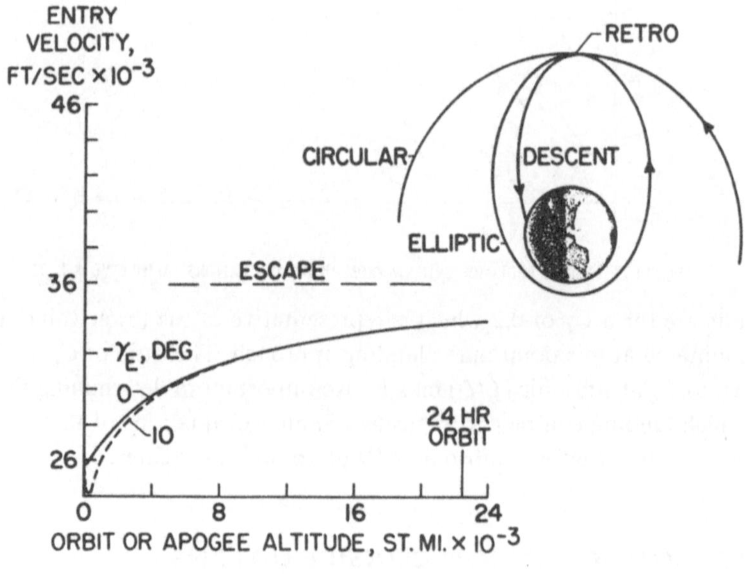


Fig. 10.16 EFFECT OF ORBIT OR APOGEE ALTITUDE ON ENTRY VELOCITY (AFTER LOVE, Ref. 34)

altitude (elliptic orbit) upon entry velocity, assumed to begin at 400,000 feet. Within the range of entry angles of interest for manned entry, about 0° to 10°, there is little effect of entry angle upon 1,000 miles or less of the earth. Once the altitude exceeds several hundred miles, there is initially a marked increase in entry velocity. The re-entry velocity for descent from an altitude of 6,000 miles exceeds 30,000 ft/sec. At the altitude for a 24-hour orbit, the entry velocity is about 34,000 ft/sec.

### [10-3.1] Deceleration

The undershoot boundary for return from orbit is usually established by a deceleration limitation or a heating limitation. For manned entry a deceleration limitation of about 12 $g$'s is generally assumed. An overshoot criterion of no-skip is also usually assumed as a typical representative entry mode in analyses. Love[32] has examined the maximum deceleration to be expected during entry and its relation to re-entry angle for the above entry corridor boundary constraints, and the following entry assumptions. Entry is considered to be initiated at 400,000 feet with the vehicle in a trimmed condition at either $(L/D)_{max}$ or $C_{L_{max}}$. A constant $L/D$ trajectory is maintained from entry to pullout. The vehicle is then rolled to maintain a constant altitude flight path, i.e., roll or lift vector modulation is assumed. This maneuver is maintained until the vehicle is unable to generate sufficient lift to sustain flight at that particular altitude. An equilibrium glide maneuver at either $(L/D)_{max}$ or $C_{L_{max}}$ is then initiated and flown to the landing point. The earth is considered to be spherical and nonrotating. Typical corridor widths are shown in Fig. 17.

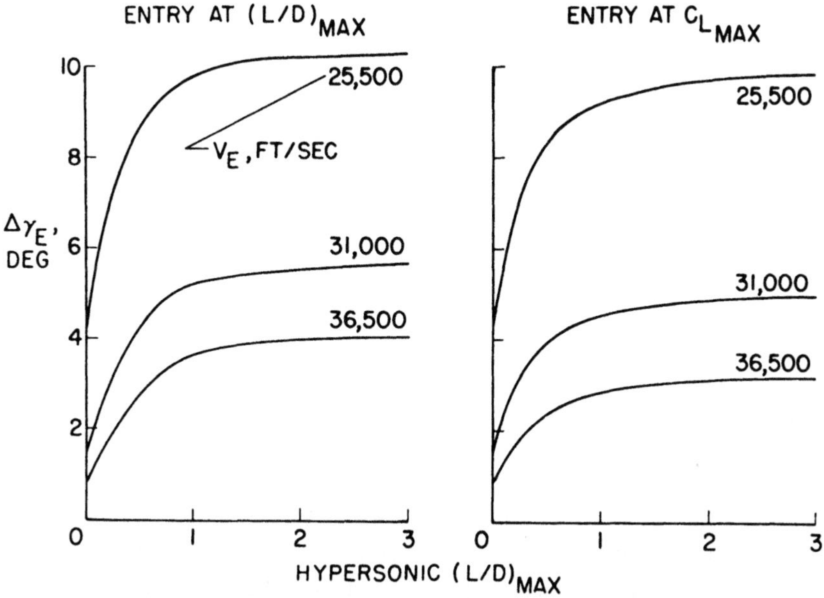


Fig. 10.17 WIDTH OF ENTRY CORRIDOR (AFTER LOVE, Ref. 34)

The maximum deceleration to be expected during entry and its relation to entry angle and entry velocity is shown in Fig. 18. The hatched areas include entry speeds from 26,000 to 36,500 ft/sec. As orbital speed increases, peak $g$ becomes increasingly sensitive to change in entry angle and for the higher velocities approaching superorbital speeds, entry angle will be limited to roughly 10° or less.

The curves outside of the hatched areas correspond to suborbital entry velocities and are included to indicate the important role that abort conditions may have upon peak $g$. It is evident that there are marked effects of the velocity at which entry is initiated following abort, and of vehicle $(L/D)_{max}$. Other factors such as altitude at abort-entry, retro application, etc., also have important bearing on abort peak $g$. With a capability to withstand the order of $12g$, a lifting vehicle would be able to handle a wide range of abort conditions. Similar results are obtained for entry at $C_{L_{max}}$.

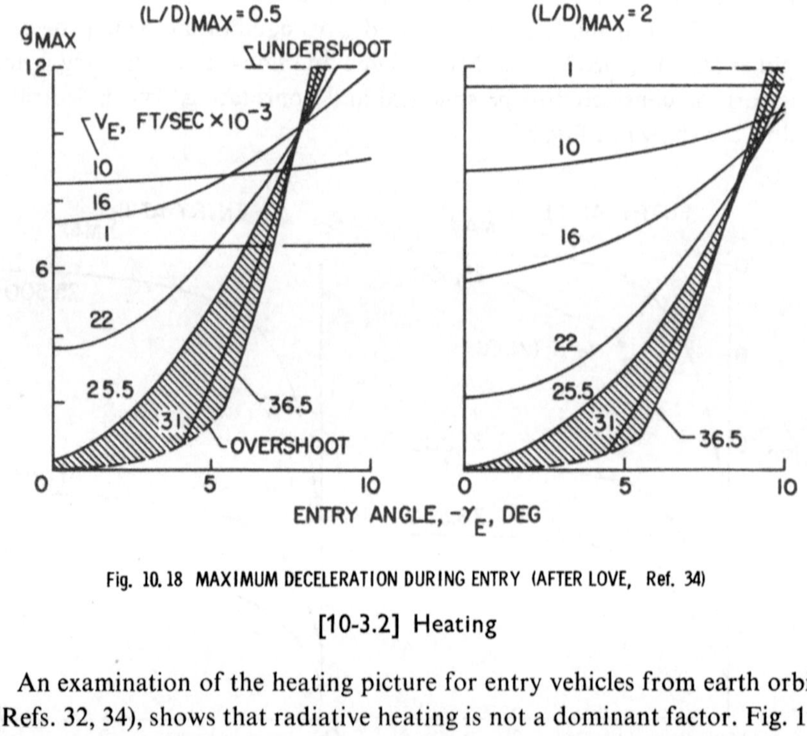


Fig. 10.18 MAXIMUM DECELERATION DURING ENTRY (AFTER LOVE, Ref. 34)

## [10-3.2] Heating

An examination of the heating picture for entry vehicles from earth orbit (Refs. 32, 34), shows that radiative heating is not a dominant factor. Fig. 19 shows in a qualitative way the heat load input to an entry vehicle in terms of Btu's per pound of vehicle weight as a function of entry velocity for several types of bodies. At entry velocities in the orbital regime, the advantage lies with the blunt body since the heating is essentially all convective and the nose radius is large. As entry velocity increases the radiative contribution first becomes significant on the blunt body, but over a narrow range of velocities the sum of the radiative and convective heat load still remains less than that for the conical body which is still deriving nearly all of its heat load from convective heating.

Figure 20 summarizes the maximum stagnation point convective heat rates

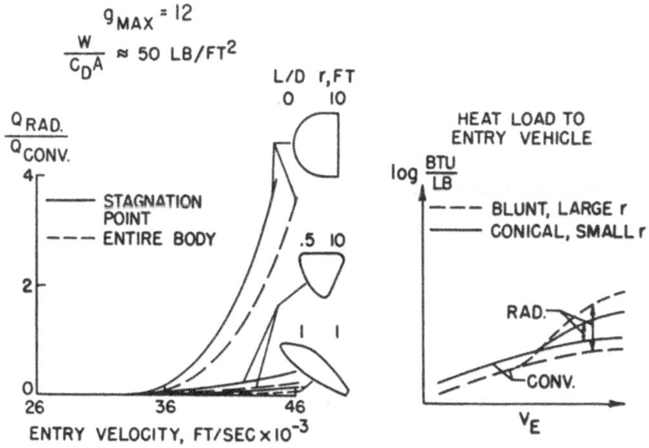


Fig. 10.19 ROLE OF VEHICLE CONFIGURATION IN RADIATIVE HEATING (AFTER LOVE, Ref. 34)

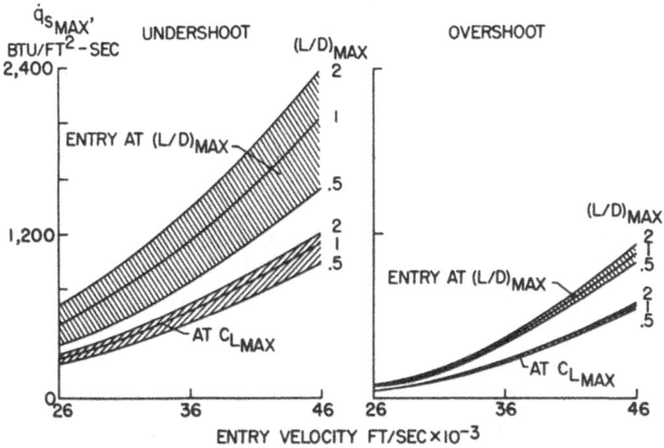


Fig. 10.20 MAXIMUM STAGNATION POINT CONVECTIVE HEAT RATES (AFTER LOVE, Ref. 34)

as a function of entry velocity. The heating results are based on an idealized vehicle with a loading $W/A$ of 35, a nose radius of 1 foot and a shape similar to that of a highly swept, delta-planform lifting body ($L/D=1$). Whether entry is made near overshoot, undershoot, at $(L/D)_{\max}$, or at $C_{L_{\max}}$, entry velocity has a major effect on the heat rates. Undershoot produces maximum heat rates much larger than those at overshoot. Entry at $C_{L_{max}}$, as contrasted with entry at $(L/D)_{max}$, has a decided advantage in reducing maximum heat rates. The maximum heat rates for entry at $C_{L_{max}}$ are less sensitive to the value of vehicle $(L/D)_{max}$. There is the overall indication that in considering heat rates, it is advantageous to keep $(L/D)_{max}$ small.

Radiation equilibrium temperatures corresponding to these heat rates, with an assumed emissivity of 0.85, are shown in Fig. 21. For entry at $C_{L_{max}}$,

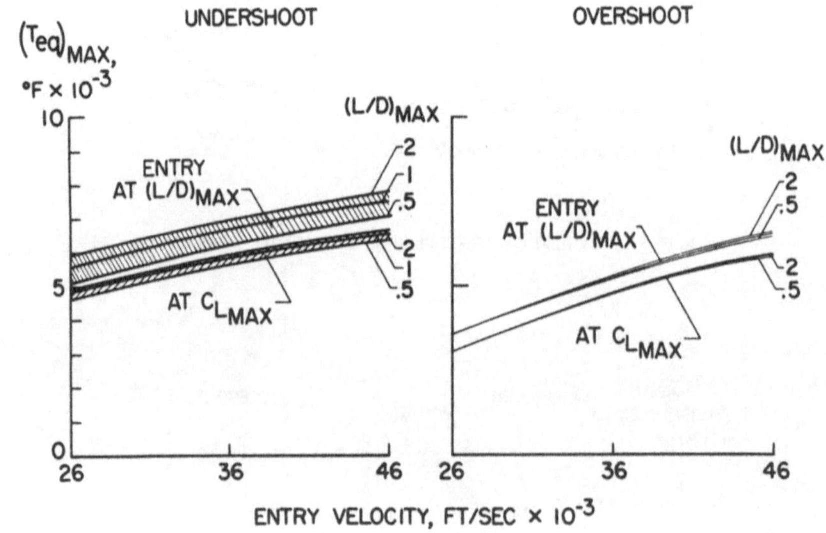


Fig. 10.21 MAXIMUM STAGNATION POINT RADIATION EQUILIBRIUM TEMPERATURES ($\epsilon$ =0.85) (AFTER LOVE, Ref. 34)

the maximum radiation equilibrium temperatures would range from about 3000°F to 6500°F, and to higher temperatures for entry at $(L/D)_{max}$. Corresponding stagnation point heat loads are shown in Fig. 22. The hatched band for any value of $(L/D)_{max}$ covers the heat loads from undershoot at the bottom of the band to overshoot. The large reduction in heat load that is realized by entry at $C_{L_{max}}$ rather than $(L/D)_{max}$ is evident; and, as in the heat rate picture, there is strong justification for keeping $(L/D)_{max}$ small, especially for entry at $(L/D)_{max}$. These factors suggest a design that gives $(L/D)_{max}$ near $C_{L_{max}}$, and that enters near $C_{L_{max}}$ and employs roll modulation to govern ranging. For entry at $C_{L_{max}}$, the increase in stagnation point heat load with entry velocity is not overly large over the velocity range.

Love[32,34] also discusses the effect that use of pitch modulation on the pullout would have on the heating results for modulation restricted to the regime between $C_{L_{max}}$ and $(L/D)_{max}$ with both lift and drag being modulated.

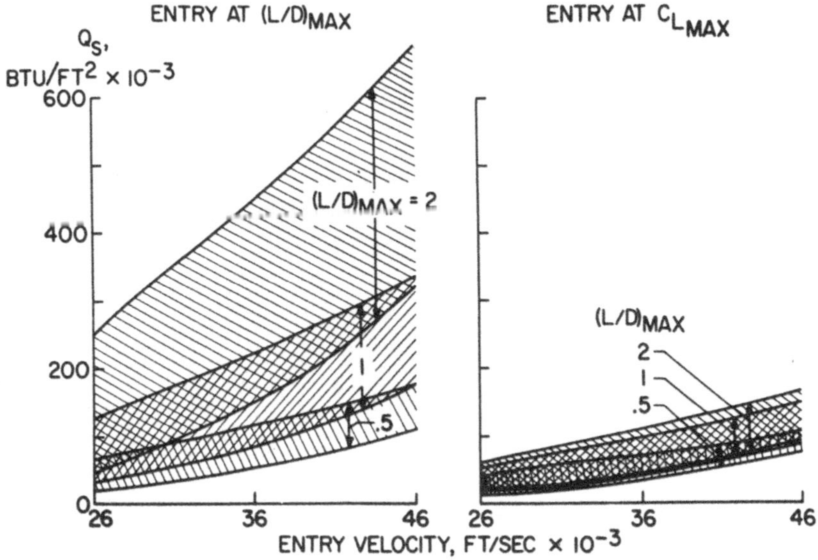


Fig. 10.22 STAGNATION POINT CONVECTIVE HEAT LOADS (AFTER LOVE, Ref. 34)

The justification for pitch modulation in pullout is reduction of deceleration and increase in corridor width. Pitch modulation, as compared with pullout at $C_{L_{max}}$, increases both heat rates and heat loads. Estimates indicate that for $L/D = 1$, the penalty in heat load is relatively moderate, but by no means insignificant. He suggests that, in light of the desirability of entering at high $C_L$ with subsequent roll modulation, the addition of some pitch modulation capability for use during and after pullout—with roll modulation remaining the primary mode—might offer a fuller realization of the aerodynamic performance potential without a prohibitive heating penalty, provided $L/D$ does not exceed about one and the pitch modulation is sufficiently restricted.

## [10-3.3] Heat Protection

Love also presents a summary of the current heat protection picture in Fig. 23 in the form of heat load versus heat rate so as to establish in a general way the relation of the stagnation point heating to materials protection concepts. The left end of each curve corresponds to entry at 26,000 ft/sec and the right end to 46,000 ft/sec. The hatched boundary suggests the limits to which metallic shields can operate; for example, the refractory metals can be expected to cope with some 40 to 50 Btu/ft$^2$/sec. A review of the current state of heat protection technology indicates that barring rapid developments in ceramics, transpiration cooling, and one or two other ap-

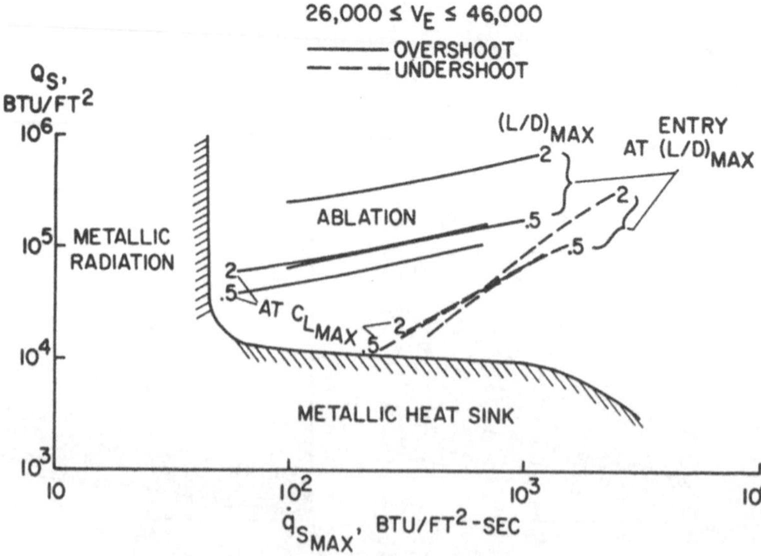


Fig. 10.23 STAGNATION POINT CONVECTIVE HEATING IN RELATION TO HEAT PROTECTION (AFTER LOVE, Ref. 34)

proaches, ablation materials will be the most likely choice for the stagnation region of vehicles in the indicated $L/D$ range, (Refs. 32-37).

Of greater concern, perhaps, than the heating to the nose or stagnation region is the heating of the major surface areas of the vehicle. An estimate of the maximum radiation equilibrium temperatures that would exist along the streamwise centerline of a delta-planform lifting body with $(L/D)_{\max}=1$ and entering at $C_{L_{\max}}$ is shown in the left-hand portion of Fig. 24. The assumed emissivity is 0.85. The hatched bands indicate the range of temperatures to be expected on the lower surface between undershoot at the top of the band and overshoot at the bottom for entry velocities of 26,000 and 36,000 ft/sec. A drop of only a few hundred degrees is indicated in progressing 20 feet rearward from the tangency point of the surface with a hemispherical nose. The curves showing the rapid decay in temperature with distance rearward for the upper surface centerline and the condition of overshoot are subject to considerable uncertainty because of insufficient knowledge concerning the expansion phenomena over the upper surface.

The status of the life of coated refractory metal sheet as summarized recently by Mathauser[38] is shown in the right-hand portion of the figure. The different curves represent different refractory metals. They include tungsten, tantalum, molybdenum, and columbium. Present-day coatings appear to be able to provide protection under continuous exposure of at least 1 hour at 3000 °F to 100 hours at 2500 °F. An order of magnitude or greater decrease in coating life is obtained under cyclic exposure conditions.

This degradation under cyclic temperature exposure reflects directly on the reuseability of refractory metal components in entry vehicles. A band is also shown indicating probable future improvement in coatings and/or ceramics as indicated by current studies although this gain has not yet been realized for sheet-type application. Current work indicates that its achievement will likely be accompanied by short material life indicating the need for refurbishment after each entry flight, and in this respect would require a refurbishing technique similar to that for a surface protected by ablation material.

Fig. 24 suggests that while current refractory metals are adequate for manned lifting vehicles entering at orbital speeds, other modes of heat protection are required over much of the vehicle if it is to have a high growth potential in entry velocity. A refurbishable ablation shield appears to offer the best approach for growth. Refractory metals might be satisfactory for large areas of the upper surface for entry at $C_{L_{max}}$. Even here the use of a refurbishable ablation covering may offer the greatest growth potential in terms of entry velocity, modes of entry, and therefore mission versatility.

## [10-3.4] Lateral Ranging

The mission objectives relating to lateral ranging during re-entry may have a dominant influence on the choice of maximum $L/D$ for the re-entry vehicle (Refs. 32, 34, 35, 39). The operation of any long-term manned space station will require some type of ferry vehicle to transport men and equipment to

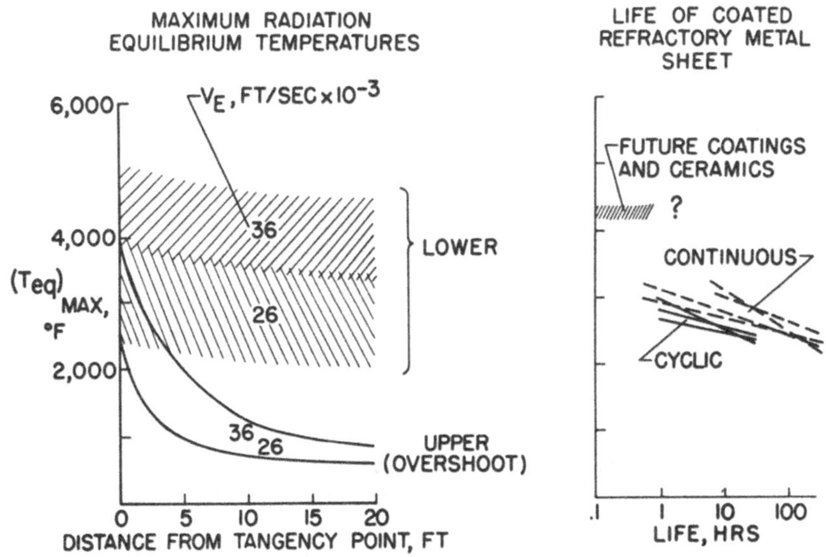


Fig. 10.24 SURFACE TEMPERATURES AND MATERIAL CAPABILITY (AFTER LOVE, Ref. 34)

and from the station with regularity and reliability. Such a vehicle, designed for entry at near-orbital speeds, could also be useful in the return from any deeper space mission if either an earth-orbit rendezvous terminal maneuver or a maneuver combining atmospheric braking and a nearearth parking orbit is used. One measure of the efficiency of a ferry vehicle between a near-earth space station and the earth is the ability to reach prechosen landing sites with some prescribed frequency.

In considering this frequency of return it is necessary to consider not only the normal mode of operation in which only infrequent returns are scheduled at desirable times, but also operation under various degrees of emergency, which dictate quick or even immediate return to earth. In extreme emergencies, when immediate return to earth is necessary, choice of landing site becomes impractical. In most cases, however, although it might be required to abandon the station quickly, the ferry vehicle could remain in orbit for some time before initiating re-entry in order to land at a prechosen site. The allowable delay time in orbit would be determined primarily by the capabilities of the ferry life-support systems.

A typical example can be considered of return from a space station which is in a circular orbit at an altitude of 200 statute miles. The results are not extremely sensitive to orbital altitude. The cross range capabilities of vehicles with various hypersonic $L/D$ ratios can be summarized as in Fig. 25. The

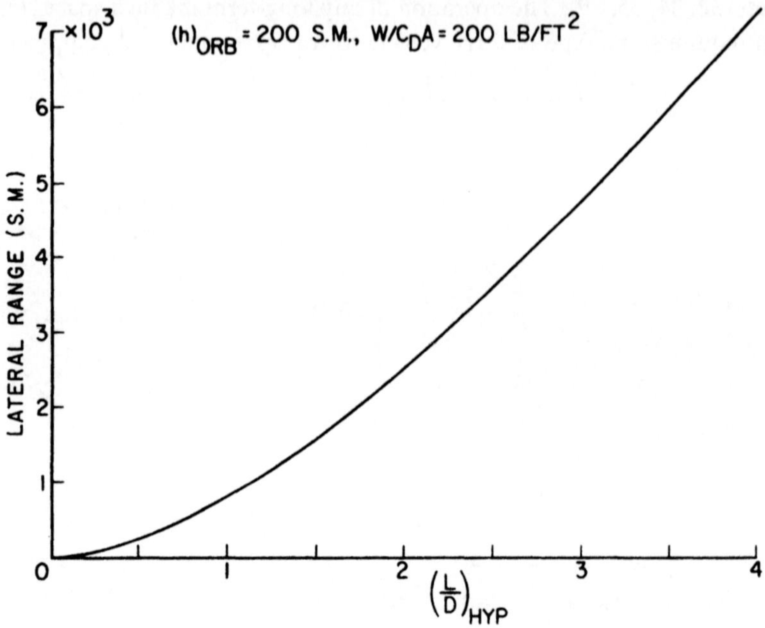


Fig. 10.25 LATERAL RANGE OBTAINABLE BY AERODYNAMICS (AFTER BARADELL, Ref. 39)

maximum lateral range achievable is presented in terms of the vehicle maximum hypersonic $L/D$. For this investigation a value of $W/C_DA$ of 200 lb/ft$^2$ was chosen, (Ref. 39). At the lower values of $L/D$ a value of 75 lb/ft$^2$ was also considered and was not found to have a significant effect on the values presented. The calculations were carried out assuming the vehicle performs a shallow re-entry and holds a constant altitude at the bottom of pullup by rolling about the wind vector until an appropriate equilibrium glide path can be followed.

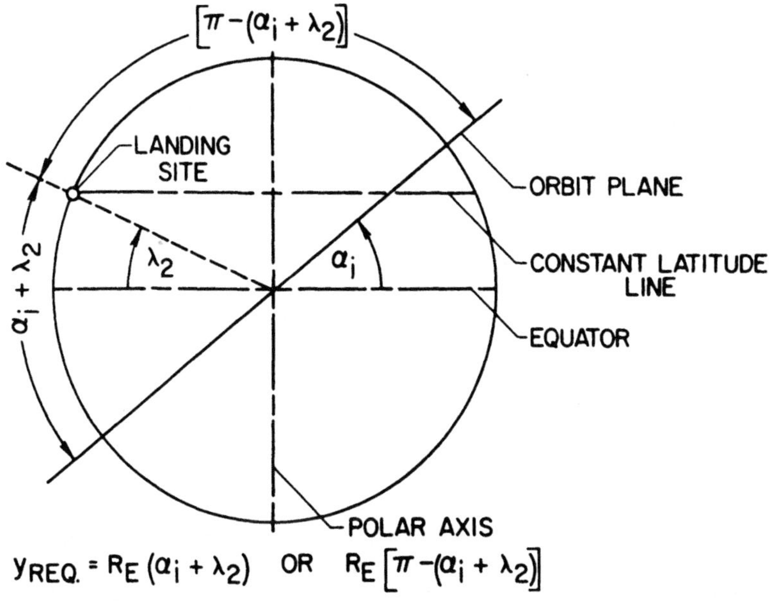


Fig. 10.26 MAXIMUM LATERAL RANGE REQUIRED FOR "QUICK" RETURN (AFTER BARADELL, Ref. 39)

The maximum lateral range that could be required for quick return from a given orbit to a landing site at a given latitude is determined, as shown in Fig. 26, to be given by

$$y_{\text{req}} = R_E(\alpha_i + \lambda_2) \quad \text{or} \quad R_E[\pi - (\alpha_i + \lambda_2)]$$

whichever is the lesser. "Delay time" is defined as the time between decision to re-enter and initiation of re-entry, and a "quick" return is defined as a return with a delay times of less than one orbit. For a quick return, therefore, the vehicle initiates re-entry some time in the first orbit after the decision to re-enter. Here $\alpha_i$ is the orbital inclination to the equator, $\lambda_2$ is the latitude of the landing site, and $R_E$ is the radius of the earth. These range requirements are shown in Fig. 27. Also shown are the values of hypersonic $L/D$

required. Several geographic points are shown to indicate the limits of the United States and of the U.S. Mainland.

The ability to reach any point on the globe once each orbit from any orbit inclination requires a hypersonic $L/D$ of about 3.6. For equatorial or polar orbits no lateral range is required to reach points on the equator or the poles, respectively, once each orbit. The minimum lateral ranges required

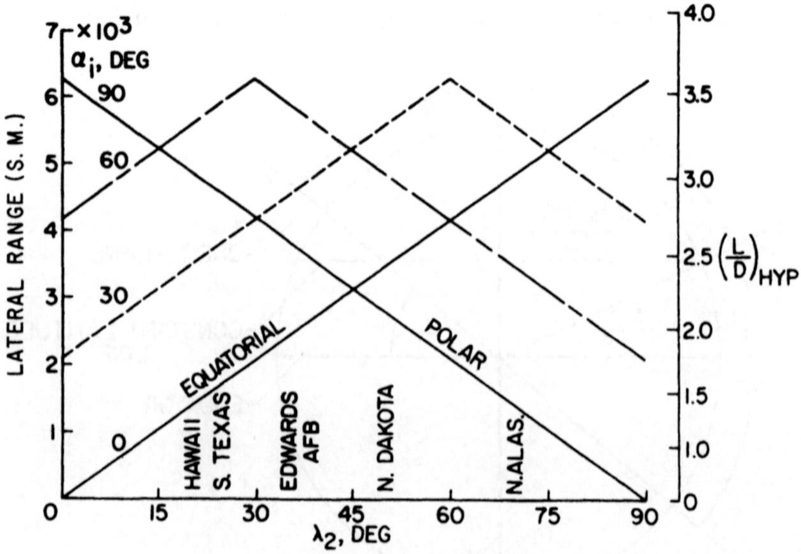


Fig. 10.27 LATERAL RANGE REQUIRED FOR "QUICK" RETURN FROM SPACE STATION ORBIT (AFTER BARADELL, Ref. 39)

for quick return from equatorial or polar orbits approach zero as the location of the landing sites approaches these points. For other inclinations no choice of landing site can assure quick returns with zero lateral range capability.

The effect of delay orbits on lateral range and associated $L/D$ requirements is illustrated in Fig. 28, which presents the maximum holding time required to reach a landing site near Edwards AFB from orbits of various inclinations and for several values of the hypersonic $L/D$ (Ref. 39). The solid curves represent boundaries or discontinuities in the picture, while the dashed curves are representative of the members of the family. As is indicated, a vehicle with an $L/D$ of 3.6 could reach Edwards AFB within one orbit for any orbital inclination. This quick return is possible for a vehicle with $L/D=2.7$ from orbits inclined at less than 22° to the equator or from a polar orbit. With less $L/D$ some holding time is required for a polar orbit and the orbits for which quick return is possible are limited to low inclination orbits as illustrated by the $L/D=2$ curve. With an $L/D$ of greater than 1.9 quick return from an equatorial orbit is possible, while with less $L/D$ the landing site cannot be reached from an equatorial orbit. The curve for $L/D = 1$ shows that at least daily return is possible to a base such as Edwards for all orbits which pass over the continental U.S. As $L/D$ is decreased further, the holding times for all inclinations increase and the band of low inclinations, for which no return is possible, grows wider. With values of $L/D$ greater than 0.84,

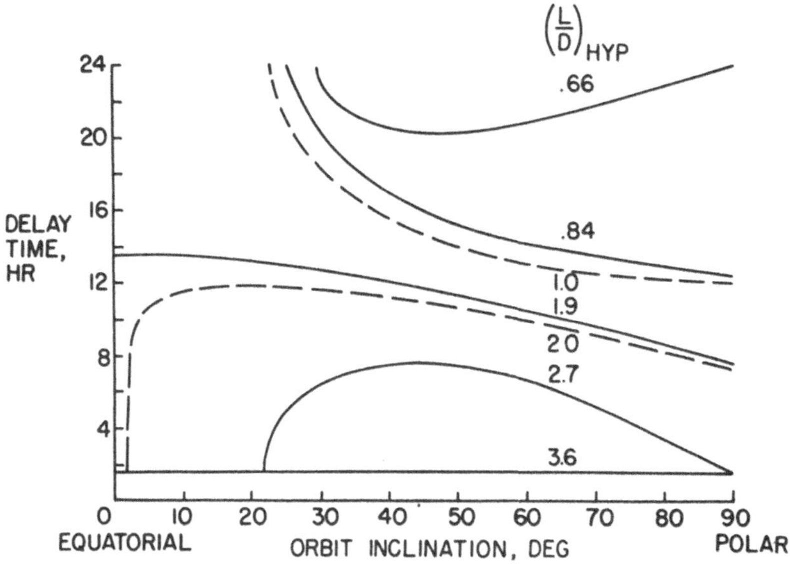


Fig. 10.28 MAXIMUM DELAY TIME FOR RETURN TO EDWARDS AFB (AFTER BARADELL, Ref. 39)

holding times of less than 12 hours are required to reach the site from a polar orbit, but for values of $L/D$ less than 0.84, the delay time for a polar orbit jumps to 24 hours and remains there for values of $L/D$ down to 0.66 after which no regular return to Edwards from a polar orbit can be counted on. A reduction in lateral range requirements can be obtained by the use of multiple landing sites.[39]

In Fig. 29, the maximum lateral range and $L/D$ required to return from a polar orbit to a spot within the continental U.S. (excluding Alaska) is indicated. An $L/D$ of about 0.7 will guarantee at least once-a-day return to any spot in the continental U.S. An $L/D$ of about 0.9 will assure twice-a-day return to the U.S. An $L/D$ of 1 will give at least once-a-day return to the U.S. from an orbit of any inclination that passes over the U.S., (i.e., for the lowly inclined orbits any spot within the southern half of the U.S. would be accessible, and as orbit inclination increases the accessible area increases until the entire U.S. is accessible for orbits inclined greater than about 37°).

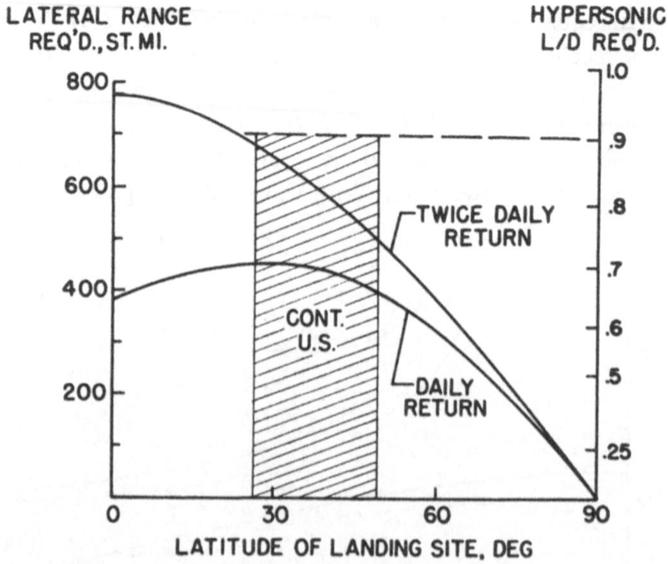


Fig. 10.29 MAXIMUM LATERAL RANGE AND L/D REQUIRED FOR SPOT RETURN FROM POLAR ORBIT (AFTER LOVE, Ref. 34)

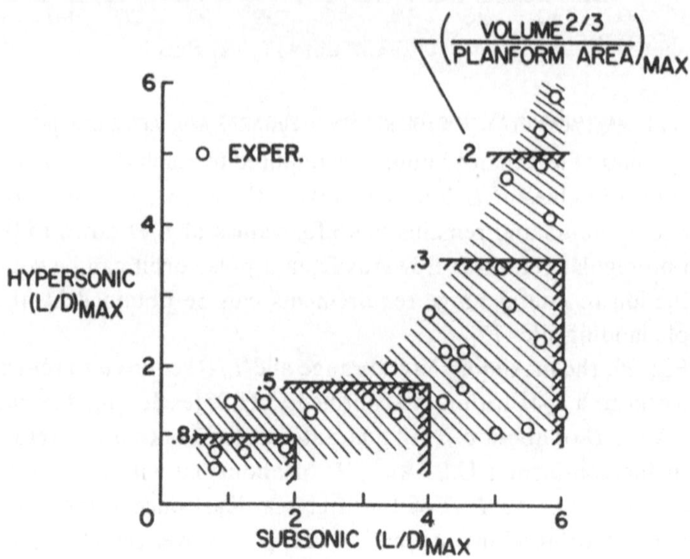


Fig. 10.30 RELATION OF HYPERSONIC TO SUBSONIC $(L/D)_{max}$ FOR ENTRY VEHICLES (AFTER LOVE, Ref. 34)

## [10-3.5] Landing

Low $g$ impact at landing is desirable for a vehicle intended for reuse, such as a multipurpose entry vehicle. For an entry vehicle with a hypersonic $L/D$ of about 1 and a wing loading in the neighborhood of 35, a subsonic $L/D$ of about 4 or more would be a desirable but not an essential goal.

Fig. 30 relates hypersonic $(L/D)_{max}$ to subsonic $(L/D)_{max}$, with their relation as a function of volumetric efficiency. The experimental points as compiled by Love[34] show that the higher the subsonic $(L/D)_{max}$, the greater the possible spread in hypersonic $(L/D)_{max}$. The boundaries denoting values of the ratio of volume to the two-thirds power to planform area were derived from experimental data. These boundaries are maxima in the sense that a given valued boundary could move down or to the left, but not up or to the right. The overall results show that hypersonic $(L/D)_{max}$ comes at greater expense to volumetric efficiency than does subsonic $(L/D)_{max}$. Of particular interest, is the indication that a fixed geometry entry vehicle with a hypersonic $(L/D)_{max}$ of the order of 1 is capable of achieving subsonic $(L/D)_{max}$ well in excess of 4 while retaining relatively good volumetric efficiency. Good volume distribution can also be realized in a vehicle that has these characteristics.

## [10-3.6] Comparison of Medium and High $L/D$ Vehicles

A reusable manned entry vehicle suitable for entry at circular to hyperbolic velocities appears practicable without incurring large penalties for aerodynamic performance. The general class of vehicle that shows considerable merit has a hypersonic $(L/D)_{max}$ of about one. Entry at high $C_L$, coupled with a design that tends toward the realization of $(L/D)_{max}$ near $C_{L_{max}}$, appears desirable from consideration of heating and lateral ranging. The alleviation of the heating problem by entry at high $C_L$ is sizeable over the entire velocity range, and entry at high $C_L$ produces no significant detrimental effect upon peak deceleration or corridor width. With entry at high $C_L$, indications are that the increase in weight of an ablation-protected vehicle in this class that is brought about by increasing entry velocity is within practical consideration. For this class of vehicles Love[32, 34] recommends refurbishable ablation; however, the use of a radiating metallic approach over significant surface areas is not ruled out for the highest entry velocity. Use of roll modulation only as the entry mode is attractive from the heating viewpoint; however, it appears that the addition of some restricted pitch modulation capability, with roll modulation as the primary mode, might offer a fuller realization of the aerodynamic performance potential without prohibitive increase in heating. Conventional landing appears within reach of such a vehicle without excessive penalties in total weight or useable volume, and with good volume distribution.

A typical example of a medium $L/D$ vehicle is the lifting body developed by the NASA/Langley Research Center. Figure 31 shows a modified version described by Eulberg.[35] Estimates of the hypersonic $L/D$ for this configu-

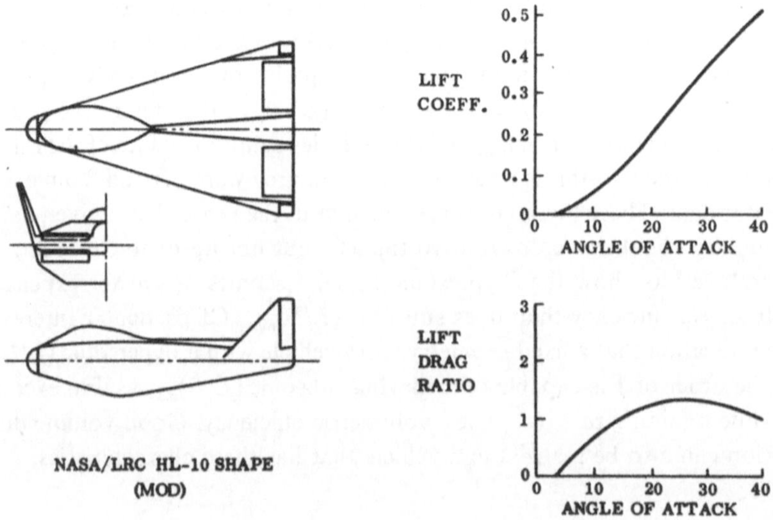


Fig. 10.31 MEDIUM L/D CONFIGURATION (AFTER EULBERG, Ref. 35)

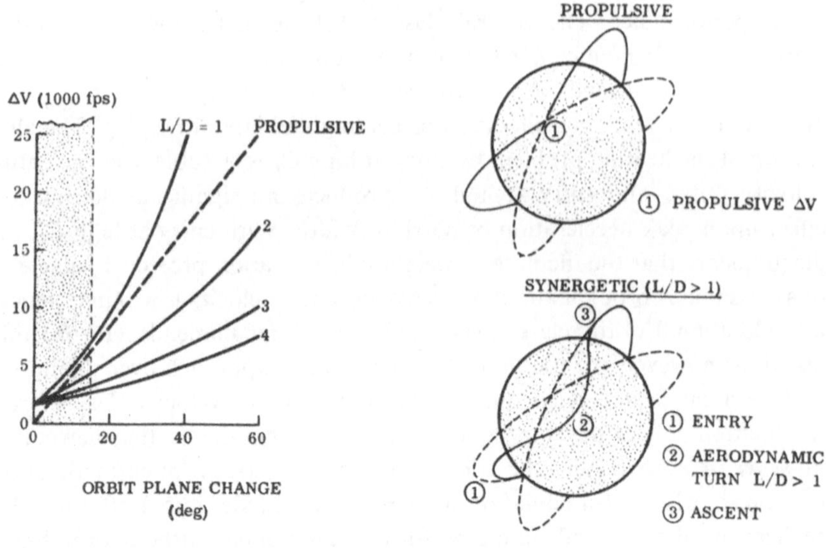


Fig. 10.32 PLANE CHANGE PARAMETERS (AFTER EULBERG, Ref. 35)

ration indicate a maximum $L/D$ of about 1.3 at an angle of attack of 26°. The lift coefficient at this angle of attack is about 0.27.

Military missions may require orbital plane change capability as well as a quick return capability, possibly for crew rescue purposes. The orbital plane change can be accomplished propulsively by a rocket impulse applied normal to the flight path heading. This is shown schematically in Fig. 32. If the orbital vehicle has sufficient aerodynamic capability (i.e., $L/D > 1$) the orbit plane could be changed by entry into the atmosphere, aerodynamic turn or heading change and re-ascent to orbit.

Recent studies of the velocity losses of the "synergetic" or aerodynamic plane change technique have been carried out by Nyland[40] and Bruce.[41] Nyland's data is shown in terms of required $\Delta V$ versus plane change angle for both a pure propulsive maneuver and with various aerodynamic shapes or $L/D$'s. For large plane changes, the $\Delta V$ savings of $L/D$ above 2 are quite apparent. However, when this is translated into relative vehicle weights, some of the benefit is lost due to the increased inert weight of high $L/D$ vehicles and the problem of efficiently housing the required propellants during the turn. Also, the aerodynamic heating during the entry and turn tend to increase the thermal protection weights of the aerodynamic vehicles.

A typical total required plane change might be 15° or less, and possibly could consist of several smaller plane changes during one mission. Therefore small plane changes can probably be made most efficiently using propulsion. In the analysis of Eulberg[35] a typical $\Delta V$ of 6500 fps is assumed for maximum plane changes of the order of 15°.

If further, a 3800 n.m. lateral range requirement is assumed from a 30° inclined orbit, Fig. 33 indicates that an aerodynamic $L/D$ capability of 3 is required while a lifting vehicle with an $L/D$ of 1 would require a $\Delta V$ of about 22,000 fps.

The Air Force Flight Dynamics Laboratory has been studying high $L/D$ configurations, similar to that shown in Fig. 34. This shape represents a typical compromise of slopes and sweeps to obtain internal volume with minimized surface area. Also shown are estimates of hypersonic aerodynamic performance of the shape. A maximum $L/D$ of about 3 is shown occurring at an angle of attack of 12°, with a resulting lift coefficient of 0.9. The vehicle possesses a subsonic $L/D$ of approximately 4.5. These data are based on calculations made at General Dynamics/Convair for a similar vehicle, with small leading edge radii (Ref. 35).

This high $L/D$ configuration has been compared with a medium $L/D$ configuration by Eulberg.[35] A minimum crew size oı two was selected, based on current military aircraft experience, where a pilot and mission equipment operator appears to be minimum. The final vehicle concept was assumed to

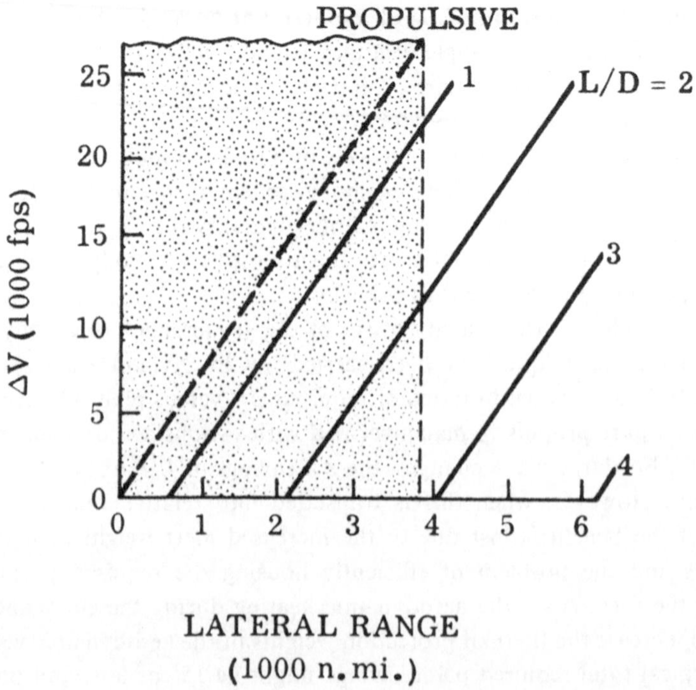


Fig. 10.33 QUICK RETURN PARAMETERS (AFTER EULBERG, Ref. 35)

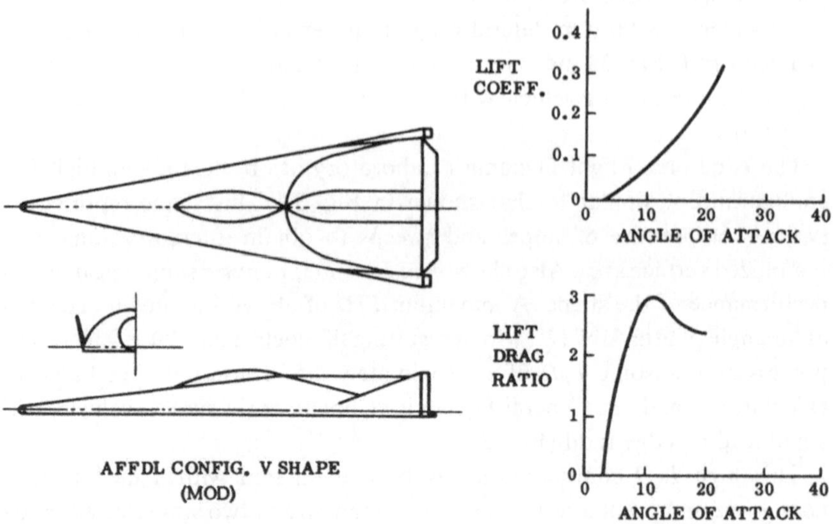


Fig. 10.34 HIGH L/D CONFIGURATION (AFTER EULBERG, Ref. 35)

have multi-mission flexibility. Equipment weight of 4500 lbs was assumed. Three passengers and 3000 lbs of cargo were considered to be adequate as an alternate for the support mission.

The vehicle shapes selected were assumed to be geometrically variable at some fixed gross weight, to determine preferable wing loading characteristics. Typical constraints on wing loading, such as landing, abort and aerodynamic heating effects were explored. From this, a compromise wing loading for each shape was arrived at, on a minimum weight basis.

The starting point of the wing loading investigation was an estimation of the maximum permissible based on internal vehicle volume requirements.

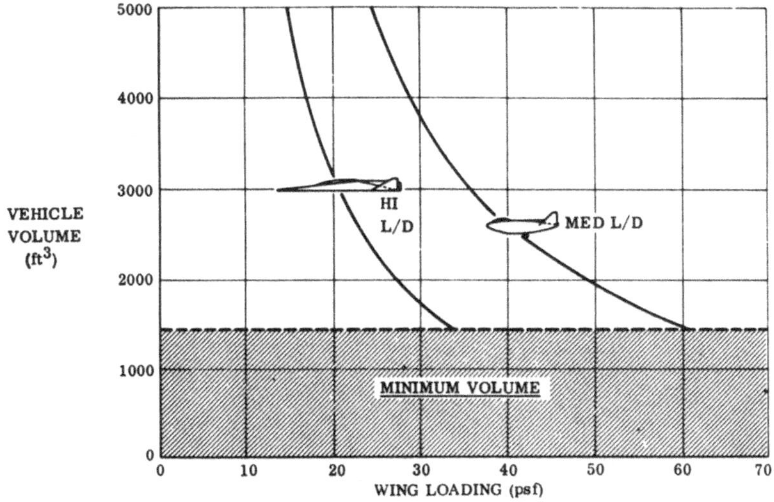


Fig. 10.35 VOLUME vs WING LOADING (AFTER EULBERG, Ref. 35)

Fig. 35 plots this volume versus wing loading for the two configurations. As shown, the planform area to volume relationships of the shapes result in separate curves. These curves are shown intersecting a minimum volume estimate to house the crew, mission equipment, vehicle subsystems, structure and unusable space, which was taken to be essentially equal for either shape. This information indicated a maximum wing loading for the high $L/D$ shape of about 34 psf and around 60 psf for the medium $L/D$.

Several effects of wing loading on aborted launch conditions were also examined. Fig. 36 plots maximum normal load factor and peak dynamic pressure versus wing loading for a typical sub-orbital abort pullout. These data were for the medium $L/D$ vehicle aborting from a Saturn Ib trajectory at the initial conditions of 12,000 fps velocity, 660,000 ft altitude and 0 degrees flight path angle. The load factor variation with wind loading was found to be small. Although dynamic pressure is highly sensitive to wing loading, as shown, the abort peak of around 900 psf (for a 60 psf wing loading) is only about 10% above the normal launch requirement. Typical VTO rocket launch vehicles generate peak dynamic pressures approaching 800 psf at about 30,000 ft, on normal launch trajectories.

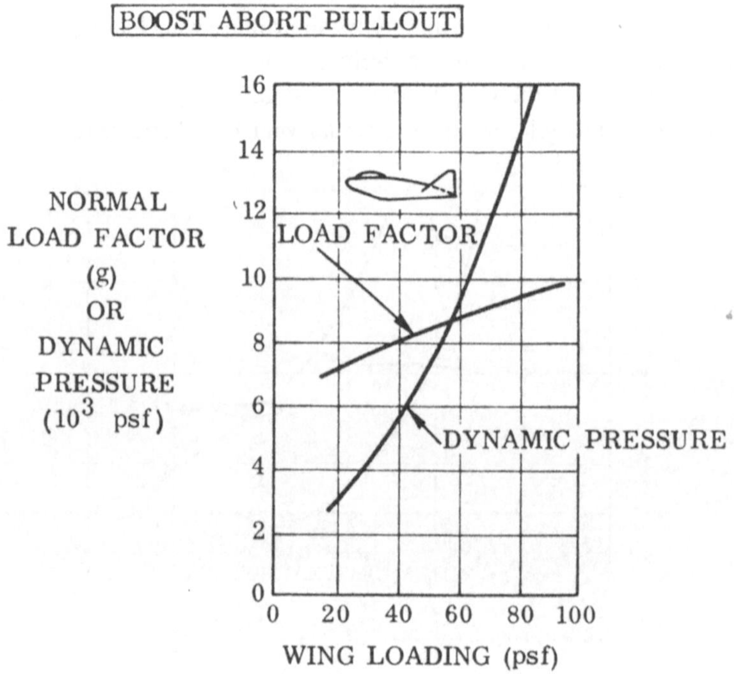


Fig. 10.36 ABORT PARAMETERS (AFTER EULBERG. Ref. 35)

The effect of wing loading on radiation equilibrium temperatures at a point on the lower surface of the two vehicle configurations flying at their respective maximum *L/D* attitudes is shown in Fig. 37. From these estimates, it is shown that the significantly better $C_L$ of the medium *L/D* vehicles is overpowered by the effect of the local angle of attack on the heating rates. In other words, although the high *L/D* vehicle flies at a lower altitude than the medium *L/D*, at the same velocity, the high *L/D* temperatures are lower than the other at all wing loadings, due to the smaller local angle of attack.

Also, shown on Fig. 37 is an assumed material temperature limit for a radiation cooled insulation thermal protection system. This was arbitrarily based on coated columbium cover panel performance degradations above 2500 °F. Ablating type thermal protection systems would not exhibit a similar heating rate or temperature limitation. It appears that if temperature limited radiation cooled thermal protection is used, a wing loading limitation would exist for the lower surface of the vehicles.

A major portion of the inert weight of these vehicles is invested in the thermal protection system. Two basic thermal protection systems have received the most attention for manned entry vehicles, namely radiation cooled, "insulation" systems and ablation cooled systems. In this study, a typical

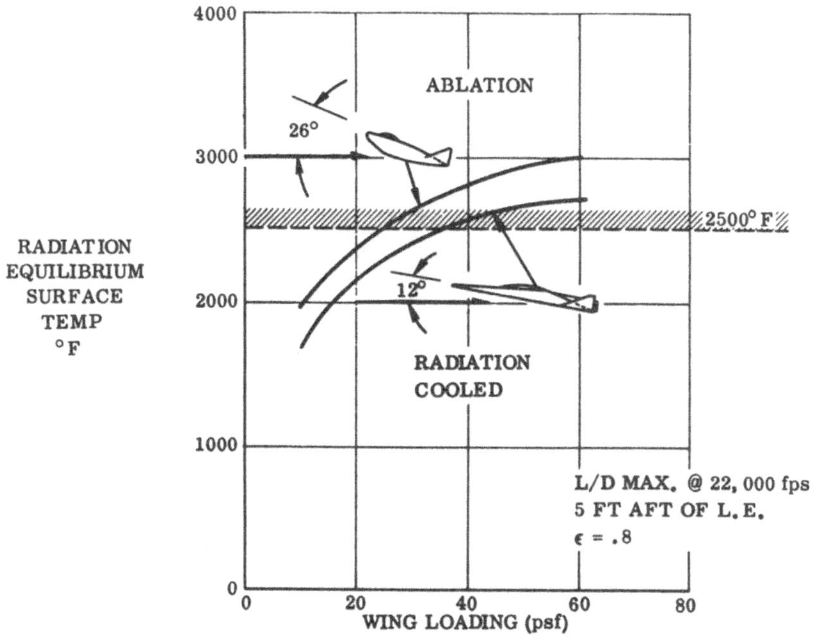


Fig. 10.37 TEMPERATURE vs WING LOADING (AFTER EULBERG, Ref. 35)

concept of each type was assumed for more detailed comparison. The insulation system was based on a coated columbium cover panel on flexible supports, with a silica fibre insulant of 4.5 lbs per cu.ft. density between the cover panel and basic vehicle structure. The ablation system assumed a silicone elastomer ablation material of 45 lbs per cu.ft. density mounted on an aluminum substrate which was in turn mechanically attached to the basic structure. Both systems were considered to be reusable, with minimum refurbishment.

Both vehicles were assumed to use either ablation or insulation on the lower surface and sides. Insulation only was presumed on the upper surfaces. The fins were based on hot nickel alloy structure, without thermal protection as such. The nose caps for both vehicles were assumed to be ceramic, radiation cooled structures. The high $L/D$ vehicle was assumed to have segmented ceramic leading edges, while the medium $L/D$ vehicle utilized the lower surface thermal protection for its sides or "leading edges."

The basic structure was assumed as a conventional aircraft aluminum sheet/stringer/frame semi-monocoque shell. This shell contained the internal pressures and supported the crew and equipment for all launch, entry, and landing loads. The thermal protection system was sized to limit the structural temperature to 200 °F at landing.

Fig. 38 presents the results of the structure plus thermal protection weights variation vs. wing loading for the above assumptions. It can be seen that

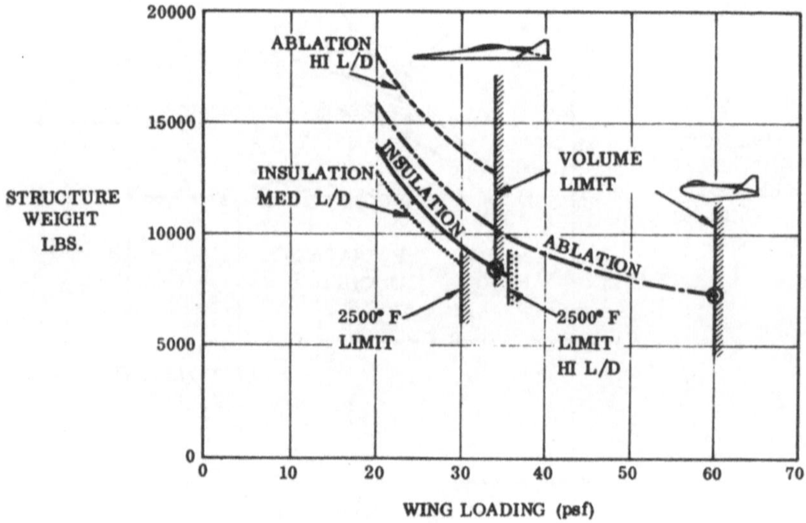


Fig. 10.38 STRUCTURE WEIGHT vs WING LOADING (AFTER EULBERG, Ref. 35)

the high $L/D$ vehicle weight is minimum with insulation at its volume limited wing loading of 34 psf, which is acceptable for the temperature limitation used. The high $L/D$ vehicle weights with ablation are considerably higher due to the longer heating times which yield greater total heat fluxes. The medium $L/D$ vehicle weights with ablation are considerably higher due to the longer heating times which yield greater total heat fluxes. The medium $L/D$ vehicle with insulation would offer attractively low weights; however, the 2500 °F surface temperature limit of the refractory cover panel demands wing loadings of less than about 30 psf. As shown, an improvement in weight is realized for the medium $L/D$ vehicle by using the ablation system at its volume limited wing loading of 60 psf.

The results of the structural weight comparisons indicated a minimum structure weight for the high $L/D$ vehicle of roughly 8600 lbs of which some

60% consisted of thermal protection. The medium $L/D$ vehicle minimum structure weight was 7200 lbs, of which nearly 67% was thermal protection.

Fig. 39 presents a bar chart comparison of the entry vehicle gross weights. The crew and equipment weights are the same for both vehicles, two 225 lb crew members and 4500 lbs of mission equipment. The subsystem weights

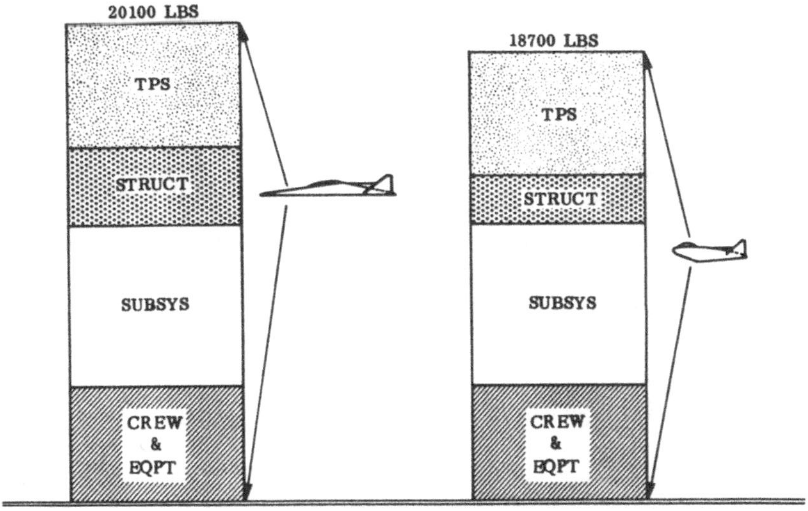


Fig. 10.39 ENTRY VEHICLE WEIGHT COMPARISON (AFTER EULBERG, Ref. 35)

are nearly identical for both vehicles, consisting of items such as auxiliary power, environmental control, furnishings, displays, communications, flight control, guidance and navigation, reaction controls, hydraulics, electrical, landing gear and abort parachute systems. The retro and abort separation systems were assumed to be part of the booster adapter system, with these weights being accounted for there. As shown in the figure, the resulting entry vehicle gross weights are 20,100 lbs for the high $L/D$ and 18,700 lbs for the medium $L/D$.

The design mission, as discussed previously, included quick return to the continental U.S. The high $L/D$ vehicle accomplished the mission with a launched gross weight estimated at 34,700 lbs, while the medium $L/D$ vehicle required nearly 9 times that gross weight because of the excessively large propulsion module required to achieve the cross range.

Several tentative conclusions were indicated, and are worth considering. First, if quick return is a tactical necessity, high $L/D$ vehicle concepts offer the least weight. However, manned military orbital vehicles, with maneuvering and quick return capability, tend to exceed current military expendable booster performance.

## SYMBOLS

$A$ reference area
$C_D$ drag coefficient $=D/\frac{1}{2}\varrho Av^2$
$C_L$ lift coefficient $=L/\frac{1}{2}\varrho Av^2$
$D$ drag force
$g$ acceleration of gravity
$g_0$ conversion from weight to mass
$h$ altitude above the earth's surface
$L$ lift force
$m$ vehicle mass
$N$ distance normal to orbital plane
$r$ radial distance to re-entry vehicle measured from earth's center
$r_0$ earth's radius assuming spherical earth
$S$ lineal distance in the orbital plane
$t$ time
$T$ temperature
$u$ horizontal component of velocity
$v$ velocity
$v_c$ local circular satellite velocity $=\sqrt{gr}$
$v_s$ satellite velocity in elliptical orbit
$w$ vertical component of velocity
$W$ vehicle weight
$y$ lateral range
$\alpha$ angle of attack
$\alpha_i$ orbital inclination to equator
$\beta$ density lapse rate
$\gamma$ flight path angle referred to local horizontal, positive for descending trajectories
$\delta$ $v/v_c^*$
$\Delta v$ velocity increment
$\lambda$ latitude
$\zeta$ $1-r/r^*$
$\eta$ deboost orientation with respect to local horizontal referred to the direction of motion
$\theta$ angular range referred to apogee
$\varrho$ atmospheric density
$\varrho_0$ reference density for atmospheric atmosphere
$\sigma$ density ratio $=e^{-\beta h}$
$\phi$ roll angle
$\psi$ turn angle out of the orbital plane

| | |
|---|---|
| $\omega$ | deboost orientation with respect to orbital plane |
| * | refers to conditions at point of deboost |
| $(\bar{\ })$ | variable normalized with respect to the local circular satellite velocity |
| $(\dot{\ })$ | differentiation with respect to time |

SUBSCRIPTS

| | |
|---|---|
| $A$ | refers to conditions at orbital apogee |
| $E$ | refers to conditions at re-entry |
| eq | equilibrium |
| $f$ | refers to final conditions |
| $i$ | refers to initial conditions |
| $V$ | refers to vertical component |
| $N$ | refers to lateral component |
| $st$ | stall speed |

## REFERENCES


1. Becker, J. V.: Entry vehicles, *Aeronautics and Aerospace Engineering*, **1** (1963).
2. Clapp, R. T.: A Small State-of-the-Art Maneuverable Lifting Re-entry Vehicle, AIAA Paper 65-492, AIAA Second Annual Meeting, San Francisco, California (July 1965).
3. Love, E. S.: Manned lifting entry, *Astronautics and Aeronautics*, **5** (1966).
4. Eulberg, A. C.: A Translation of Manned Military Missions Into an Entry Vehicle Concept, AIAA Paper 65-488, AIAA Second Annual Meeting, San Francisco, California (July 1965).
5. Friedenthal, M. J.: Control of Re-entry From Orbit, *Transactions of the 7th Symposium on Ballistic Missile and Space Technology*, vol. **II** (1962).
6. Galman, B. A.: Some Fundamental Considerations For Lifting Vehicles in Return From Satellite Orbit, *Proceedings of the 4th AFBMD/STL Symposium on Ballistic Missile and Space Technology*, vol. II (Pergamon Press, New York, 1961).
7. Low, G. M.: Nearly Circular Transfer Trajectories for Descending Satellites, NASA TR R-3 (1959).
8. Baker, J. M. and B. E. Baxter: Optimum deboost altitude for specified atmospheric entry angle, *AIAA J.* **1** (1963).
9. Detra, R. W., F. R. Riddell, and P. H. Rose: Controlled recovery of nonlifting satellites, *ARS J.* **30** (1960).
10. Galman, B. A.: Retrorocket alignment for maximum entry angle, *ARS J.* **32** (1962).
11. Esses, H.: Maximum ejection velocity for return from satellite orbits, *ARS J.* **29** (1959).
12. Royce, W.: The Altitude Oscillations of a Vehicle Re-entering the Atmosphere at a Constant Lift Coefficient, Space Technology Laboratories TN STL/TN-60-0000-09006 (14 January 1960).
13. Campbell, G. S.: Long period oscillations during atmospheric re-entry, *ARS J.* **29** (July 1959).
14. Slye, R. E.: An Analytical Method for Studying the Lateral Motion of Atmospheric Entry Vehicles, NASA TN D-325 (September 1960).
15. Love, J. A.: A Drag Monitoring Guidance Scheme for a Lifting Re-entry Vehicle, Space Technology Laboratories TM STL/TM-60-0000-09047 (23 May 1960).
16. Assadourian, A. and D. C. Cheatham: Longitudinal Range Control During the Atmospheric Phase of a Manned Satellite Re-entry, NASA TN D-253 (May 1960).
17. Cheatham, D. C., J. W. Young, and J. M. Eggleston: The Variation and Control of Range Traveled in the Atmosphere by a High-Drag Variable-Lift Entry Vehicle, NASA TN D-230 (March 1960).
18. Eggleston, J. M. and J. W. Young: Trajectory Control for Vehicles Entering the Earth's Atmosphere at Small Flight Path Angles, NASA Memorandum 1-19-59L (February 1959).

19. Eggers, A. J., J. H. Allen, and S. E. Neice, A Comparative Analysis of the Performance of Long-Range Hypersonic Vehicles, NACA TN-4046 (October 1957).
20. Wang, H. E. and R. S. Skulsky: Characteristics of lateral range during constant altitude glide, AIAA J. **1** (1963).
21. Haig, C. R. and L. D. Perlmutter: Use of lateral maneuver to extend the period of daylight landings from orbit, *J. SR*, **2** (1965).
22. Wallace, R. A. and W. A. Gray: Minimum lift-drag ratio required for global landing coverage, *AIAA J.* **1** (1963).
23. Wagner, W. E.: Roll modulation for maximum re-entry lateral range, *J. SR*, **2** (1965).
24. Wang, H. E. and S. T. Chu: Variable lift re-entry at superorbital and orbital speeds, *AIAA J.* **1** (1963).
25. Wang, H. E.: Motion of re-entry vehicles during constant-altitude glide, *AIAA J.* **3** (1965).
26. Loh, W. H. T.: *Dynamics and Thermodynamics of Planetary Entry*, Prentice-Hall, Inc., Englewood Cliffs, New Jersey, 1963.
27. Volgenan, E.: *Boost Glide and Re-entry Guidance and Control*, C. T. Leondes, Ed., McGraw-Hill Book Company, Chap. 10.
24. Boyle, E. J.: Recall and Return of a Manned Vehicle From Satellite Orbit, *Proceedings of Symposium on Space Rendezvous Rescue and Recovery*, vol. **16**, (September, 1963 Advances in the Astronautical Sciences, Publication of American Astronautical Society).
29. Eggers, A. J. and T. J. Wong: Motion and heating of lifting vehicles during atmospheric entry, *ARS J.* **31** (1961).
30. Shen, Y. C.: Series solution of equations of re-entry vehicles with variable lift and drag coefficients, *AIAA J.* **1** (1963).
31. Allen, H. J. and A. J. Eggers: A study of the Motion and Aerodynamic Heating of Missiles Entering the Earth's Atmosphere at High Supersonic Speeds, NACA TN-4047 (October 1957).
32. Love, E. S.: Factors influencing configuration and performance of multi-purpose manned entry vehicles, *J. SR.*, **1** (1964).
33. Spielberg, I. N. and C. B. Cohen: Janus: A manned orbital spacecraft with staged re-entry, *J. SR.*, **2** (1965).
34. Love, E. S. and E. B. Pritchard: A Look at Manned Entry at Circular to Hyperbolic Velocities, *Proceedings of the Second Manned Space Flight Meeting*, Dallas, Texas (1963).
35. Eulberg, A. C.: A Translation of Manned Military Missions Into an Entry Vehicle Concept, AIAA Paper 65-488, AIAA Second Annual Meeting, San Francisco, California (July 1965).
36. Grahane, W. E.: Determination of Re-entry Performance of the M-2 Lifting Body Based on Aerodynamic Heating Considerations, *Transactions of the 9th Symposium on Ballistic Missile and Space Technology*, vol. II (1964).
37. Syvertson, C. A., et. al.: Some Considerations of the Performance of a Maneuverable, Lifting-Body Vehicle, *Proceedings of Symposium on Space*

*Rendezvous Rescue and Recovery*, vol. 16, (September 1963 *Advances in the Astronautical Sciences*, Publication of American Astronautical Society).

38. Mathauser, E. E.: Research, Design Considerations, and Technological Problems of Structure for Winged Aerospace Vehicles, *Proceedings of the NASA-University Conf. on the Science and Technology of Space Exploration*, vol. 2 (Chicago, Illinois, 1962).
39. Baradell, D. L. and C. H. McLellan: Lateral Range and Hypersonic Lift Drag Ratio Requirements for Efficient Ferry Service from a Near Earth Manned Space Station. Paper presented at AIAA Second Manned Space Flight Meeting, Dallas, Texas (1963).
40. Nyland, F. S.: The Synergetic Plane Change for Orbiting Spacecraft, RAND Memorandum RM-3231-PR, Santa Monica, California (August 1962).
41. Bruce, R. W.: The Combined Aerodynamic Propulsive Orbital Plane Change Maneuver, Aerospace Corporation Report SSD-TDR-64-98, El Segundo, California (June 1964).

# 11

# *Entry from Lunar and Planetary Missions**

JOHN F. McCARTHY, Jr., Vice President
*Research, Engineering and Test*
*Space Division*
*North American Rockwell Corp.*

## [11-1] INTRODUCTION

In recent years, space technology has progressed from that required for small unmanned earth-orbital satellites and manned earth-orbital spacecraft to unmanned planetary probes and unmanned lunar-landing devices. Firm national commitments have been made for a manned lunar landing within the next several years. With the completion of the design, ground test, and preliminary flight test of the Apollo spacecraft, new emphasis is being placed on missions leading to greater utility of earth-orbital systems. Of particular interest are increased payload, extended mission duration, and more operational flexibility. Concurrently, extensive effort is being expended to examine the factors involved in extending the initial lunar-landing capability of early Apollo flights to lunar exploration, planetary flyby, and, finally, planetary landing. It is with the latter missions that this chapter is concerned, since these missions may require the use of new design concepts. In defining the problems associated with flights to the planets and developing their solutions, it is useful to exploit the technology already gained from contemporary earth-orbital entry vehicles.

### [11-1.1] Entry-Vehicle Evolution

One of the best indicators of the flight performance of an entry vehicle is the ratio of the aerodynamic lift to drag ($L/D$), since longitudinal range is almost directly proportional to $L/D$, lateral range is approximately propor-

* The author gratefully acknowledges the assistance of Mr. G. M. Hanley, whose many discussions and critiques contributed significantly to the preparation of this chapter.

tional to $(L/D)^2$, and entry corridor depth, which will be discussed later, is significantly influenced by $L/D$. The evolution of entry-vehicle performance, as characterized by $L/D$, is shown in Fig. 11-1 as a function of time.

In the early era of ballistic nose-cone development, the main emphasis was on survival of the nose cone when subjected to what was then considered a severe thermal environment. Investigation of this survival problem led to a comprehensive understanding of the high-speed convective heat-transfer process and to the development of effective ablative heat-protection schemes.

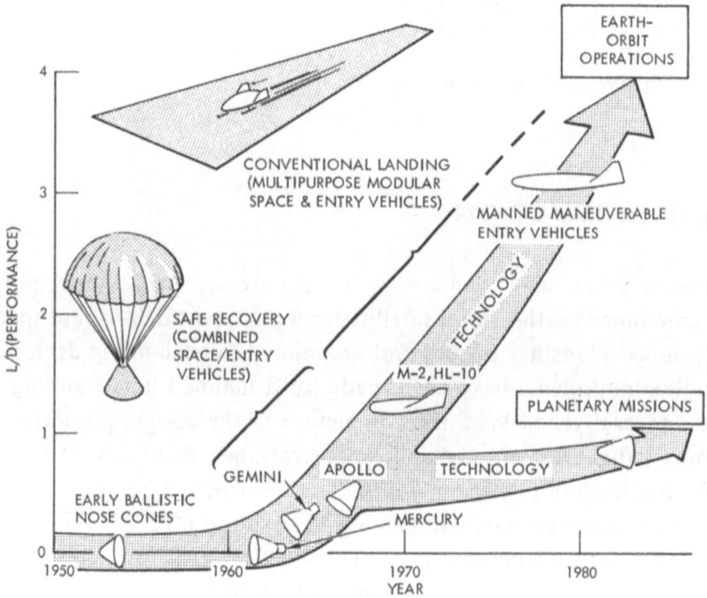


**Fig. 11-1.** Entry-vehicle evolution.

Nose-cone technology has played a significant role in the development of the Mercury and Gemini manned earth-entry systems, and is making a substantial impact on the development of the Apollo entry system. For these relatively low-performance manned systems, which feature a combined space and entry vehicle in capsule-like form, the major operational emphasis is on safe recovery, with little consideration given to reusability. Beyond Apollo, the trend in performance is expanding in two directions: (1) toward higher $L/D$ (greater than 0.5) for increased flexibility of earth-orbital operations; (2) toward higher entry velocities (greater than 36,000 feet per second) for return from planetary missions (initially from Mars and Venus).

For earth-orbital operations, an increasing mission frequency may lead to the development of higher-performance systems which achieve operational flexibility by their ability to land in a more conventional sense, by design which allows for multipurpose operation, and through use of the concept of reusability. Such vehicles will be used in operations related, for example, to the resupply of space stations, space rescue, and satellite repair. These missions allow for separate entry and mission modules, so that the design of the entry vehicle need no longer be penalized by mission considerations. One of the major problems encountered in this development area is the identification of realistic performance requirements for future systems and the design of a basic reusable entry vehicle. In the design of entry vehicles for these missions, unlike those for the planetary mission branch of expanding performance (Fig. 11-1), emphasis is being placed on the refinement of current concepts. Steps in this direction are characterized by the M-2 and HL-10 lifting-body entry vehicles, which possess hypersonic lift-to-drag ratios of about 1.2 and improved horizontal landing characteristics.

Since earth-entry configurations for lunar and planetary missions are not currently constrained by a high operational frequency, the emphasis is on safe recovery following entry through an extremely severe, and yet uncharted, thermodynamic environment. Furthermore, there is no need for lift-to-drag ratios greater than approximately 0.5, since the space trajectory can be modified with very little expenditure of propellant to reach virtually any desired landing site. In the subsequent sections, the problems related to safe recovery of these systems are discussed in detail, and means are illustrated by which the current technology may be successfully employed to solve these problems.

### [11-1.2] Design Interaction

Major considerations that influence the final design of entry vehicles and their mission profiles are illustrated in Fig. 11-2, which has been separated into two major segments – the mission and the system, both of which interact in the development of mutual compatibility. Since any space mission is subject to certain inherent restrictions imposed by the type of mission and the technology available, it is useful to identify and categorize some of the more significant constraints that affect the design and selection of the entry system. Of these, two broad categories can be identified, each having a direct effect upon one or more elements of design. These categories are (1) operational constraints which primarily influence the overall flight planning through the trajectory and performance parameters, and are generally specified by the customer at the time the mission is defined; and (2) systems constraints which primarily arise from state-of-the-art limitations, cost, or requirements for compatibility among subsystems and which tend to directly influence the vehicle design parameters.

Fig. 11-2 shows the interaction of the two classes of operational and systems constraints, with trade-off considerations between design and performance indicated. The design parameters are those quantities related to the vehicle itself and subject to some control by the designer. Although only the major aerodynamic characteristics are shown in the illustration, many other parameters become involved as design progresses. The trajectory and performance parameters include those quantities which characterize the flight profile of the vehicle, and are generally obtained from a knowledge of the design parameters through solutions of the equations of motion.

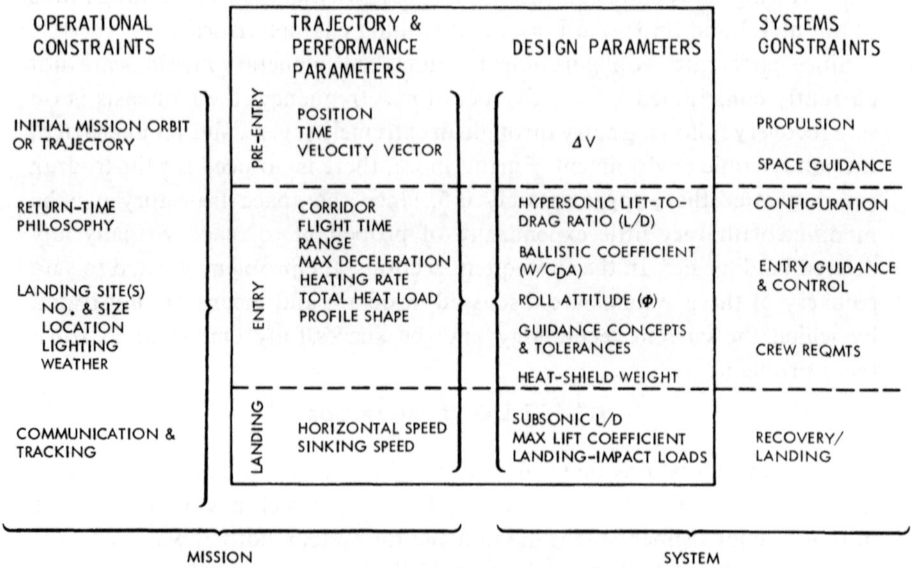


**Fig. 11-2.** Manned entry vehicle design interactions.

Three general flight phases are also identified in Fig. 11-2: (1) pre-entry, or space flight; (2) entry, or atmospheric flight; and (3) landing. While many published parametric studies have been concerned with only one of these phases in considering the influences of operational and system constraints, all three phases must be integrated to ensure overall mission and system compatibility. This amalgamation is particularly true for the lunar mission, where the integration of the pre-entry and entry flight phases dominates the entire selection process for the entry trajectory.

The interaction among flight phases can be better understood by considering some of the elements of a lunar or planetary mission. The entire entry phase of flight may be regarded as a two-point boundary-value problem for which the initial conditions are set by the original mission trajectory plus any propulsive velocity change used to alter it and for which the terminal conditions are dictated by the location of the selected landing site on earth. In fact, the entry sequence begins upon departure from the initial mission trajectory on a conic which is designed to produce an atmospheric penetration with selected position and velocity vectors relative to the earth at a point in time. There are, therefore, trade-offs between the pre-entry space trajectory and post-entry atmospheric flight profile, as well as separate trade-offs within each phase. From an operational point of view, therefore, it is not surprising that a detailed consideration of those factors constraining the entry phase should reveal items which influence the initial and terminal points. The initial mission trajectory being flown and the time of departure from this trajectory relate to the initial conditions, while the earth-landing site and earth-based communications and tracking requirements relate to the terminal conditions. Operational considerations associated with return-time philosophy, such as mission duration, could also constrain the end points.

The systems constraints, which are primarily dictated by the state of the art, cost, and subsystems compatibility, have a major impact on design of the entry configuration, and, consequently, have an impact upon the achievable performance. Some of the important considerations are shown in Fig. 11-2. For example, current state of the art requires the use of storable propellants for mission durations (about two weeks) characterized by the manned lunar-landing mission, although the use of cryogenic propellants to improve the future performance of space propulsion systems is being considered. The ability of the space guidance system to accurately determine position and velocity also influences the amount of propulsive correction necessary to meet the requirements for position accuracy at the interface of space flight and atmospheric entry.

The entry portion of flight is constrained by vehicle configuration, entry guidance and control, and crew requirements. As used herein, configuration implies more than the superficial constraints on shape; in this context, the broader implications of materials, structural elements, and manufacturing capabilities are examples of configuration constraints. In the usual situation, design constraints on the configuration would include specifications in terms of maximum weights and/or dimensions for booster compatibility and required volumes for specific payloads. Although extremely complex, the design process in which these constraints are satisfied while the entire system is optimized is receiving extensive attention, particularly for high $L/D$ entry vehicles. Perhaps the most critical element in this process is the manner in which the geometrical variables of the configuration are manipulated to produce the desired aerodynamic performance and stability characteristics while providing efficient payload packaging and maintaining structural and heat-protection integrity.

The magnitude of the entry velocity, combined with expected trajectory deviations prior to entry and limitations on crew acceleration during entry, necessitate some form of flight-path control. Further considerations of crew safety, rescue, and retrieval problems lead to the requirement that the entry vehicle be capable of controlling the trajectory to a preselected landing site. The guidance mechanism senses parameters which determine the flight path, such as acceleration and velocity, and compares them with pre-established tolerances. For Apollo-shaped vehicles, trajectory control can be accomplished by modulating the vehicle's rotation, and consequently the lift vector, with respect to the local vertical. This control technique, commonly known as roll modulation, requires a vehicle which is neutrally stable in roll and which changes and holds roll position with reaction control. Although this method of control is effective and conceptually simple, the sensitivity to center-of-gravity location and the relative lack of mission and payload flexibility may suggest that future, higher $L/D$ vehicles will employ techniques of aerodynamic control approximating those used on conventional aircraft.

The presence of a crew imposes an additional constraint on the entire trade-off process, particularly because of tolerance limits on human deceleration. Peak deceleration not exceeding 10 $g$'s and sustained deceleration not exceeding 5 $g$'s are typical limits. These peak and sustained deceleration limits define, respectively, the ultimate limit of the undershoot boundary of the entry corridor and the minimum entry range. Providing proper restraining devices, space, and visibility for the crew within the vehicle is also an important constraining factor on certain configurations where the requirements to accommodate men dictate the minimum dimensions of the vehicle.

Although extensive work is in progress to develop entry vehicles capable of horizontal landing in a more conventional aircraft sense, current entry vehicles use separate recovery and landing subsystems. Compromises in the hypersonic performance of many high $L/D$ vehicles necessary to obtain adequate low-speed flying qualities both typify the magnitude of constraint that this factor represents and suggest that the use of separate recovery or landing subsystems may continue to have a place in the design of entry systems. For Apollo, the terminal descent is controlled by parachute, and an internal shock-attenuation system is used to control the landing-impact loads. This system was selected on the basis of development schedule constraints and to satisfy low weight and volume requirements. The landing system influences the entry flight phase because of the need to determine areas on the earth's surface compatible with the technique chosen for terminal descent.

In the following discussion, attention will be focused on those interacting performance and design parameters which most significantly influence successful earth entry from lunar and Mars and Venus missions. These parameters are the entry corridor, range, thermodynamic environment (heating rate and heat load), and the weight and material characteristics of the heat shield. The influence of $L/D$ and the ballistic coefficient $(W/C_DA)$ on entry performance will also be discussed.

### [11-1.3] Launch-Vehicle Constraints

Historically, the manned space program has been paced by the development of launch vehicles. However, with the advent of the Saturn series of launch vehicles and the utilization of multiple launches and spacecraft assembly in earth orbit, sufficient payloads can be launched to allow feasible manned planetary missions. Fig. 11-3a shows the injected payload of the Saturn launch vehicle for three versions of the booster. The injected payload is plotted versus the injection-velocity increment from a 100-nautical-mile altitude in earth orbit. The bar graph in Fig. 11-3b shows the number of launches required to deliver sufficient payload and propellants into orbit to

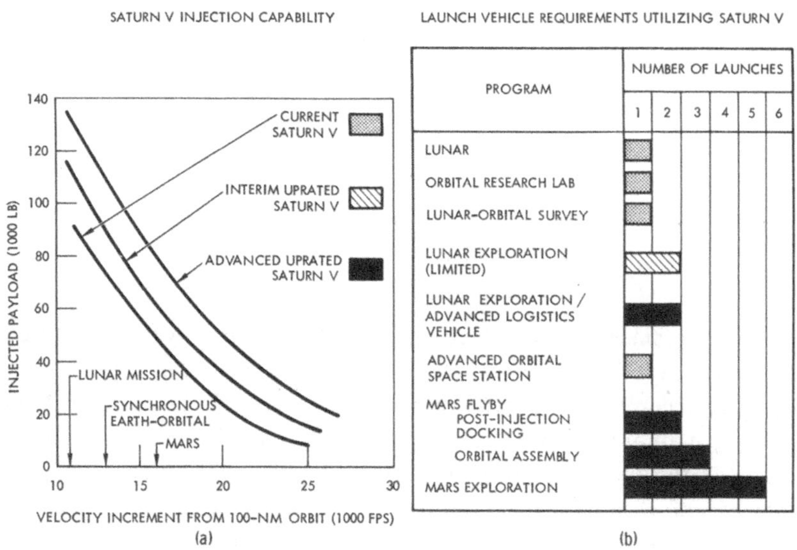


**Fig. 11-3.** Launch vehicles.

accomplish the indicated missions. For example, a Mars flyby mission, assuming assembly in earth orbit and using the Saturn V launch vehicle, requires three launches of an advanced uprated Saturn V. Since the current Saturn V launch vehicle is designed specifically for the Apollo lunar mission, only one launch is needed to fulfill mission requirements.

### [11-1.4] Earth-Entry Velocity Constraints

The expected increase in launch-vehicle capability tends to shift the emphasis to other developmental problems, one of which is the earth-entry vehicle. Fig. 11-4 shows the earth-entry velocities associated with a return from Mars for coasting-type trajectories. The entry velocity becomes a function of the

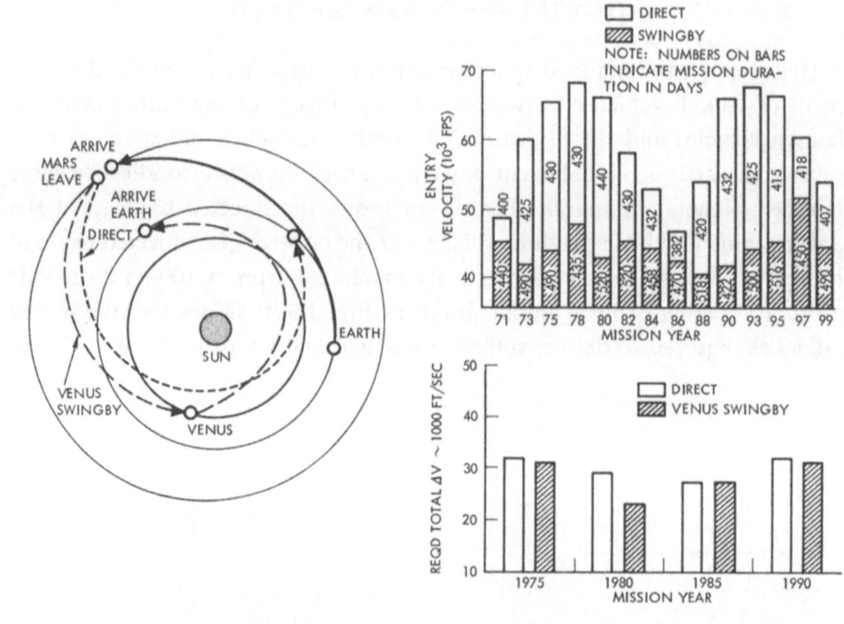


**Fig. 11-4.** Mars-Venus swingby (Ref. 1).

return mode, since substantial reductions can be realized by utilizing a Venus encounter (Ref. 1). In this maneuver, the gravitational field of Venus is used to alter the direction of the velocity vector of the spacecraft as it passes by Venus, resulting in a more tangential approach to earth. The advantages of the Venus-swingby return mode can be demonstrated for the 1975 mission, which is a typically unfavorable year for a direct return. In the direct-return mode, the spacecraft passes inside the orbit of Venus to effect the rendezvous with earth at a resultant entry velocity of 65,000 feet per second. By adjusting the return trajectory slightly, the spacecraft can be made to rendezvous with Venus; and if a darkside passage is made at an altitude of about 2000 miles, the heliocentric velocity of the spacecraft is reduced by about 15,000 feet per second. The resultant Venus-to-earth trajectory approaches earth at a lower heliocentric angle, so that the entry velocity is reduced to 44,000 feet per second. During those opportunities when the spacecraft would not normally pass close to Venus during the return trip, the spacecraft could be parked in Mars orbit for relatively short periods of time to await a favorable rendezvous with Venus. Since the angular rate of travel of Venus is high compared with that of Mars, the waiting time would normally be small. With these modes, a comparison of mission time, in days, can be obtained by referring to the numbers inside the bars that indicate the entry velocity. A nominal ten-day stay on Mars has been assumed in the calculations. In addition to reducing the earth-entry velocity, use of the Venus swingby can, in some cases, also reduce the propulsive velocity increment required to return to earth from a Mars orbit.

As can be seen from Fig. 11-4, the earth-entry velocities associated with Mars missions, even if a Venus encounter were used, range up to 50,000 feet per second. The earth-entry velocity corresponding to a lunar mission is approximately 36,000 feet per second. As a result of the large increase in entry velocity for planetary missions, compared with earth-orbital and lunar missions, it may be anticipated that the thermodynamic environment and heat-protection problems during entry will be more severe and that the problems associated with entry flight mechanics will increase significantly.

The following sections are devoted to a discussion of the entry flight profile, thermodynamic environment, and characteristics of heat-protection systems for the Apollo lunar mission, and comparisons are made with similar characteristics associated with earth entry from planetary missions. In particular, means are developed by which the existing technology may be employed to adequately accomplish earth entry for these missions.

## [11-2] ENTRY FLIGHT PROFILE CHARACTERISTICS

In general, the entry sequence is characterized by three distinct phases, each with its own set of constraints: initial entry, range control, and recovery/landing. Initial entry is assumed to occur at an altitude of 400,000 feet, where aerodynamic forces are just beginning to play a significant role in the motion of the vehicle. There are two major constraints on the flight-path angle at this altitude for a given entry velocity: ability to stay within the atmosphere, on the one hand; and the maximum deceleration that can be tolerated by the crew, on the other. These limits define the depth of the entry corridor.

During the range-control portion of the trajectory, the aerodynamic forces are modulated to effect the lateral and longitudinal maneuvering required to reach the recovery area within reasonable limits of deceleration and heating. The final phase, recovery or landing, is initiated at slightly supersonic Mach numbers after major range control has been accomplished. Major problems encountered during this phase are the dynamics caused by activation of the landing system and the impact at the recovery area. In this section, emphasis will be placed on the initial entry and range-control phases.

### [11-2.1] Equations of Motion

During entry, motion of the vehicle is governed by the forces resulting from the earth's gravitational attraction and the aerodynamic forces arising from flight in an atmosphere. Fig. 11-5 illustrates the forces acting on an entry vehicle for motion in the pitch plane, which is the plane passing through the center of the earth and containing the velocity vector. If a spherical, non-rotating earth is assumed, a summation of forces in the directions normal and tangential to the flight path results in the following two equations of

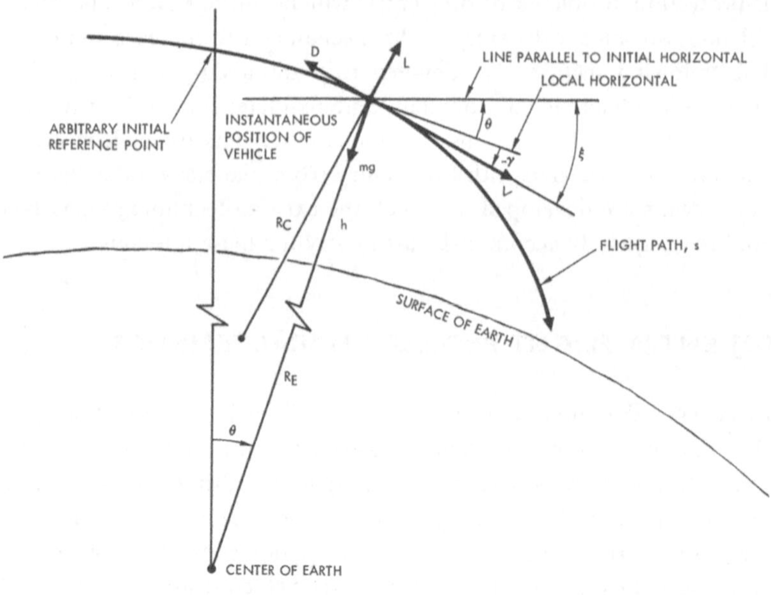


**Fig. 11-5.** Force diagram.

motion:

$$L - mg\cos\gamma = -\frac{mV^2}{R_C} \tag{11–1}$$

$$D + mg\sin\gamma = -m\frac{dV}{dt} \tag{11–2}$$

where

$L$ = lift
$D$ = drag
$V$ = velocity
$m$ = mass
$g$ = gravity
$\gamma$ = flight-path angle
$t$ = time
$R_C$ = radius of curvature

Since

$$\text{flight path curvature} = \frac{1}{R_C} = \frac{d\xi}{ds} = \frac{d(\theta - \gamma)}{ds}$$

$$\frac{d\theta}{dt} = \frac{V\cos\gamma}{R_E + h}$$

and

$$\frac{d\theta}{ds} = \frac{d\theta}{V\,dt} = \frac{\cos\gamma}{R_E + h}$$

where

$s$ = flight-path distance
$R_E$ = earth radius
$h$ = altitude

the radius of curvature can be eliminated from Eq. (11–1) to yield

$$L - mg\cos\gamma = -mV^2\left(\frac{\cos\gamma}{r} - \frac{1}{V}\frac{d\gamma}{dt}\right) \tag{11–3}$$

where

$$r = R_E + h$$

By employing the usual aerodynamic terms for the lift and drag forces, i.e.,

$$L \equiv C_L\frac{\varrho V^2}{2}A$$

$$D \equiv C_D\frac{\varrho V^2}{2}A$$

where

$C_L$ = lift coefficient
$C_D$ = drag coefficient
$\varrho$ = atmospheric density
$A$ = reference area (usually vehicle planform or base area)

Eqs. (11–3) and (11–2), respectively, can be expressed as

$$\frac{L}{D}\frac{C_D A}{W}\frac{\varrho V^2}{2} - \cos\gamma = -\frac{V^2}{g}\left(\frac{\cos\gamma}{r} - \frac{1}{V}\frac{d\gamma}{dt}\right) \tag{11–4}$$

$$\frac{C_D A}{W}\frac{\varrho V^2}{2} + \sin\gamma = -\frac{dV}{g\,dt} \tag{11–5}$$

after division by $W = mg$

For first-order computations, it is sufficient to assume that $g$ is constant and $h$ is negligible when compared with the radius of the earth, so that $r \cong R_E$. Solution of Eqs. (11–4) and (11–5) requires two additional relationships involving $\varrho$ and $h$. In general, for spherically symmetric atmospheres

$$\varrho = f(h) \tag{11–6}$$

and the altitude can be related to the flight-path angle, time, and velocity by

$$\frac{dh}{dt} = V\sin\gamma \tag{11–7}$$

Eqs. (11–4) through (11–7) constitute a set of simultaneous differential equations which, in general, are amenable only to computer solutions. The aerodynamic parameters, i.e., lift-to-drag ratio $L/D$ and ballistic coefficient $W/C_D A$, may be considered as forcing functions in these equations, whose values may be specified as a function of variables such as altitude, velocity, Mach number, and time.

As part of the early efforts in ballistic research, Allen and Eggers (Ref. 2) achieved a simple, but very significant solution to the equations of motion by assuming that the earth's atmosphere is isothermal and the density can be approximated by the expression $\varrho = \varrho_{SL} e^{-\beta h}$, that $L/D = 0$ (ballistic flight), that the flight-path angle, $\gamma$, is constant over the portions of the trajectory where significant heating and deceleration occur, and that $|\sin\gamma| \ll C_D A \varrho V^2/2W$. As a result of these assumptions, Eq. (11–4) is eliminated and Eq. (11-5) may be solved directly with the aid of Eq. (11–7).

The resultant solution is

$$\ln\frac{V^2}{V_i^2} = \frac{g(\varrho - \varrho_i)}{\left(\frac{W}{C_D A}\right)\beta\sin\gamma_i} \tag{11–8}$$

where

$\varrho \quad = \varrho_{SL} e^{-\beta h}$

$\varrho_i \quad = \varrho_{SL} e^{-\beta h_i}$

$\varrho_{SL}$ = density at sea level

$(\ )_i$ = refers to initial condition

$\beta \quad$ = scale height

Eq. (11–8) may be used to calculate the altitudes where maximum deceleration and heating occur. It is interesting to note that the ballistic coefficient, $W/C_D A$, is a scaling parameter that causes a proportional change in density at which these events occur. The ballistic coefficient, however, does not influence the peak deceleration, which is determined only as a function of the initial velocity and flight-path angle.

Another meaningful approximate solution to the equations of motion, called equilibrium glide, can be obtained by assuming that

$$|\gamma| \ll 1$$

$$\left|\frac{d\gamma}{dt}\right| \ll 1$$

Under these assumptions, $\cos\gamma \cong 1$, $\sin\gamma \cong 0$, and $d\gamma/dt \cong 0$, and Eq. (11–4) and (11–5) can be written as

$$\frac{L}{D} \frac{C_D A}{W} \frac{\varrho V^2}{2} = \left(1 - \frac{V^2}{V_C^2}\right) \tag{11–9}$$

$$\frac{C_D A}{W} \frac{\varrho V^2}{2} = -\frac{1}{g} \frac{dV}{dt} \tag{11–10}$$

where $V_C = \sqrt{g R_E}$ is the earth circular-orbit velocity. These assumptions result in a flight condition in which the vehicle instantaneously is assumed essentially to be in a circular orbit around the earth, and the defect in centrifugal force is compensated by the aerodynamic lift, as implied in Eq. (11–9).

Eq. (11–9) can be immediately solved for density, giving

$$\varrho = \frac{2D}{V^2 L} \frac{W}{C_D A} \left(1 - \frac{V^2}{V_C^2}\right) \tag{11–11}$$

As in the ballistic case, the density is proportional to $W/C_D A$ for a given $L/D$.

An expression for determining the flight time during equilibrium glide can be derived by combining Eqs. (11–9) and (11–10) to obtain

$$\frac{D}{L}\left(1 - \frac{V^2}{V_C^2}\right) = -\frac{1}{g} \frac{dV}{dt} \tag{11–12}$$

Solving Eq. (11–12) for the flight time results in

$$t = \frac{V_C L}{2gD} \ln \left\{ \left( \frac{1 - \frac{V}{V_C}}{1 + \frac{V}{V_C}} \right) \left( \frac{1 + \frac{V_i}{V_C}}{1 - \frac{V_i}{V_C}} \right) \right\} \tag{11–13}$$

Note that the flight time is directly proportional to $L/D$.

The range angle, $\theta$ (Fig. 11-5), can be determined by substituting $\mathrm{d}t$ from Eq. (11–12) into

$$\frac{\mathrm{d}\theta}{\mathrm{d}t} = \frac{V \cos\gamma}{R_E} \cong \frac{V}{R_E}$$

to obtain the expression

$$\mathrm{d}\theta = \frac{L}{2D} \frac{\mathrm{d}\left(\frac{V^2}{V_C^2}\right)}{\left(1 - \frac{V^2}{V_C^2}\right)} \tag{11–14}$$

Integration of Eq. (11–14) yields

$$\theta = \frac{L}{2D} \ln \left\{ \frac{\left(1 - \frac{V^2}{V_C^2}\right)}{\left(1 - \frac{V_i^2}{V_C^2}\right)} \right\}$$

where $\theta = 0$ when $V = V_i$. Like the flight time, the range angle is also proportional to $L/D$.

A point of interest is the influence, in an uncoupled manner, of $W/C_D A$ and $L/D$ on the flight of the entry vehicle. The influence on performance due to $W/C_D A$ is to scale the altitude of flight, whereas the range and flight time are affected by $L/D$.

Consider now the case of flight at constant altitude ($\varrho = 0$, $\gamma$ = constant, and $\mathrm{d}\gamma/\mathrm{d}t = 0$). Of interest are expressions giving the requirements for roll-angle modulation to maintain the desired flight profile, and the flight time. The assumption of a constant-altitude deceleration with $L/D$ variable and $W/C_D A$ constant may be realized when roll modulation of the lift vector is employed, which, as indicated earlier, is the flight mode used for the Apollo command module. In this case, the vehicle angle of attack is held constant, and the lift in the pitch plane is modulated by rolling about the velocity vector, resulting in a constant value of $C_D$.

Under this restriction, $L/D$ varies as

$$\frac{L}{D} = \left(\frac{L}{D}\right)_{\phi=0} \cos\phi \tag{11–15}$$

where

$\phi$ = roll angle

The required roll angle, $\phi$, is determined from Eqs. (11–4) and (11–15):

$$\cos\phi = \frac{2}{\dfrac{C_D A}{W}\left(\dfrac{L}{D}\right)_{\phi=0} \varrho V^2}\left(1 - \frac{V^2}{V_C^2}\right) \tag{11–16}$$

Eq. (11–5) can be rewritten as

$$\frac{\mathrm{d}\left(\dfrac{V}{V_C}\right)}{\dfrac{V^2}{V_C^2}} = -\frac{C_D A g \varrho V_C}{2W}\,\mathrm{d}t \tag{11–17}$$

Integration of Eq. (11–17) yields the following expression for the time of flight, $\Delta t$, in the constant-altitude mode:

$$\Delta t = \frac{2W}{V_C C_D A g \varrho}\left(\frac{V_C}{V} - \frac{V_C}{V_i}\right) \tag{11–18}$$

The manner in which the above set of relationships can be employed to obtain significant results related to earth entry from lunar and plantary missions will be discussed in the following sections.

### [11-2.2] Entry Corridor

The concept of an entry corridor, established by Chapman (Ref. 3), provides the link between guidance requirements and entry-vehicle capability. The entry-corridor depth, illustrated in Fig. 11-6, is defined as the distance between the perigees of two vacuum trajectories (conics) that are established by placing certain constraints on the entry trajectories resulting from the vacuum trajectories. The conic having the highest perigee altitude is defined as the overshoot conic, and is usually determined by the constraint that the entry vehicle be just capable of staying in the atmosphere (escape-limited) when utilizing full negative lift. The conic having the lowest perigee is defined as the undershoot conic, and is determined by a maximum $g$-limitation during atmospheric entry with full positive lift.

The implications of these constraints are illustrated on the altitude-velocity

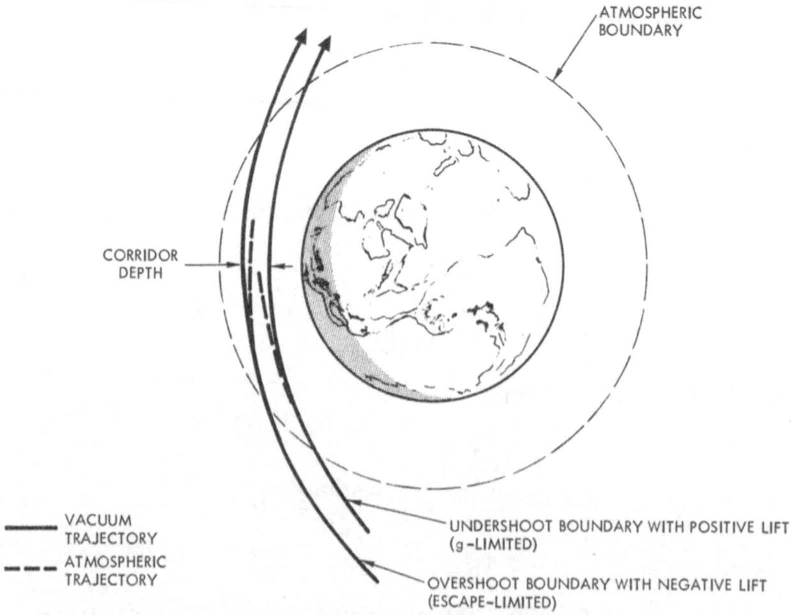


**Fig. 11-6.** Definition of entry corridor.

plot shown in Fig. 11-7a. A vehicle entering the atmosphere at supercircular velocities, $V/V_C > 1$, along an overshoot trajectory is required to achieve a flight condition yielding tangency with the line $L/W = [1 - (V^2/V_C^2)]$. Negative lift is required when $V/V_C > 1$ in order to satisfy the no-escape condition at sufficiently high supercircular velocities. Entry along the undershoot boundary requires that the entry trajectory achieve a condition of tangency

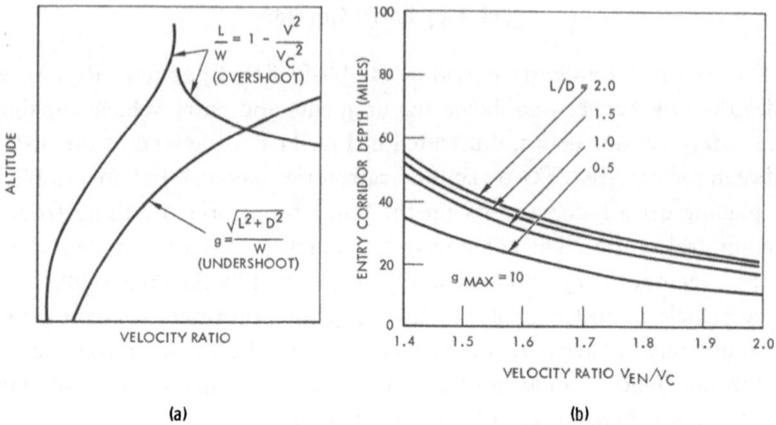


**Fig. 11-7.** Effect of entry velocity on corridor.

with the imposed maximum load-factor line defined by

$$\frac{\sqrt{L^2 + D^2}}{W} = g$$

The result of applying these constraints on the corridor depth is illustrated in Fig. 11-7b for a 10-$g$ load factor as a function of $L/D$ and entry velocity relative to the earth-orbital velocity. As shown in Fig. 11-7b, the corridor depth decreases with increasing entry velocity. Increasing the maximum $L/D$ of the entry vehicle from 0.5 to 1.0 significantly increases the entry corridor depth. Beyond this point, increases in $L/D$ have an insignificant influence on corridor depth. The Apollo command module, with a maximum usable $L/D$ of 0.5, has a corridor depth of about 35 miles for entry from lunar missions ($V_{EN}/V_C = \sqrt{2}$). For a 52,000-foot-per-second entry velocity ($V_{EN}/V_C = 2.0$), the corridor depth is reduced to 10 miles. Results of recent planetary mission studies indicate that a 10-mile corridor is achievable with currently available guidance concepts.

### [11-2.3] Range Control

Once safe capture by the atmosphere is achieved by entry within the corridor limits, the aerodynamic forces must be modulated to allow landing and recovery at some predesignated area. A typical entry trajectory, shown in Fig. 11-8, indicates some of the possible range-modulation schemes. The

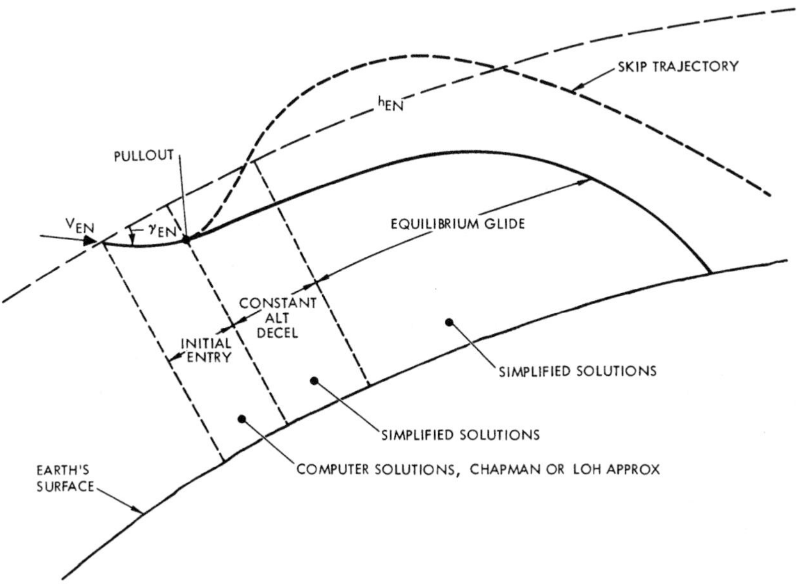


**Fig. 11-8.** Typical entry trajectory.

first portion of the entry trajectory, initial entry, is initiated at an altitude (400,000 feet) where aerodynamic forces begin to influence vehicle motion and terminates when pullout ($\gamma=0$) occurs. This portion of the trajectory, which is related to the entry corridor previously discussed, requires the use of a computer to obtain accurate solutions, although the approximate solutions of Chapman (Ref. 4) or Loh (Ref. 5) may be employed to obtain preliminary flight characteristics. After pullout, the range-modulation portion of the entry trajectory is initiated. Possible means by which this modulation can be accomplished are illustrated in Fig. 11-8. One possibility is to employ a constant altitude deceleration until the range required can be satisfied by equilibrium gliding flight. These conditions can be readily computed by use of the relationships presented during the discussion of the equations of motion. Another possible range-modulation scheme is a skip-trajectory mode. After pullout from the initial entry portion of the trajectory and a reduction in velocity to nearly orbital velocity, sufficient lift is applied to eject the vehicle from the sensible atmosphere, thus achieving ballistic flight, followed by a subsequent re-entry. Theoretically, very large longitudinal ranges may be achieved by use of this technique for entry from planetary or lunar missions. However, since these ranges are sensitive to the velocity and flight-path angle at ejection, use of the very long-range skip trajectories is limited by the capability of the guidance and control system to sense the small ejection flight-path angles required for the long-range trajectories.

In addition to control of longitudinal range, the requirement exists for controlling lateral range in the event that the landing area is not in the initial flight plane. Lateral maneuvering techniques are illustrated in Fig. 11-9. Flight out of the initial plane is achieved by directing a portion of the lift force in the desired out-of-plane direction by rolling the vehicle about the resultant velocity vector. Maximum lateral ranges are obtained by employing large roll angles early in the trajectory, followed by a reduced roll-angle schedule as the desired heading is achieved. A mapping of the maximum achievable longitudinal and lateral range boundaries is referred to as a footprint, which represents the region where recovery can be effected.

The range potential of a vehicle entering the atmosphere at nearly escape speed from the two extreme corridor positions of overshoot and undershoot is illustrated schematically in Fig. 11-10. The figure indicates three factors: (1) the attainable range footprint depends on corridor conditions; (2) the footprints for the two extreme conditions overlap or have an area of common range potential; and (3) because a skip-out maneuver is possible for these initial entry speeds, the footprint is essentially open-ended. Since it is impractical to design systems for nearly unlimited range capability, some criteria are required to bound the footprint in order to establish design limits,

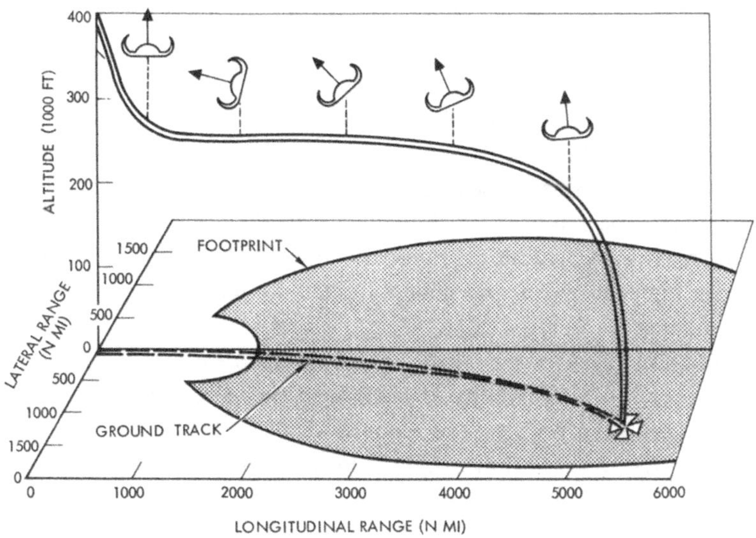


**Fig. 11-9.** Lateral maneuver technique.

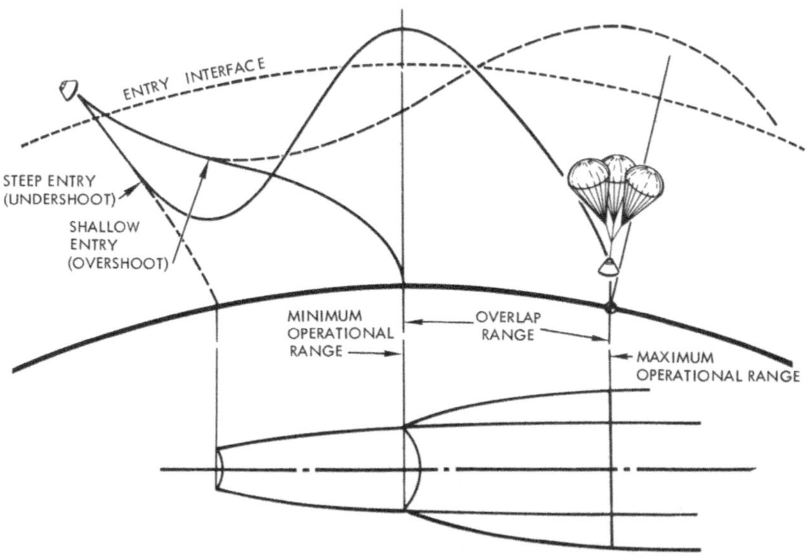


**Fig. 11-10.** Operational-range requirements.

and the concept of "operational range" provides the necessary criteria.

This concept of operational range is illustrated schematically in Fig. 11-10, where the minimum operational range is considered to be determined from entry at the overshoot boundary, while the maximum operational range is determined from entry at the undershoot boundary. The overlap range represents the range increment between these two limiting conditions. To achieve touchdown at either the primary or alternate landing sites throughout the entry corridor, the target loci relative to the initial entry must be within the operational range boundaries, and the minimum required overlap range must be at least as large as the distance separating the primary and alternate landing sites. The majority of the entry-system requirements are derived from the maximum operational range because the minimum allowable $L/D$, maximum flight duration, and maximum heating load are all derived from this parameter.

Examination of entry trajectories at parabolic speeds shows that, for a specified entry range, flexibility also exists in the selection of the shape of the entry trajectory profile. The amount of flexibility depends upon the vehicle $L/D$, the required range, and the initial position of the vehicle within the entry corridor. To achieve a maximum range with a given $L/D$, the choice of trajectory profiles multiplies as the entry-corridor condition varies from the undershoot toward the overshoot boundary. Two profiles which achieve the same longitudinal range to a given landing site are illustrated in Fig. 11-11. Profile 1 characteristically parallels the supercircular equilibrium glide

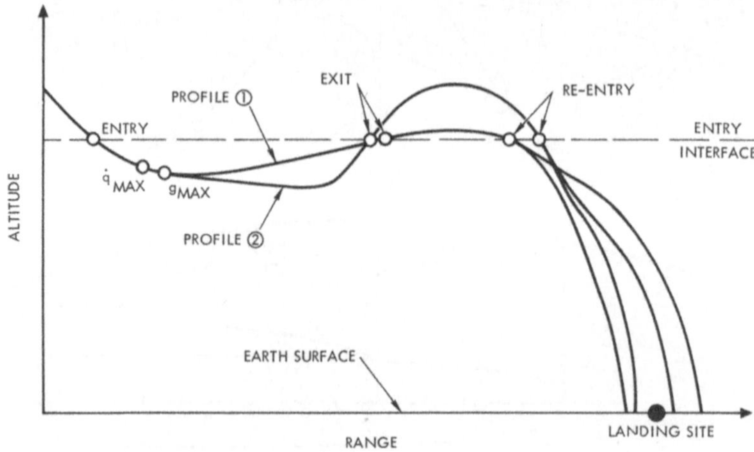


**Fig. 11-11.** Alternate entry trajectory profiles.

line and results in a short-range ballistic exit to achieve the required range, whereas profile 2 maintains a constant deceleration level throughout a portion of the supercircular flight regime that results in a relatively long-range ballistic exit to achieve the required range.

As the entry-corridor position approaches the undershoot boundary, the exit maneuver must be initiated sooner in the trajectory in order to attain the required total range. For a given $L/D$ and entry range, the undershoot boundary defines the limiting cases in which all available energy must be used to achieve the required range. Under this situation, the flexibility of profile selection is reduced to zero.

Considerable differences, however, exist between the two classes of trajectory design for entries from corridor positions other than on the undershoot boundary. To examine these differences for Apollo as they affect the design of vehicle systems, the total stagnation-point heating load was determined for a typical entry-corridor position and $L/D$. The heating-load variation as a function of range and profile shape is shown in Fig. 11-12, which indicates that significant variations in heating load occur for all ranges as a function of trajectory shape.

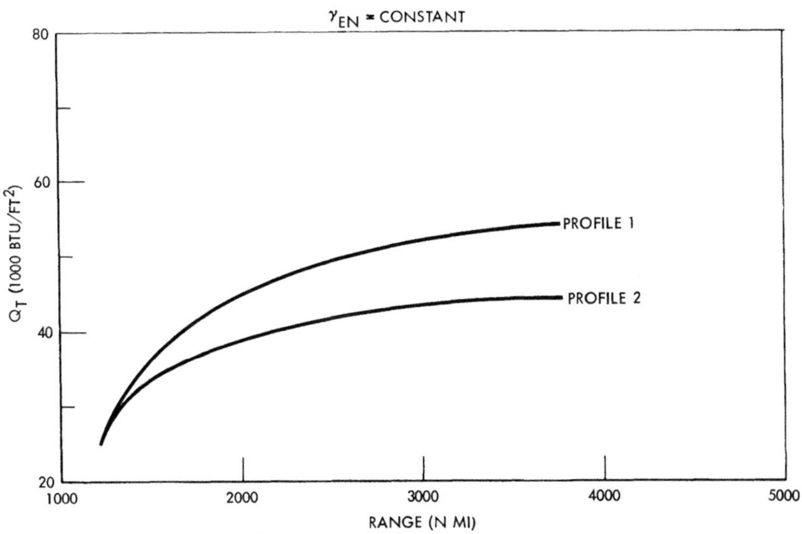


**Fig. 11-12.** Influence of trajectory shape on stagnation-point heat load.

## [11-3] THERMODYNAMIC ENVIRONMENT

In order to establish design conditions for an entry vehicle, the total thermal environment to be encountered must be considered. As shown in Fig. 11-13, heating contributions are encountered during the boost/abort, earth-orbital, deep-space, lunar-orbital, and earth-entry phases of the mission. The total boost/abort heating is usually small compared with total entry heating, but can be significant for regions of a configuration which are sensitive to heating rate, depending on the shape of the vehicle and boost

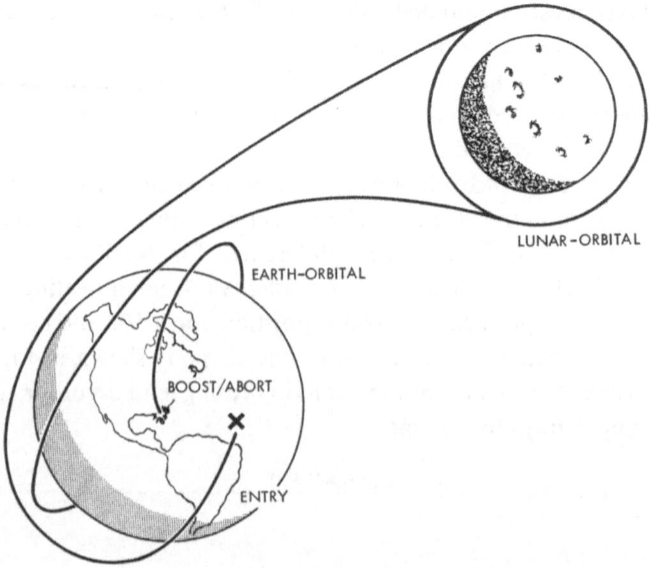


**Fig. 11-13.** Thermodynamic environments.

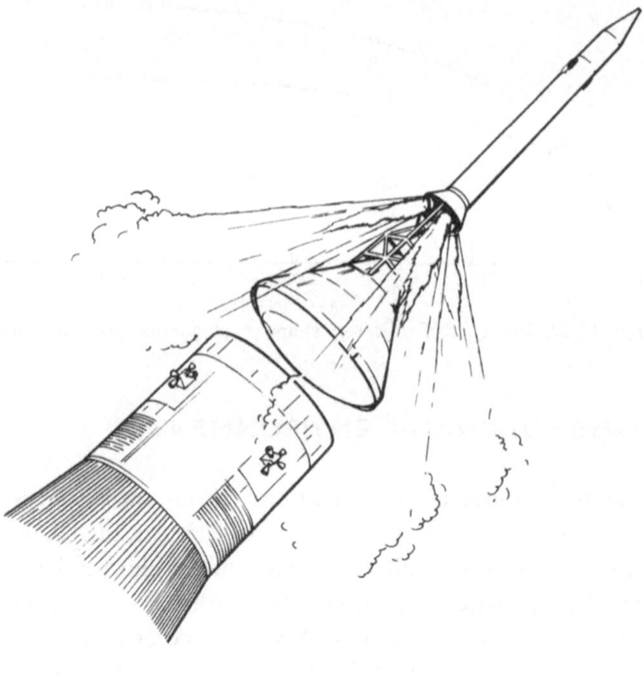

**Fig. 11-14.** Emergency abort (120 Btu/Ft$^2$-Sec).

orientation. On the Apollo mission, for example, an abort heating rate of 120 Btu/ft$^2$-sec is encountered on the leeward surfaces due to impingement of the abort rocket (Fig. 11-14). In the earth-orbital and space phases of flight, heat transfer is due to radiation, and the temperature of the radiation source may vary considerably depending on whether the vehicle is exposed to solar radiation or is in deep space behind the earth's shadow. These phases are illustrated in Fig. 11-15. For the Apollo mission, the spacecraft

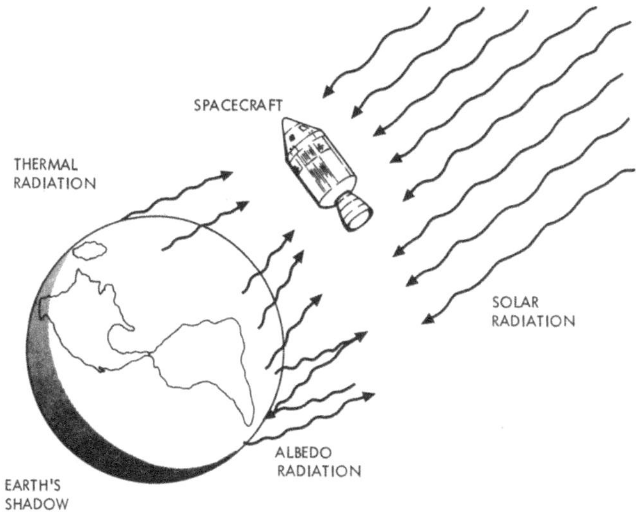


**Fig. 11-15.** Space flight thermal environment.

must be "barbecued" (slowly rotated) during the space phase to maintain an acceptable temperature. The most critical heating encountered by an entry vehicle occurs during the earth-entry phase, and this phase will be given emphasis in the following sections.

The heating experienced by a vehicle during entry is composed of both radiation and convection, as shown in Fig. 11-16. Radiative heating is due to thermal excitation of the air particles passing through the shock wave in front of the vehicle, while convective heating is caused by friction as the gas in the shock layer passes over the vehicle. Of the six heating effects itemized in Fig. 11-16, nonequilibrium effects on heat transfer will be discussed briefly; nonadiabatic effects will be treated in more detail because of their significance in determining the radiative heating. The remaining four effects, listed for completeness, are of significance only in a detailed design, and will not be considered here. These effects are generally not amenable to theoretical analysis and are usually treated empirically.

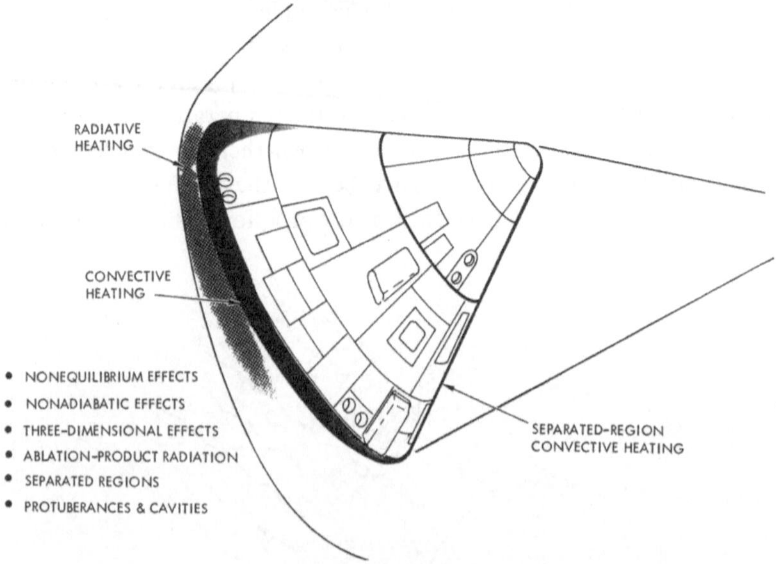


**Fig. 11-16.** Gas-dynamic heating.

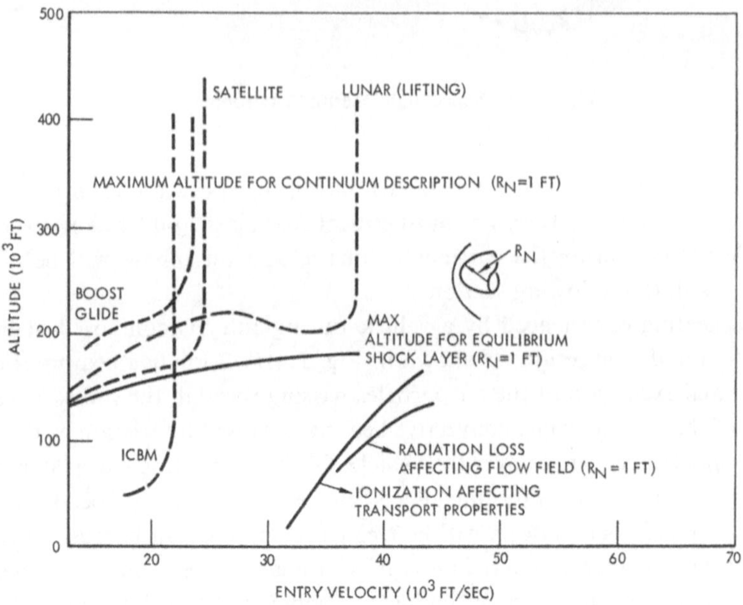


**Fig. 11-17.** Aerodynamic-environment regimes (Ref. 6).

Fig. 11-17 illustrates some typical entry trajectories and regions of significance for determining the heating of the vehicle. For a vehicle with a characteristic radius of one foot, the maximum altitude for which a continuum description of the flow is valid is about 330,000 feet. If the characteristic radius is larger, the maximum altitude would also increase, since the limiting value of the controlling parameter is an inverse function of the ratio of the mean-free path of the air particles to the characteristic radius of the vehicle.

Nonequilibrium effects become important above about 150,000 feet for a one-foot characteristic radius. Again, this boundary moves higher in altitude as vehicle size increases because the controlling parameter depends on the ratio of the thickness of the nonequilibrium zone to the shock-layer thickness. The thickness of the nonequilibrium zone is inversely proportional to density, while shock thickness is proportional to radius. Ionization effects do not depend on vehicle size but simply on the pressure and temperature in the shock layer. The effect of radiation on the flow field is another size-dependent phenomenon, and the boundary shown would again move up with increasing vehicle size. For entry velocities of approximately 50,000 feet per second, such as would be encountered during return from a planetary mission, radiation losses become significant. Consequently, radiation losses will be discussed in some detail.

Fig. 11-18 shows some typical heating results for entries from lunar and

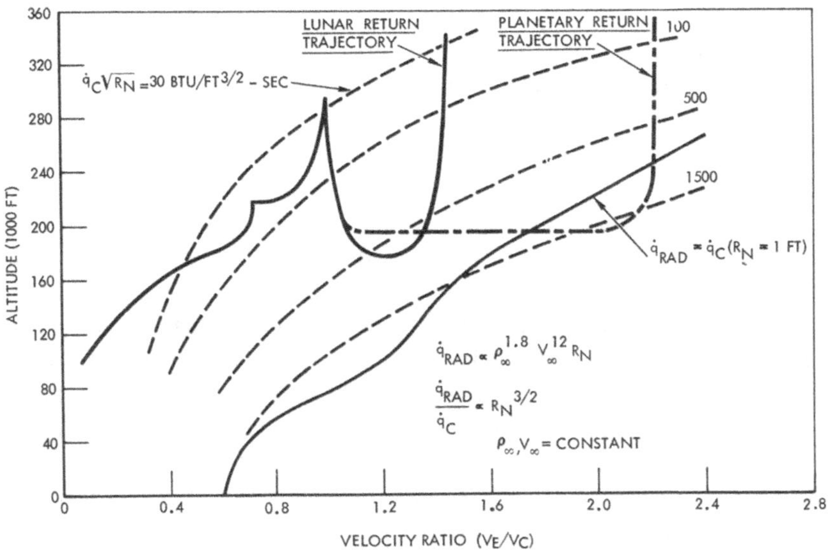


**Fig. 11-18.** Total entry thermal environment.

planetary missions. The parameter $\dot{q}_c\sqrt{R_N}$ is the laminar convective heating rate times the square root of the characteristic radius. Radiative heating is proportional to the radius, which causes a blunt vehicle to encounter more radiative heating than a fine vehicle. For a one-foot radius, conditions are indicated under which radiative heating and convective heating are the same. This line would move upward as the vehicle size increases since the radiative-to-convective heating ratio is proportional to the characteristic radius raised to the $\frac{3}{2}$ power.

### [11-3.1] Convective Heating

As the shock-heated gases flow over the surface of a vehicle, the viscosity of the gas results in a boundary layer at the surface which has a velocity profile ranging from zero at the wall to a value normally computed for the inviscid case. This gradient in velocity, illustrated in Fig. 11-19, assumes a

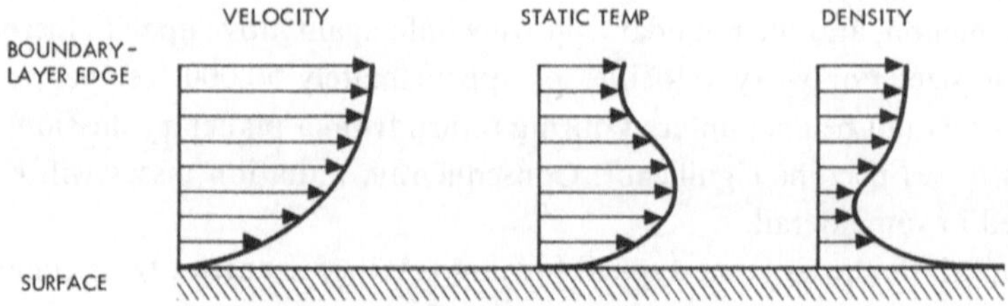


**Fig. 11-19.** Convective heating.

no-slip condition, or adhesion, of the molecules at the body surface. This shear layer near the surface introduces large temperature gradients through the boundary layer. Because of the temperature gradient and chemical reactions occurring at the surface, heat transfer occurs at the surface by a process called convection. Theoretically, the heat-transfer rate is computed by establishing the rate of heat conduction due to the gas conductivity and temperature gradient at the surface plus the heat released or absorbed due to chemical reactions at the surface.

Since the problem of computing the temperature gradient at the surface depends upon the velocity profile, it is necessary in computing convective heat-transfer rates to solve the boundary-layer equations of momentum, continuity, and energy conservation, which are derived from the general Navier-Stokes equations for a viscous fluid, assuming that the viscous influence is restricted to a thin layer near the surface which has a negligible pressure gradient in the direction normal to the surface. Because of the complex interrelation of the transport properties (viscosity and conductivity, for example) and the thermodynamic properties (temperature and density, for example) at hypersonic velocities, the set of coupled differential equations are solved on high-speed computers, subject to the boundary conditions at the edge of the boundary layer, i.e., velocity ($V_\delta$), pressure ($p_\delta$), temperature ($T_\delta$), and the gradient in edge velocity in a direction parallel to the surface ($dV_\delta/dx$). In addition, boundary conditions at the surface must be imposed: wall temperature ($T_W$) and the condition of no-slip $V_W = 0$.

Because computer solutions of these equations are too expensive to be repeatably employed for preliminary design analyses, available computer results are correlated in terms of the significant variables. Examples of correlation equations for stagnation-point and flat plate heating are presented below:

*Laminar Flow*

Stagnation Point (Hoshizaki, Ref. 7)

$$\dot{q}_0 = 2.59 \times 10^4 \sqrt{\frac{\varrho_\infty (2\varrho_\infty/\varrho_0)^{\frac{1}{2}}}{R_S}} \left(\frac{V_\infty}{10^4}\right)^{3.2} \left(1 - \frac{h_W}{h_0}\right)$$

Flat Plate (Hanley, Ref. 8)

$$\dot{q}_{FP} = 1.17 \times 10^4 \sqrt{\frac{\varrho_\infty (V_\delta/V_\infty)(p_\delta/p_0)}{X}} \times$$

$$\times \left(\frac{T_W}{900}\right)^{-0.051} \left(\frac{V_\infty}{10^4}\right)^{3.2} \left[1 - 1.205\left(\frac{h_W}{h_0}\right)\right]$$

*Turbulent Flat Plate* (Eckert, Ref. 9)

$$\dot{q}_{FP} = \lambda^* \times 10^6 \frac{[\varrho_\infty (V_\delta/V_\infty)(p_\delta/p_0)]^{0.8}}{X^{0.2}} \left(\frac{V_\infty}{10^4}\right)^{4.4} \left(1 - \frac{h_W}{h_R}\right)$$

$$\lambda^* = \frac{285{,}000(\mu^*)^{0.2}}{(Z^* R T^*)^{0.8}} = f(h^*, p)$$

$$h^* = \tfrac{1}{2}(h_\delta + h_W) + 0.0387\left(\frac{V_\delta}{100}\right)^2$$

where

$\dot{q}$ = heating rate (Btu/ft$^2$-sec)
$\varrho$ = density (slugs/ft$^3$)
$V$ = velocity (ft/sec)
$p$ = pressure (lb/ft$^2$)
$h$ = gas enthalpy (Btu/lb)
$T$ = temperature in degrees Rankine
$X$ = distance from leading edge of flat plate (ft)
$R_S$ = shock-wave radius of curvature (ft)
$R$ = gas constant

and the subscripts or superscripts are defined as follows:

$FP$ = flat plate
$\infty$ = free-stream conditions
$\delta$ = conditions at edge of boundary layer
$0$ = conditions at stagnation point
$W$ = wall or surface conditions
$R$ = recovery conditions
$*$ = reference conditions

If the wall temperature is assumed to be small compared with the flow stagnation temperature (a good assumption for hypersonic flight) and if $\varrho_\infty/\varrho_0$ is assumed to vary slowly with velocity and altitude, laminar heat transfer for a given configuration varies approximately as $V_\infty^{3.2}\sqrt{\varrho_\infty}$ and turbulent heat transfer varies approximately as $V_\infty^{4.4}\varrho_\infty^{0.8}$. Geometrical variations enter the equations as $R_S^{-1/2}$ and $X^{-1/2}$ for laminar flow, and $X^{-0.2}$ for turbulent flow. A matrix which illustrates the free-stream density and velocity dependency of gasdynamic convective and radiative heat transfer is presented below. The specified values listed for radiation do not take into account the flow-field interactions caused by potentially large energy losses due to radiation at high velocities. This problem is discussed in the following section.

Table **11-1**

| | Coefficient* | | |
|---|---|---|---|
| Heating Source | $c_1$ | $c_2$ | $c_3$ |
| *Convection* | | | |
| Stagnation Point | 0.5 | 3.2 | —0.5 |
| Laminar Flat Plate | 0.5 | 3.2 | —0.5 |
| Turbulent Flat Plate | 0.8 | 4.4 | —0.2 |
| *Radiation* | | | |
| $V_\infty < 28{,}000$ ft/sec | 1.7 | 7.4 | 1.0 |
| $28{,}000 < V_\infty < 38{,}000$ ft/sec | 1.4 | 20.0 | 1.0 |
| $V_\infty > 38{,}000$ ft/sec | 1.3 | 8.0 | 1.0 |

* coefficients are defined by:

$$\dot{q} \propto \rho_\infty{}^{c_1} V_\infty{}^{c_2} X^{c_3}$$

The laminar equations presented above can be employed to obtain a reasonable estimate of convective heat transfer. For turbulent flow, current theory is inaccurate, and the relationship presented above for a flat plate, which is empirically derived, represents what is believed to be a conservative estimate of turbulent heat-transfer rates.

### [11-3.2] Radiative Heating

At earth-orbital entry velocities (26,000 feet per second), aerodynamic heating is caused almost entirely by convective heat transfer. As entry velocity is increased to 36,000 feet per second (earth entry from lunar missions), convective heating is still predominant, but heat transfer due to hot-gas radiation is a significant factor in defining heat-protection requirements. During earth entry from planetary missions, the velocity is greater than 45,000 feet per second, and, under these circumstances, radiative heating becomes a predominant heat-transfer mode.

Fig. 11-20 illustrates the phenomena related to radiation heat transfer. As

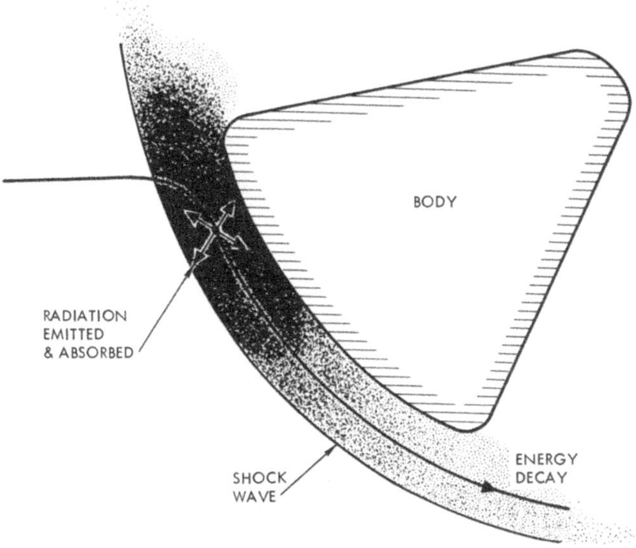


**Fig. 11-20.** Radiation flow-field phenomena.

the cold free-stream air passes through the shock-wave system in front of the vehicle, the temperature of the air rises sharply, approaching the total temperature of the free-stream near the stagnation region. Because of the high temperatures, the gas emits radiation and partially absorbs radiation emitted from other portions of the flow field. Due to this interchange, or transport, of energy from one portion of the flow field to another and ultimately to the vehicle surface, the flow field surrounding the body no longer exhibits the adiabatic nature assumed in classical inviscid theory.

Early attempts at computing the radiant energy transported to an entry vehicle were made assuming an adiabatic, nonabsorbing media. Under certain conditions, such assumptions result in a reasonable approximation of the radiant heating rate. These conditions are approximately satisfied when the radiative heating rate is small compared with both the black-body radiation ($\sigma T^4$) and the free-stream kinetic energy rate ($\varrho_\infty V_\infty^3/2$). Both conditions place an upper limit on the energy that can be radiated to an entry vehicle per unit time. When $\dot{q}$ is on the order of $\sigma T^4$, significant gaseous self-absorption of radiation occurs. Conversely, when $\dot{q}$ is on the order of $\varrho_\infty V_\infty^3/2$, "decay" of the flow energy occurs, i.e., a significant portion of the total flow energy is dissipated as radiant energy.

Results of radiation studies by Yoshikawa and Chapman (Ref. 10), shown in Fig. 11-21, illustrate these flow regimes as a function of stagnation-point

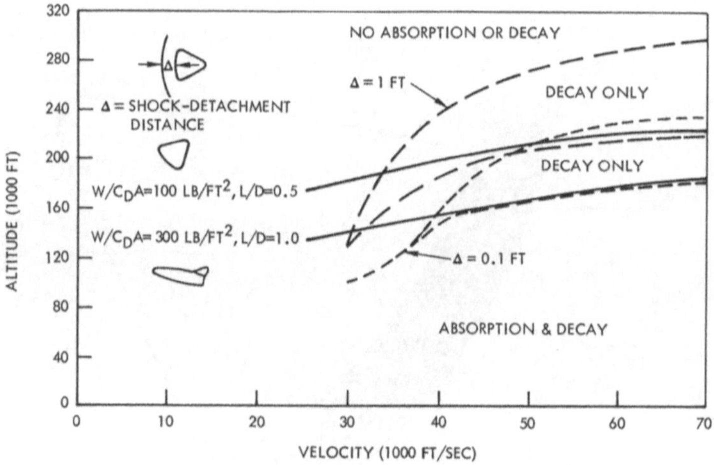


**Fig. 11-21.** Radiation flight regimes (Ref. 10).

shock-detachment distance, altitude, and velocity of flight in the earth's atmosphere. Also shown are lines of 10-g constant total load factor for

$$\frac{W}{C_D A} = 100\ \text{lb/ft}^2, \quad \frac{L}{D} = 0.5, \quad \text{and} \quad \frac{W}{C_D A} = 300\ \text{lb/ft}^3, \quad \frac{L}{D} = 1.0.$$

The 10-g load factor is typical of a maximum limit for safe manned entry. The values of the ballistic coefficient and the lift-to-drag ratio are also typical of the range expected for high-speed manned entry. Characteristic of vehicles with low ballistic and $L/D$ parameters are large radii in the stagnation region (for example, $R_N \cong 10$ feet), whereas, for vehicles having the higher range of values for these parameters, the nose radii in the stagnation region are characteristically much smaller (for example, $R_N \cong 1$ foot). The dimensions of the nose radii yield shock-detachment distances which are smaller by about an order of magnitude. For example, a nose radius of 10 feet would give rise to a shock-detachment distance of approximately 1 foot. It can be noted in Fig. 11-21 that for manned earth entry constrained to less than a 10-g load factor, the flow fields lie in the "decay only" flight regime, and the effects of gaseous self-absorption are not of critical importance.

### [11-3.2.1] Stagnation-Point Heating

The solution of an inviscid flow field that emits radiation without absorption is accomplished by the simultaneous solution of the conservation equations of momentum, mass, and energy. The basic difference between this solution and the adiabatic solution is the inclusion of a radiation-energy-loss term in the energy equation.

In Ref. 11, Hanley and Korkan derived equations for the stagnation region by employing the nomenclature shown in Fig. 11-22 and assuming that the

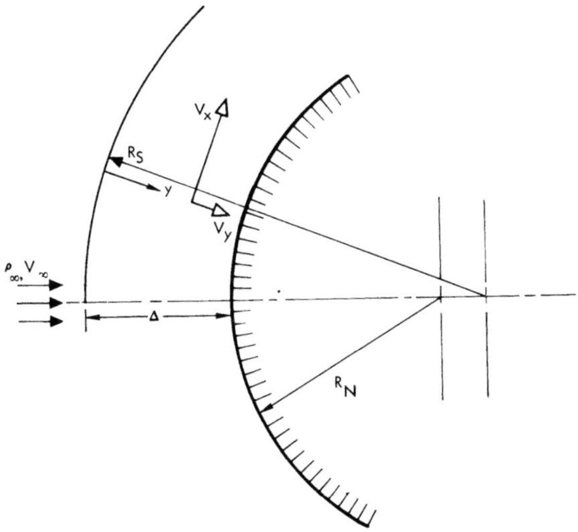


**Fig. 11-22.** Nonadiabatic stagnation-point radiative heating (Ref. 11).

shock-detachment distance from the body is small compared with the stagnation-point radius of curvature (a usual hypersonic flow-field assumption). The resultant relationships are as follows:

Continuity

$$(j + 1)f' + V_y\left(\frac{\varrho'}{\varrho}\right) + V_y' = 0$$

Momentum

$$\left(\frac{\varrho}{\varrho_i}\right)(f'^2 + V_y f'') = 2k(1 - k)\left(\frac{V_\infty}{R_S}\right)^2$$

Energy

$$\varrho V_y(h' + V_y V_y') = 1$$

State

$$p = \varrho ZRT$$

where

$j$ = an integer ($j=1$ for axially symmetrical flow and $j=0$ for two-dimensional flow)
$f'$ = is defined by $V_x = xf'(y)$
$V_x$, $V_y$ = velocity component in $x$ and $y$ directions, respectively
$\varrho$ = density
$k$ = density ratio across a normal shock
$R_S$ = shock-wave radius of curvature
$h$ = gas enthalphy
$I$ = radiation intensity [$I$=function $(p, T)$]
$Z$ = compressibility factor
$(\ )_i$ refers to conditions initially behind the shock
$(\ )_\infty$ refers to free-stream conditions
and the prime and double-prime superscripts represent derivatives with respect to the $y$ coordinate.

These equations were simultaneously solved on a high-speed digital computer to obtain the velocity distribution in the stagnation region, radiant-intensity distribution, and the rate of radiant energy transferred to the stagnation-point surface. Results of many computer solutions were correlated to determine the influence of radiation on the significant flow-field variables. These results are shown in Fig. 11-23. A correlation parameter, $\bar{q}_i$, is defined as the ratio of the radiative heat-transfer rate at the stagnation point calculated from an adiabatic flow field nondimensionalized by the free-stream kinetic energy rate. That is,

$$\bar{q}_i = \frac{(\dot{q}_{RAD})_i}{(1/2)\varrho_\infty V_\infty^3} = \frac{I_i \Delta_0}{\varrho_\infty V_\infty^3}$$

Again, the subscript, $(\ )_i$, refers to conditions immediately behind the shock, $I_i$ is the gas intensity as calculated adiabatically, and $\Delta_0$ is the shock-detachment distance as calculated adiabatically. The parameter, $\bar{q}_i$, can be looked upon as the heat-transfer rate to the stagnation point with no decay in the flow field. This approach is convenient because the calculation can be made without considering the influence of radiant-energy transport on the flow field.

As shown in Figure 11-23a, even when $q_i$ is nearly 2000 times the free-stream kinetic-energy rate, $(\frac{1}{2})\varrho_\infty V_\infty^3$, the pressure across the shock layer is nearly constant and almost equal to the value just behind the normal shock

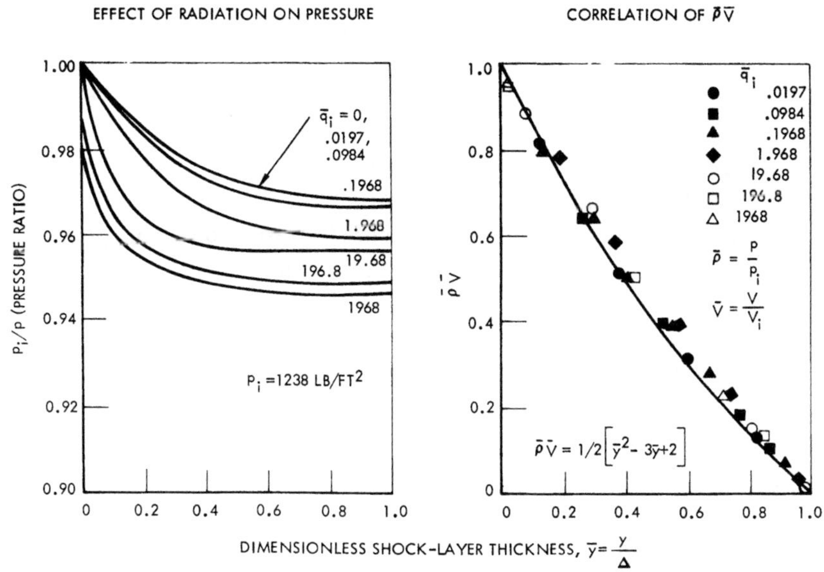


(a) (b)

**Fig. 11-23.** Numerical results for stagnation region (Ref. 11).

wave. It was found that when the dimensionless density-velocity product is plotted as a function of $\bar{y}$, the points fall on a single curve (Fig. 11-23b) having the following equation:

$$\bar{\varrho}\bar{V} = \tfrac{1}{2}(\bar{y}^2 - 3\bar{y} + 2)$$

where

$\bar{\varrho}$ = dimensionless density $(\varrho/\varrho_i)$
$\bar{V}$ = dimensionless velocity $(V_y/V_{y_i})$
$\bar{y}$ = dimensionless coordinate $(y/\Delta)$

These correlations were then employed by Hanley and Korkan (Ref. 12) to derive an approximate solution employing only the energy equation. As a result, a relationship was obtained that allows a rapid computation of the radiative heat-transfer rate to the stagnation point when the influence of energy losses in the flow field due to radiation are included. Fig. 11-24, which shows curves summarizing the results obtained from the simplified calculations, indicates that the radiant heat transfer can be readily determined when $\bar{q}_i$ and $\eta$ are known. The parameter $\eta$ is defined in Ref. 12, as

$$\eta = \left[\frac{\partial(\ln I)}{\partial(\ln h)}\right]_{p=\text{constant}}$$

The variation of $\eta$ with free-stream velocity can be calculated for any combination of gases if the radiative properties of the constituents are known. Thus, for example, a plot of $\eta$ versus pressure and free-stream velocity could be calculated for the Martian atmosphere, if its properties were known. The plot shown in Fig. 11-24, which was calculated for air, shows that $\eta \cong 2$ to 3 except in the vicinity of dissociation and initial ionization (25,000 ft/sec $< V_\infty \gtrsim 40{,}000$ ft/sec).

Calculations of $\bar{q}_i$ are readily accomplished employing tabulated radiation-intensity values such as those given in Ref. 13. A plot of $\bar{q}_i$ as a function of free-stream velocity and stagnation pressure is presented in Fig. 11-25. Consider the following example of a typical flight condition: if the radiant heat-transfer rate is desired for a velocity of 55,000 feet per second at a stagnation pressure ($p_0 \approx \varrho_\infty V_\infty^2$, assuming Newtonian flow) of 1 atmosphere at the stagnation point of a vehicle having a radius of 5 feet with $q_i \cong 0.4$ and $\eta = 3$, a dimensionless, nonadiabatic heat-transfer rate of $\bar{q} = 0.12$ can be obtained from Fig. 11-24. Multiplying this value of $\bar{q}$ by the free-stream kinetic-energy rate gives a nonadiabatic radiative heating rate of 9000 Btu/ft$^2$-sec. For the case of an adiabatic flow field, the corresponding radiative heating rate is 30,000 Btu/ft$^2$-sec. The importance of accounting for flow-field energy losses in determining radiative heat transfer for earth entry from planetary missions is readily apparent.

### [11-3.2.2] Radiative-Heating Distributions

In order to properly distribute the heat-protection system around a configuration, distribution of the heat transfer around the vehicle is required. Because of the strong interaction between radiative heat transfer and the flow field surrounding the vehicle, which was discussed in the preceding section, the analysis of radiative heat-transfer distributions requires a detailed analysis that accounts for radiation losses in a manner similar to the method described for the stagnation region.

Because of the insensitivity of pressure to radiation effects, a rather simple technique, described in Ref. 14, may be employed to calculate the radiation effects on the flow field. This method consists first of calculating the adiabatic flow field by employing standard techniques such as the method described in Ref. 15. A one-dimensional stream-tube analysis is then applied along the streamlines established from the adiabatic analysis, assuming that the pressure as a function of distance along the streamlines is not influenced by radiation in the flow field. Radiation-energy losses along the stream tubes are included in the simple stream-tube relationships given by:

Momentum

$$V \frac{\mathrm{d}V}{\mathrm{d}S} = -\frac{1}{\varrho}\frac{\mathrm{d}p}{\mathrm{d}S}$$

Continuity

$$\varrho V A = \text{constant}$$

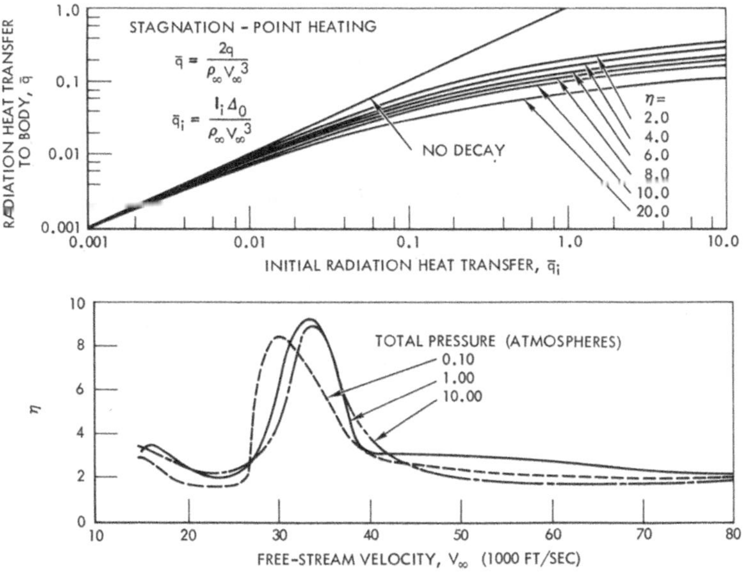


**Fig. 11-24.** Stagnation-point radiation correlation (Ref. 12).

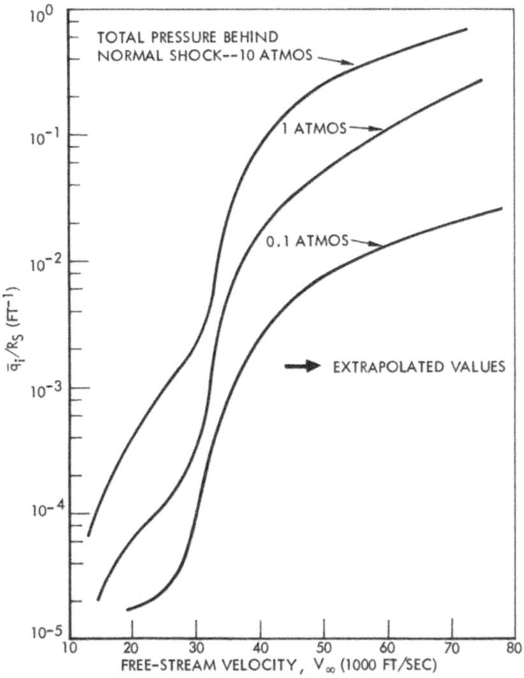


**Fig. 11-25.** Variation of $\bar{q}_i/R_S$ with free-stream velocity and total pressure behind normal shock—axisymmetric case (Ref. 11).

State

$$\frac{p}{\varrho} = ZRT$$

Energy

$$\frac{dh}{dS} + V\frac{dV}{dS} = -\frac{I}{\varrho V}$$

These equations are solved simultaneously along the defined streamlines, subject to the boundary conditions calculated immediately behind the shock wave and the condition that the static pressure as a function of distance is as determined in the adiabatic flow-field analysis. From these calculations, the new state properties, velocity, and radiative-intensity variations in the flow field are obtained.

The above method was used to compute the flow field along the pitch plane of the Apollo command module at an angle of attack of 33 degrees, assuming two-dimensional flow. The results of the analysis are shown in Fig. 11-26 and 11-27. Fig. 11-26 indicates the calculated streamline pattern

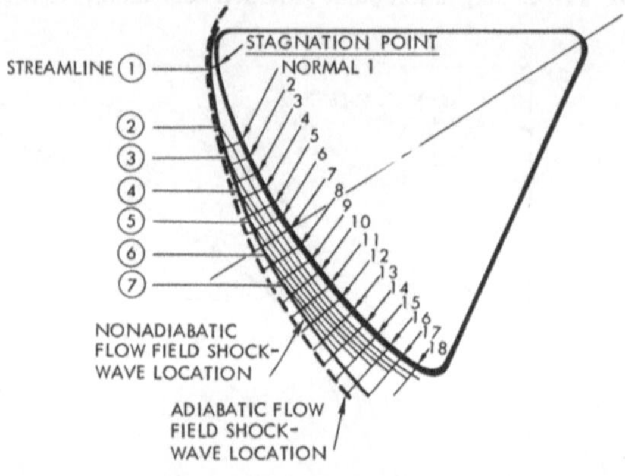


**Fig. 11-26.** Two-dimensional nonadiabatic flow field (Ref. 14).

and the shock-wave pattern resulting from the adiabatic and nonadiabatic analyses. Since radiation reduces the flow-field temperature, the density increases, causing the shock wave to move closer to the body. Fig. 11-27 shows the distributions of temperature, density, intensity, and velocity at normal 7 (the geometric body center) for assumed adiabatic and nonadiabatic flow for flight at a free-stream velocity of 52,000 feet per second and an altitude of approximately 200,000 feet. This is a condition typical of the point in flight where peak radiative heating occurs. Compared with the results obtained with adiabatic flow, examination of Fig. 11-27 indicates that the loss of energy in the nonadiabatic flow field results in a significant drop in temperature and radiation intensity near the body, an increase in density near the body, and a virtually unaffected velocity distribution. The reduced radiative intensity significantly reduces the heat transfer by radiation to blunt bodies at high velocities.

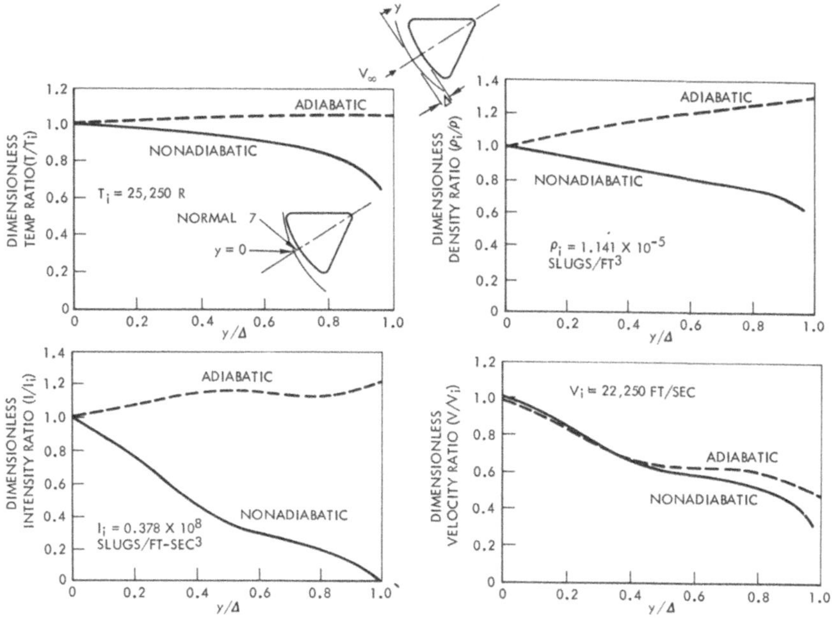


**Fig. 11-27.** Flow-field profiles (Ref. 14).

The radiative-heating distribution around the vehicle can be calculated by using the flow-field results derived above to determine the radiative intensity and by assuming the local radiating geometry to be an infinite plane of width equal to the shock-detachment distance. The resulting equation is

$$\dot{q}_{RAD} = \frac{1}{2}\int_0^{\Delta} I \, dy$$

where

$\dot{q}$ = heating rate

$I$ = radiation intensity

$y$ = coordinate locally normal to the body ($y = 0$ is located at the shock wave)

and

$\Delta$ = shock-detachment distance

Figure 11-28 shows the distribution of the heat-transfer rate, nondimensionalized by the stagnation-point heat-transfer rate, for the Apollo command module at an angle of attack of 33 degrees and for the conditions of $V_\infty =$ 52,000 feet per second and an altitude of 200,000 feet. This illustration shows that the nonadiabatic calculation leads to a peak heating rate closer to the stagnation point than that predicted by the adiabatic calculation. Since the heat-transfer rate at the stagnation point, $\dot{q}_0$, is significantly higher for the adiabatic case, the relative magnitudes are greatly distorted for the two cases in Fig. 11-28.

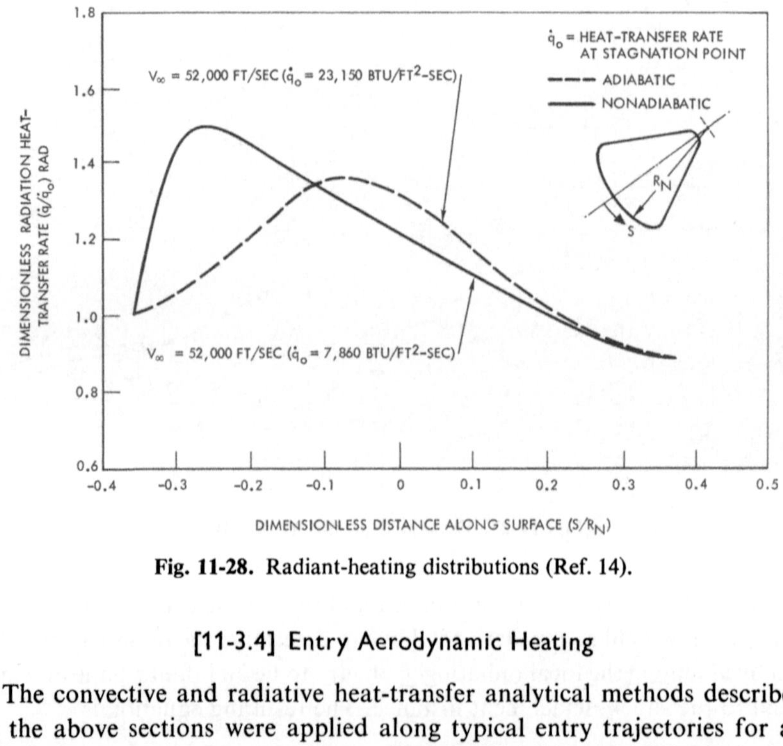


**Fig. 11-28.** Radiant-heating distributions (Ref. 14).

## [11-3.4] Entry Aerodynamic Heating

The convective and radiative heat-transfer analytical methods described in the above sections were applied along typical entry trajectories for an Apollo-type vehicle having $W/C_DA=100$ lb/ft$^2$ and $L/D=0.5$. Fig. 11-29 shows the altitude, velocity, and total load factor along an undershoot trajectory constrained to a maximum load factor of 10 $g$'s and an initial entry velocity of 55,000 feet per second. During the initial portion of entry, up to pullout, a roll angle of zero degrees (lift vector fully upright) was employed. After pullout, roll modulation of the lift vector was employed to maintain flight at a constant altitude until the lift vector was again fully upright, which occurred at about 400 seconds after initiation of entry. From this point on, a zero roll angle was maintained, resulting in an approximate equilibrium glide condition until impact.

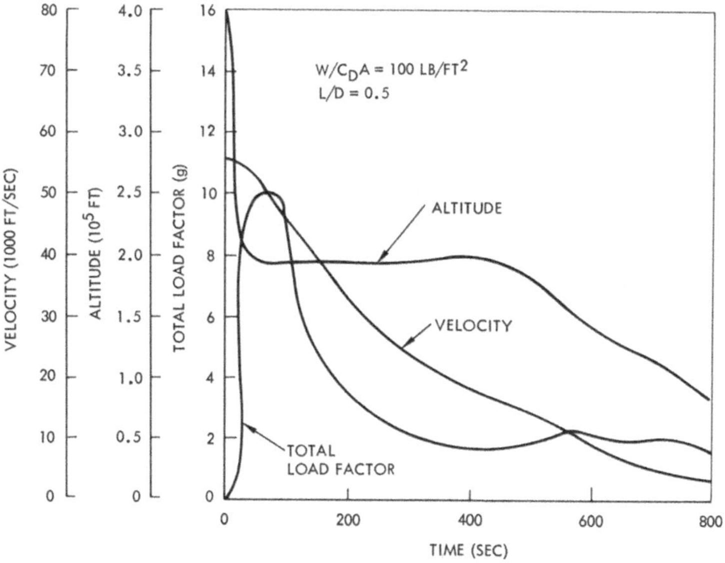


**Fig. 11-29.** Typical undershoot trajectory.

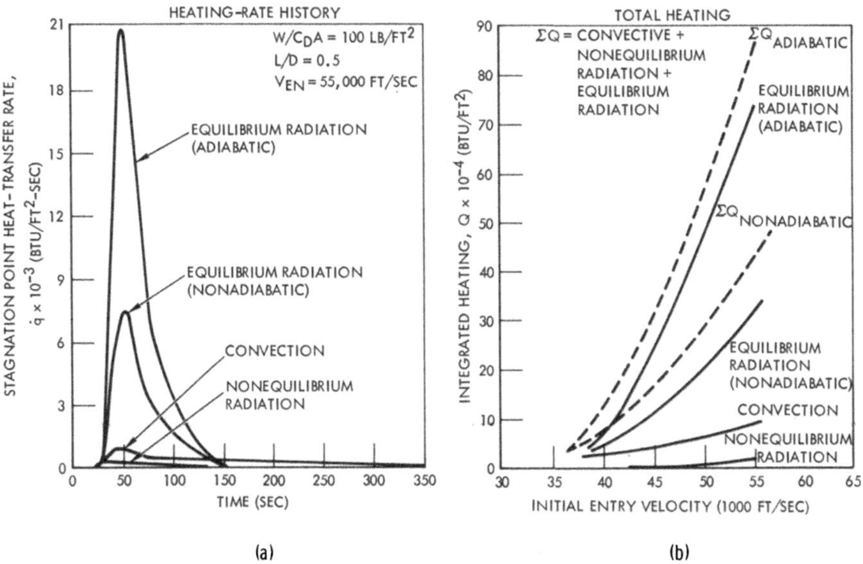


**Fig. 11-30.** Undershoot stagnation-point heating (Ref. 14).

Fig. 11-30 shows the heating-rate history along the above trajectory at the stagnation point of the Apollo vehicle. Three heating-rate contributions are indicated: equilibrium radiation, nonequilibrium radiation, and convection. As shown in this illustration, nonadiabatic equilibrium radiation results in a peak heat-transfer rate of only 7000 Btu/ft$^2$-sec; whereas, when adiabatic equilibrium radiation is assumed, the peak heat-transfer rate is 21,000 Btu/ft$^2$-sec. At an entry velocity of 55,000 feet per second, the maximum convective heat-transfer rate is only 1/7 of the maximum equilibrium radiation heat-transfer rate. Nonequilibrium radiation heat transfer results from a thin zone of gas immediately behind the shock wave that has not yet attained chemical equilibrium. Fig. 11-30a indicates that this contribution is negligible. Fig. 11-30b shows the total heating, integrated over the entire trajectory, as a function of entry velocity along 10-g undershoot trajectories. This figure illustrates the rapid increase in equilibrium radiation with increasing entry velocity and the large reduction in the calculated total equilibrium radiative heating that results from energy losses in the flow field.

The above aerodynamic heating environments will be employed in the next section to compute the requirements of a heat-protection system for earth entry from lunar and planetary missions.

## [11-4] HEAT-PROTECTION SYSTEMS

Design of heat-protection systems for entry vehicles requires an analysis of the entire thermodynamic environment, including not only entry, but also heating during boost, abort, and space flight. The degree to which the nonentry environments must be considered depends upon the overall system design. If the entry vehicle is employed as a living module during virtually the entire flight, as it is for the Apollo system, the heat load during the more than two days translunar space flight is significant for the design of portions of the heat-protection system. For other systems, where the entry vehicle is enclosed in a controlled environment, as it may be for some planetary-mission configurations, the space environment is of secondary importance in entry-vehicle design.

In addition to considerations of the thermodynamic influences of the environment, other factors influence the selection and design of the heat-protection system. These factors include ablator/structural thermal stresses and the influence of hard vacuum, space radiation, and cold soak on material properties. Because of these factors, virtually the entire environment from launch to landing must be considered in the design of heat-protection systems for entry vehicles.

### [11-4.1] Accommodation of Thermal Environment

The accommodation of the incident heat flux to an entry vehicle is illustrated in Fig. 11-31. This flux is composed of both convected and radiated

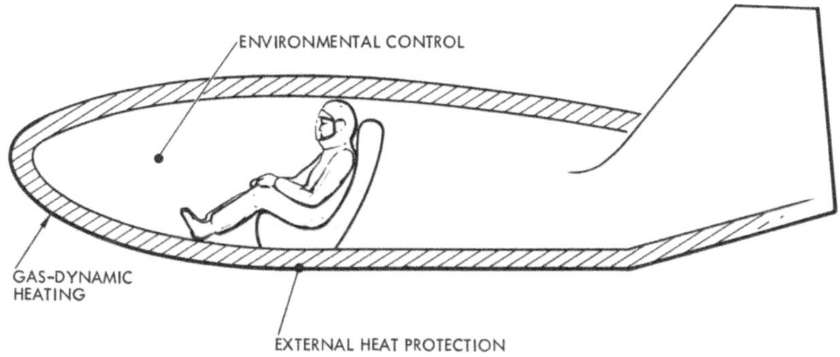


**Fig. 11-31.** Thermal-analysis problems.

energy, with the radiated energy being partially absorbed at the surface and the remainder reflected. The means by which the resultant heat flux to the surface is accommodated depends upon the type of heat-protection system employed and the rate and duration of the incident heat flux. An environmental control system is used to reject the residual heat energy.

### [11-4.1.1] Space Thermal-Environment Accommodation

The heating rates previously described for a typical Apollo environment indicate that the space thermal environment is characterized by a long-duration, low heating rate. Depending upon the type of trajectory and vehicle dynamics, the heating-rate history may vary considerably. Alternate heating and cooling of the surface elements of the space vehicle may occur as a function of two conditions: whether the elements are facing a heat source such as the earth, sun, moon, or nearby planet or whether they are exposed to the approximately $-260$ F of deep space. In all these situations, the mode of external energy transport is radiation. Another cause of high heating rate, which occurs for very short durations, is the firing of rocket engines during space maneuvering. In addition to these external heat sources, the overall heat-balance analysis must consider the internal heat generated by the crew and electronic equipment.

Since the external heat-protection system, the structural insulation, and much of the structure are typically designed by considerations of the entry environment, these portions of the vehicle configuration and their related thermodynamic characteristics represent a fixed set of conditions for determining how the space environment influences the entry vehicle and its contents. However, this does not imply that the space environment has no influence on the selection of the external heat-protection system. A discussion of this influence is presented in Subsection 11-4.2, Lunar Mission Heat-Protection System.

Control of the spacecraft temperature in a space environment may be achieved in several ways, including surface coatings for controlling the absorption and reflection of incident radiation. Use of this technique requires a knowledge of the spectral distribution of the incident radiation, which varies with the source. Since spectral distributions of solar, lunar, and earth radiation are quite different, design of surface coatings must be optimized for vehicles operating in all three environments during a mission, as is the case for a lunar mission, to minimize internal environmental control requirements.

In approaching the analysis of temperature control arising from consideration of both the external and internal heating environment, it is expedient to employ a three-dimensional thermal model of the structure that includes conduction and radiation among the structural elements. The model is then subjected to the history of the external environment and the internal sources of heat generation to determine the duty cycle of the environmental control system.

### [11-4.1.2] Entry Thermal-Environment Accommodation

The external environment imposed on a vehicle during entry is considerably different than the space environment, being characterized by a brief history of very high heating rates, composed of radiation and convection from the hot gases surrounding the entry vehicle. This environment is accommodated primarily by the external heat-protection system. Because of considerations in terms of weight and structural limitations, the external heat-protection system is not designed to be completely opaque to heat penetration, and a small percentage of the energy reaching the surface is allowed to diffuse into the structure and cabin.

Even though only a small percentage of the external energy finally reaches the cabin area, compared with the heating rates imposed by the space environment, the requirements for entry cabin cooling are usually considerably larger for a short period of time than are those of space. Four types of external heat-protection systems that have received consideration for earth entry are heat sinks, radiation cooling, transpiration cooling, and ablation. The usefulness of these heat-protection systems is primarily dependent upon two factors: the peak heating rate during entry and the duration of heating during entry. Entry duration and peak heating rate determine the total heat load for which a system must be designed.

### [11-4.1.3] Heat Sinks

The concept of a heat sink, as shown in Fig. 11-32, depends primarily upon storage of the imposed entry thermal environment in a highly con-

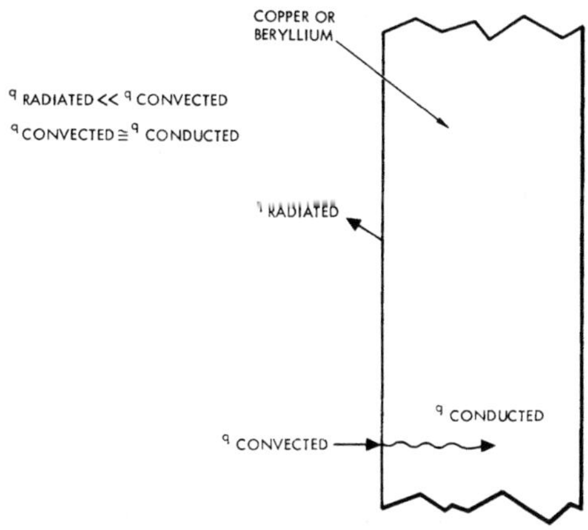


**Fig. 11-32.** Heat sink.

ductive metal. Analysis of heat sinks is accomplished by employing the familiar equation for transient heating of a solid, which for one-dimensional heat transfer is

$$\frac{\partial}{\partial x}\left(K\frac{\partial T}{\partial x}\right) = \varrho C_p \frac{\partial T}{\partial t}$$

where

$x$ = linear dimension
$K$ = thermal conductivity of material
$T$ = temperature
$\varrho$ = density of material
$C_p$ = specific heat of material
$t$ = time

and where the term on the left is the heat conducted internally due to temperature gradients in the solid, and the term on the right is the heat stored in the solid as the temperature rises. This equation is solved for the heat sink, subject to the external heat-transfer-rate boundary condition and a boundary condition at the inside surface.

At the external surface, assuming no melting, the boundary condition is given by

$$\dot{q}_W = -\left(K\frac{\partial T}{\partial x}\right)_W + \sigma\varepsilon T_W^4 \qquad (11\text{–}19)$$

where

the subscript $W$ refers to the wall or surface
$\varepsilon$ = surface emissivity
$\sigma$ = Stefan-Boltzmann constant

This boundary condition means that the heating rate at the surface is conducted into the material (the first term on the right) and reradiated from the surface (the second term on the right).

Since the external heat-protection system is the primary means of accommodating the external thermal environments, the back surface of the heat sink is normally insulated from the remainder of the vehicle; consequently, little energy is conducted through the rear surface of the heat sink. As a first approximation, the rear surface may be considered to be adiabatic, i.e.,

$$\left(\frac{\partial T}{\partial x}\right)_{\text{back surface}} = 0$$

can be taken as the rear-surface boundary condition.

In view of the high heating rates associated with earth entry from a lunar or planetary mission, a highly conductive material such as copper or beryllium would have to be employed in a heat-sink design to avoid surface melting. Thus, the amount of heat reradiated at the surface would normally be small compared with the amount conducted inward. The total amount of energy that can be stored in a heat sink per unit of surface area before melting occurs is given by

$$Q_T = \varrho \bar{C}_p (T_{\text{melt}} - T_{\text{initial}}) \tau$$

where

$Q_T$ = total heat load
$\varrho$ = material density
$\bar{C}_p$ = average specific heat of material between $T_{\text{melt}}$ and $T_{\text{initial}}$ and
$\tau$ = thickness

The efficiency of the system in accommodating the thermal environment per unit weight of the heat-protection system is given by

$$\text{Efficiency} = \bar{C}_p (T_{\text{melt}} - T_{\text{initial}})$$

If the initial temperature of the heat sink is assumed to be 500 R, the efficiency of copper is 270 Btu/lb and the efficiency of beryllium is 1500 Btu/lb.

For an Apollo lunar overshoot boundary trajectory, the total vehicle heat load is $7.76 \times 10^6$ Btu, and use of a heat sink would require a total heat-shield weight of approximately 5200 pounds, about half the weight of the Apollo entry vehicle. Obviously, use of a heat sink is not practical for Apollo. Practical application of the heat-sink design concept is limited to missions of short duration, such as those involving ballistic nose cones.

### [11-4.1.4] Radiation Cooling

The concept of radiation cooling is illustrated in Fig. 11-33. This concept

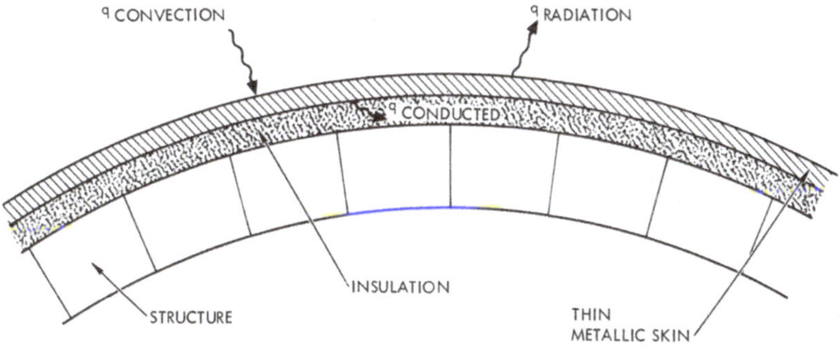


**Fig. 11-33.** Radiation cooling.

employs a thin external metal skin with insulation to protect the structure and cabin from the external environment. Compared with the heat-sink concept, the external thin metallic skin has a negligible heat-storage capacity, and the surface boundary condition given in Eq. (11–19) may be approximated by

$$\dot{q}_W = \sigma \varepsilon T_W^4$$

which implies that negligible heat is conducted inward compared with that reradiated away from the surface. As a result, this type of heat-protection system is sensitive primarily to the applied heating rate and secondarily to the total heat load because of the limitation in surface temperature imposed on the thin metallic skin. Candidate materials for reradiation heat shields over large surface areas include the superalloys such as René 41 and refractories such as columbium. The strength of these materials as a function of temperature is shown in Fig. 11-34. From this illustration, it is clear that the superalloys can be employed for surface temperatures up to 1700 F and that refractories such as columbium are limited to temperatures of about 2600 F. These temperatures, through Eq. (11–19), correspond to heating rates of 8 Btu/ft$^2$-sec and 33 Btu/ft$^2$-sec, respectively, assuming a surface emissivity of 0.8.

Since the stagnation-region peak heating rate, even along the lunar mission overshoot boundary trajectory, is about 470 Btu/ft$^2$-sec, radiative cooling is not feasible for Apollo-type entry vehicles returning from lunar or planetary missions. The radiation-cooling, or hot-structure, concept can be employed

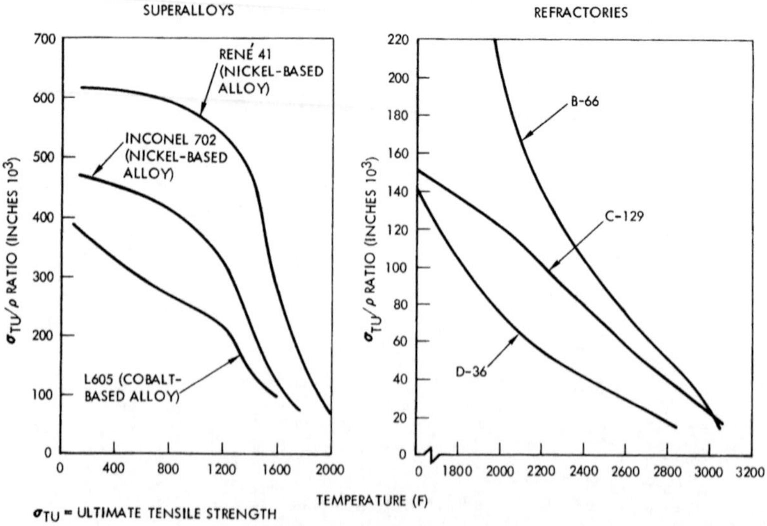


**Fig. 11-34.** Strength/density ratios.

on vehicles with missions having heating rates less than about 30 Btu/ft$^2$-sec, which is characteristic of lifting-body vehicles entering at orbital and sub-orbital velocities.

### [11-4.1.5] Transpiration Cooling

The concept of transpiration cooling, illustrated in Fig. 11-35, is based on the use of a porous material through which gases are forced. Injection of these cooling gases into the boundary layer results in the blockage of convective heat transfer. For such a system, the surface boundary condition can be written as

$$\dot{q}_{\text{combustion}} + \psi \dot{q}_{\text{convection}} + \dot{q}_{\text{radiation}} = -\left(K \frac{\partial T}{\partial x}\right)_W + \sigma \varepsilon T_W^4$$

where $\psi$, which is the ratio of convective heat transfer with injection to the convective heat transfer without injection, represents the influence of mass injection on the reduction of convective heating, and $\dot{q}_{\text{combustion}}$ is the heat added to the surface due to combustion of the injected and boundary-layer gases. If the injected gases do not absorb energy radiated to the surface from the flow field, no blockage due to coolant injection occurs for the radiative component of heat transfer.

As is usually the case with external heat-protection systems, it is desirable to reject as much heat as possible at the surface by reradiation, which implies allowing the surface temperature to rise as high as possible within the limits of the material employed. For this reason, the surface temperature is an important criterion for the rate of mass injection, since the net heating rate to the surface by combustion and convection depends upon the rate of mass injection. Efficient use of a transpiration-cooling system requires the modulation of the coolant gas flow as a function of the external heat-transfer rate so that the surface temperature is kept below the melting temperature. This concept requires a sensor to indicate the surface temperature and a system to modulate the coolant flow rate. Although such systems may prove to be dependable in the future, their use on present manned entry vehicles is impractical because of the requirements for high system reliability.

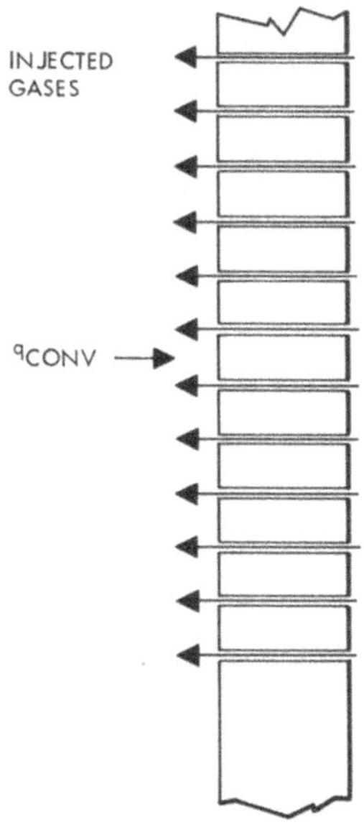


**Fig. 11-35.** Transpiration cooling.

### [11-4.1.6] Ablation

Ablative heat-protection systems are, in principle, similar to transpiration cooling heat-protection systems. The major difference between the two systems is that the transpiration of gases from an ablative system is self-controlled by the chemical kinetic behavior of the ablation material as it is heated, and no auxiliary systems are required to sense temperatures or to pump the transpiring gases into the boundary layer.

Two types of ablative heat-protection systems are illustrated in Fig. 11-36: subliming and charring. When the material of a subliming ablator reaches a given temperature, it sublimes and gases are injected into the boundary layer at a rate dependent upon the recession rate, $\dot{s}$. Assuming only convective

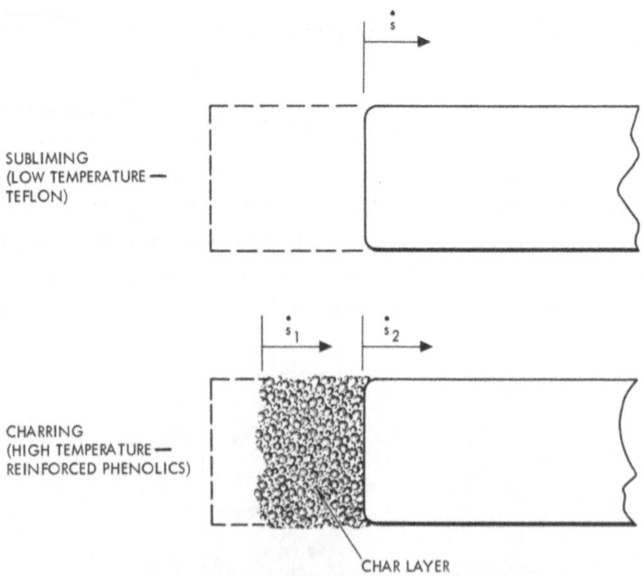


**Fig. 11-36.** Ablation heat-protection systems.

heating, the surface boundary condition can be written as

$$\psi \dot{q}_{\text{convective}} + \dot{q}_{\text{combustion}} = \dot{m}(\Delta h)_{\text{sublimation}} - \left(K \frac{\partial T}{\partial x}\right)_W + \sigma\varepsilon T_W^4 \qquad (11\text{–}20)$$

where

$\dot{m}$ = mass rate of material lost

$\Delta h$ = change in material enthalpy due to chemical reaction

In this relationship, the net heat flux to the surface by convection ($\psi \dot{q}_{\text{convective}}$) and combustion is accommodated by sublimation of the material, conduction into the interior, and reradiation from the surface. With subliming ablators, such as Teflon, the sublimation temperature is low (about 1000 R), and, consequently, the reradiated energy and the conducted energy are small compared with the energy absorbed by sublimation ($\dot{m}\{\Delta h\}_{\text{sublimation}}$).

By defining the "effective heat of ablation" as

$$h_{\text{eff}} = \frac{\dot{q}_{\text{convective}}}{\dot{m}}$$

the relative efficiency of such materials for heat protection can be readily determined. Usually, $\dot{q}_{\text{combustion}}$ in Eq. (11–20) is small, and $h_{\text{eff}}$ can be obtained by neglecting $\dot{q}_{\text{combustion}}$, $K(\partial T/\partial x)$, and $\sigma\varepsilon T_W^4$, giving

$$h_{\text{eff}} = \frac{(\Delta h)_{\text{sublimation}}}{\psi}$$

However, since $\psi$ is a function of $\dot{m}$, other material, and aerothermodynamic parameters, the effective heat of ablation is coupled to the material characteristics and the thermal environment. Experimental results for Teflon, shown in Fig. 11-37 for laminar and turbulent convective heating, display a linear dependence of $h_{eff}$ on the difference between total enthalpy and static enthalpy of the gas at the wall. The flight velocities corresponding to the enthalpy-difference parameter are also shown in Fig. 11-37.

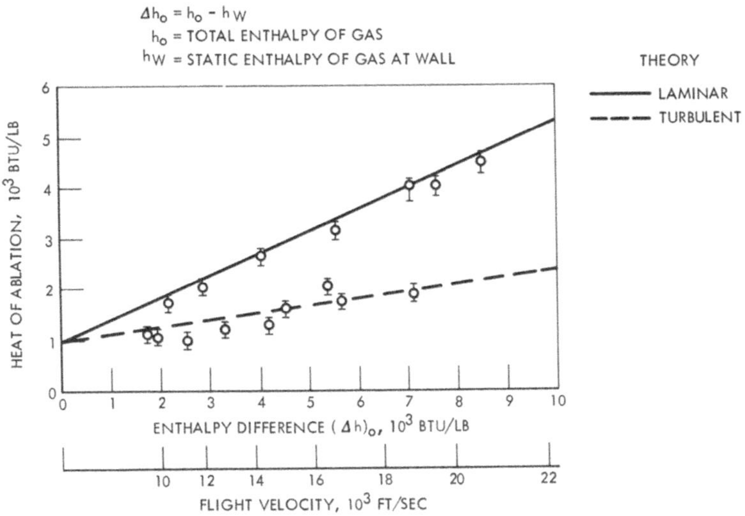


**Fig. 11-37.** Teflon's effective heat of ablation (Ref. 16).

Charring-ablation systems, shown in Fig. 11-38, exhibit a complex behavior in accommodating the entry heating environment. These systems are composed of an initial mixture of plastics which, upon being heated, undergo a series of complex chemical reactions that result in the formation of gases and a solid residue which is referred to as char. The char layer is normally composed of carbonaceous material that may contain reinforcing organic or inorganic fibers to provide the strength required to withstand thermal and shear stresses.

Since the char is composed of a carbonaceous material, a chemical reaction occurs between the char surface and the oxygen and nitrogen gases which diffuse to the surface through the boundary layer, forming primarily CO and CN gases. For this reason, charring ablation is characterized by dual moving boundaries: an outer-surface boundary which lies between the boundary layer and the surface, and a boundary between the virgin plastic and the plastic undergoing decomposition. More precisely, three zones may be defined (Fig. 11-38): char layer, decomposition region, and virgin material.

The virgin material is that region in which no chemical changes have occurred in the original solid plastic. The decomposition region is characterized by a chemical reaction that results in the production of gases and the solid char residue. The char-layer region is characterized by a mature char that no longer undergoes significant chemical decomposition.

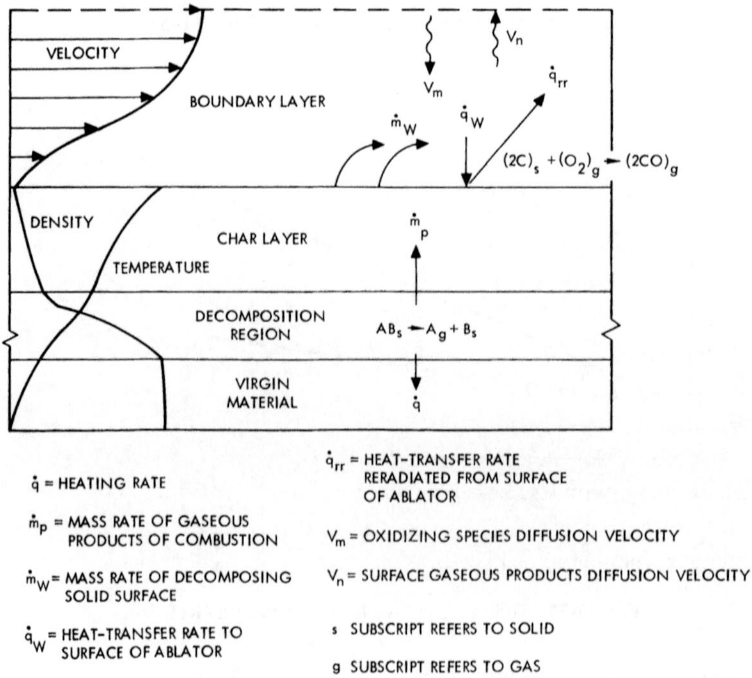


**Fig. 11-38.** Charring abaltion.

Several modes of energy accommodation are employed in a charring-ablation system. Examples include (1) reradiation of energy at the surface, (2) blockage of convective heating by surface and decomposition-region gases, (3) decomposition of the virgin plastic, (4) absorption of energy by the gas products of decomposition as they transpire through the char layer, and (5) conduction and storage of heat in the solid material. All these processes of energy accommodation are closely coupled, and the analysis of charring-ablation heat-protection systems requires a numerical solution of the complex energy equation and chemical kinetic relationships, including a coupled analysis of these mechanisms. Sophisticated models of this process have been programmed for machine computation, and a more detailed discussion of the theoretical approach and typical results for a phenolic-nylon charring ablator may be obtained from (Ref. 17).

Useful results can be obtained parametrically for charring-ablation systems by imposing a quasi steady-state assumption on the ablation process. This assumption implies a fixed temperature distribution through the ablator measured from the receding surface, which is satisfied only in the limit as time goes to infinity for a semi-infinite slab subjected to a constant heat flux. This approximation is validly applied for transient situations in which the rate of change of thickness of the thermally affected zone is small compared with the rate of surface recession.

The results of a quasi steady-state analysis of a phenolic-nylon charring ablator, derived from (Ref. 17), are shown in Fig. 11-39, where the rate of

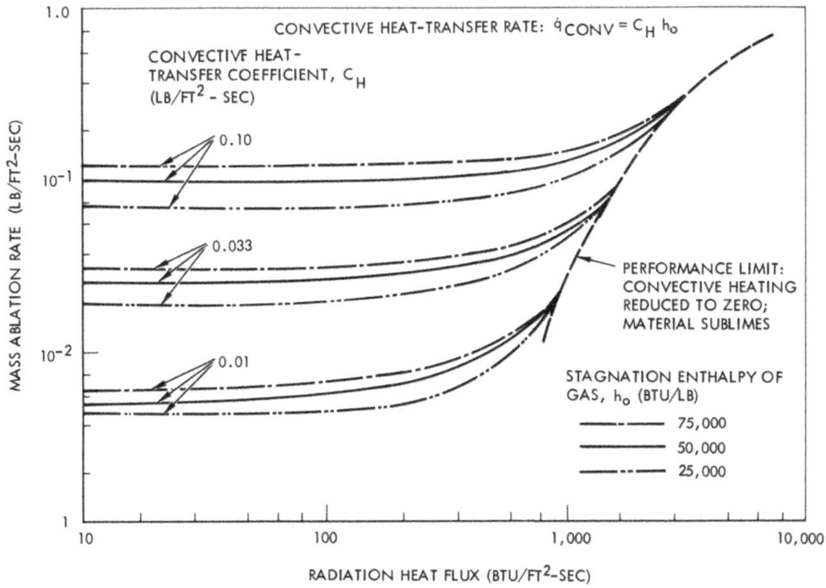


**Fig. 11-39.** Performance of phenolic-nylon ablator (Ref. 17).

mass loss is shown as a function of the convective and radiative heating rates. This illustration shows that even in the limit, when the radiative heating rates are high enough to cause sublimation of the material, the effective heat of ablation defined by $h_{eff} = \dot{q}/\dot{m}$ is about 15,000 Btu/lb. The total performance of the charring ablator is not indicated solely by the effective heat of ablation, which only predicts instantaneous mass-loss characteristics. Conduction of heat through the virgin plastic to the inside surface of the ablator is also significant in determining charring-ablator thickness requirements. For relatively long-duration flight (such as 1000 seconds), the use of low-density ablators, which contain phenolic microballoons and have densities of about

30 to 40 lb/ft$^3$, helps to minimize heat conduction because low-density ablators are fairly effective insulators.

A summary of the thermodynamic environment to which classes of entry systems are exposed is shown in Fig. 11-40 in terms of the peak lifting-surface heating rate, peak radiation equilibrium temperature corresponding

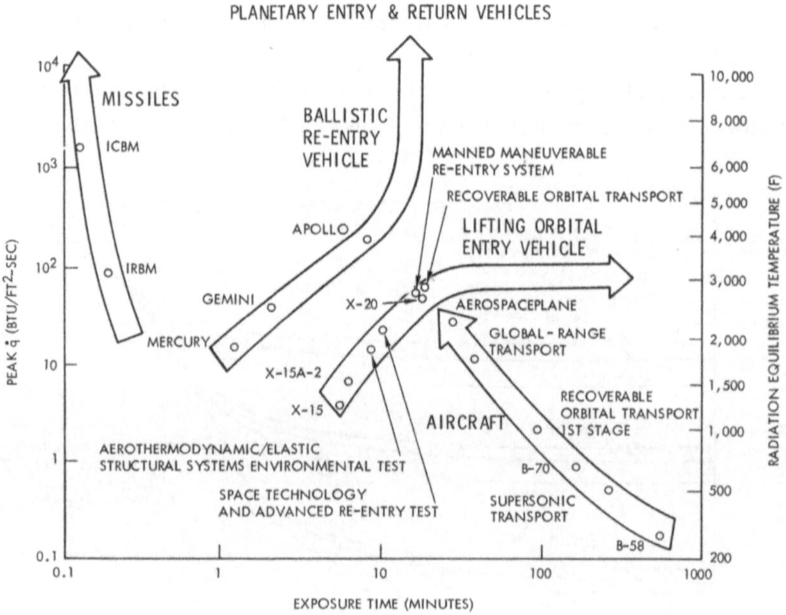


**Fig. 11-40.** Trends in thermal environment.

to this heating rate, and exposure time. This illustration indicates that ballistic missile environments are characterized by high heating rates and low exposure time; lifting entry-vehicle earth-orbital environments are characterized by moderate heating rates and long exposure times; and cruise-type vehicle environments vary from low heating rates and long exposure times for low-speed aircraft to moderate heating rates and moderate exposure times for high-speed (suborbital) flight vehicles. The advanced earth-entry vehicles for lunar and planetary missions are exposed to moderate and high heating rates, respectively, for moderate exposure times.

The influence of these environments on types of heat-protection systems is illustrated in Fig. 11-41. Because of the high thermal efficiency of charring-ablation systems over a large range of heating rates and total heat loads, this heat-protection system is considered useful for applications ranging from ballistic missiles to entry into the earth's atmosphere at hyperbolic velocities. This type of system is currently employed on the Gemini entry vehicle, is

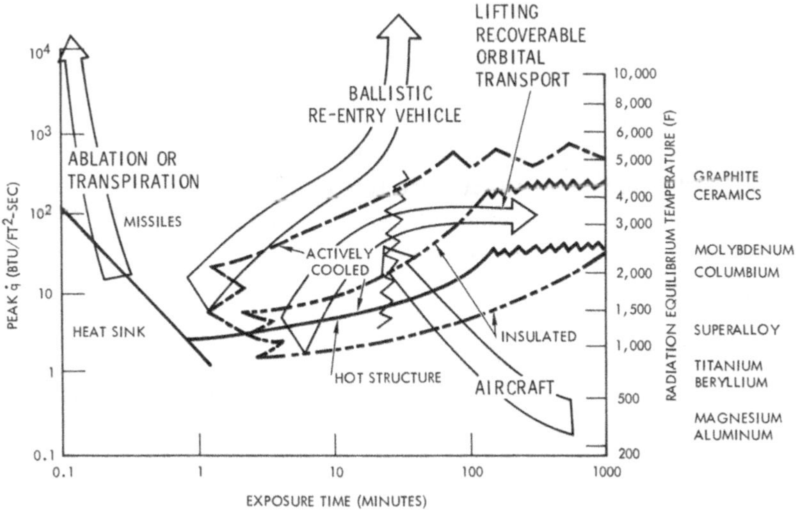


**Fig. 11-41.** Thermal-protection concepts.

planned for use on the Apollo command module for the lunar mission, and is considered highly competitive for future intermediate lift-to-drag-ratio ($L/D \approx 1.0$ to 1.5) earth-orbital entry vehicles.

## [11-4.2] Lunar Mission Heat-Protection System

Selection and development of a heat-protection system for the Apollo lunar mission required initial theoretical and experimental studies to select appropriate materials, further detailed theoretical and experimental studies to design and develop the heat-protection system, and unmanned flight testing of the heat-protection system to qualify the design concept for manned application during the lunar mission.

### [11-4.2.1] Charring-Ablation Material Selection

Selection of heat-protection materials required analyses and tests simulating the entire thermodynamic environment to which the Apollo command module is subjected, including boost and abort, earth-orbital flight, translunar flight, lunar-orbital operations, and earth entry. This environment, which was discussed earlier, results in temperatures on the blunt end of the vehicle that vary from −150 F in space to about 5000 F during entry. During the entire flight, the interioı of the Apollo command module must be maintained at temperatures required for crew comfort. During the conditions of extreme hot and cold exterior temperature, severe thermal gradienst exist across the heat-protection system. Since the ablator is exposed to the severest portions of the environment, thorough testing and analysis of several types of charring ablators were required to select the final system.

Criteria for selecting the charring-ablation system included: (1) Thermodynamic efficiency in the entry environment, (2) Compatibility in the space environment, (3) Fabricability and (4) Structural compatibility. As a result of these tests, an ablator was selected for Apollo that performed well in all phases of the test program. Many of the materials that showed good thermodynamic performance during entry did not perform well during other phases of the test program.

### [11-4.2.2] Heat-Protection System Design

After materials were selected, a detailed design of the heat-protection system was conducted within the constraints of the overall design philosophy of the Apollo command module and the mission thermodynamic environment. A cross section of the Apollo command module, shown in Fig. 11-42,

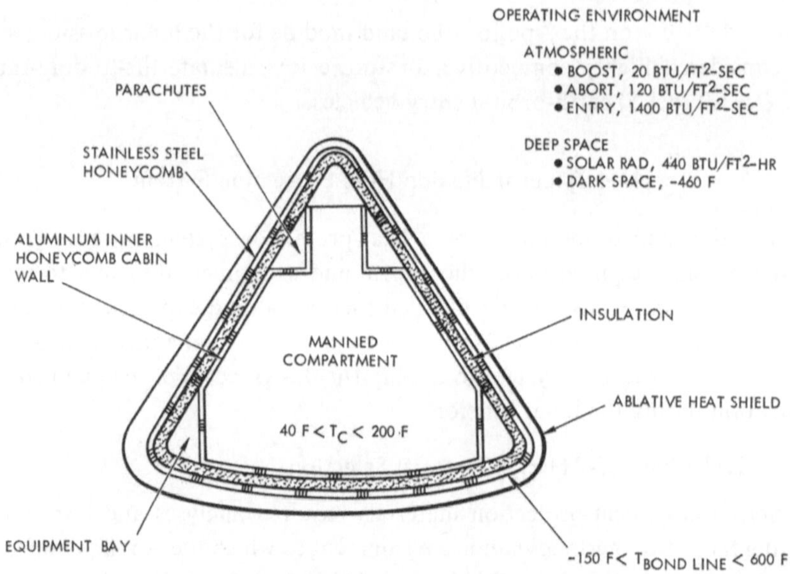


**Fig. 11-42.** Apollo thermal-protection system.

indicates the general approach for the heat-protection design. The ablative material is injected into a fiber glass honeycomb matrix which, prior to material injection, has been bonded to a stainless steel honeycomb substructure. A superinsulation (TG-15000) is located between the stainless steel honeycomb and the aluminum honeycomb pressure shell that houses the crew.

[11-4.2.3] Design Criteria

Design of the thermal-protection system is constrained by temperature limitations imposed on the manned compartment and the steel honeycomb structure to which the ablator is attached. The imposed design conditions are:

Temperature of the Manned Compartment (1) Cabin walls not to exceed 200 F and (2) Cabin walls not less than 40 F.

Temperature of Steel Structure (1) Maximum temperature not to exceed 600 F and (2) Minimum temperature greater than −150 F.

The manned-compartment limitations were dictated by the strength requirements of aluminum, condensation considerations, and the capability of the environmental control system to provide a comfortable environment. The maximum temperature of the stainless steel substructure was dictated by the strength of the bond between the ablator and substructure, whereas the minimum temperature of the steel substructure was imposed in order to limit thermal stresses between the ablator, steel substructure, and aluminum pressure shell.

The maximum temperature level at the bond line between the ablator and stainless steel honeycomb substructure is determined primarily by the entry heating environment, although the space environment has an influence on the initial temperature of the ablator and the steel substructure just prior to entry. Because of the thermal "inertia" of the ablator, this peak interface temperature does not occur until the spacecraft nears earth touchdown. Minimum bond-line temperature, which is dictated by space-cooling conditions, is controlled by proper orientation of the vehicle. Occasionally, however, when it is necessary to take sightings or fire rocket engines for course corrections, a preferential orientation is required for a short period of time. Because periods are short, local drops in temperature are not severe enough to drive the bond-line temperature below the specified limit.

Limitations on the cabin wall temperature are dependent on the thermodynamic characteristics of the entire heat shield, all the way from the outer edge of the ablator to the inside of the aluminum cabin wall, as well as on the characteristics of the environmental control system. Although the entry environment normally determines the insulation requirements for protection against a maximum temperature of 200 F, consideration must also be given to solar radiant heating encountered in space. The minimum temperature requirements of the cabin are determined only by heat losses that occur in the space environment.

[11-4.2.4] Ablator Design

Since the major design conditions for the ablator occur during earth entry, it is necessary to survey the thermodynamic conditions that exist within the entry flight profile to determine the trajectories that design the ablative heat shield. The environment along the limits of the flight envelope are illustrated in Fig. 11-43. Ablative heat-protection systems are characteristically de-

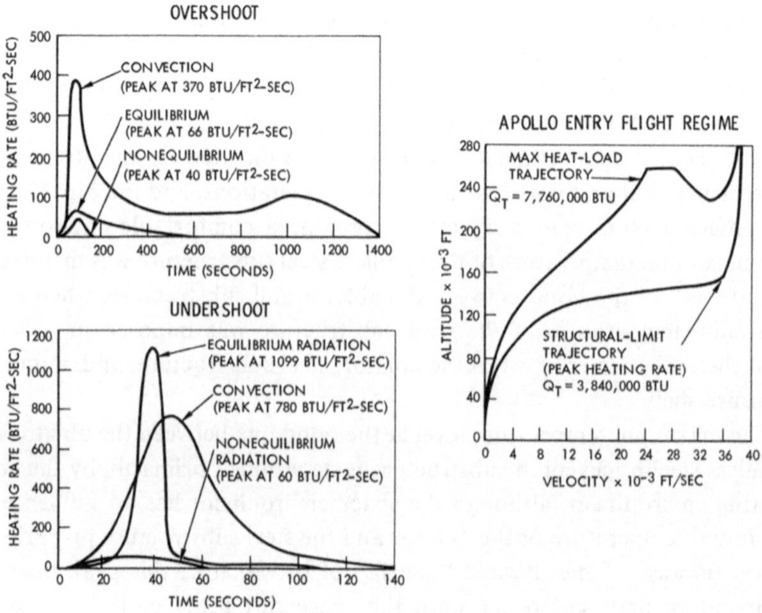


**Fig. 11-43.** Apollo entry design heat loads.

signed by the maximum heat load, which corresponds to the overshoot trajectory. Along this trajectory, the total heat load to the entire vehicle, including radiation and convection, is $7.76 \times 10^6$ Btu. Although a much higher peak heating rate occurs along the undershoot trajectory, the total heat load is approximately one-half the load along the overshoot trajectory. Since some elements of the command module, such as windows, are limited by the heating rate rather than by heat load, consideration must also be given the undershoot trajectory for local thermodynamic design problems.

Design of the ablative heat-protection system requires a detailed knowledge of the chemical kinetic behavior and thermodynamic properties of the ablator, which are obtained experimentally. These characteristics are employed in a complex charring-ablation computer program to determine the detailed behavior of the ablator as a function of time during entry.

### [11-4.2.5] Insulation Design

To determine the critical conditions, including solar heating, heat loss to dark space, and entry heating, insulation thicknesses must be obtained at several points on the entry vehicle for virtually the entire mission profile. These studies also require the use of automatic computation techniques, but the thermodynamic characteristics of the insulation are much simpler to analyze than are the ablator characteristics. The insulation thicknesses obtained at each point to satisfy each of the design criteria are then compared with the greatest thickness satisfying all mission criteria.

### [11-4.2.6] Heat-Shield Weight

The heat-shield weights for the Apollo and Mercury entry vehicles are summarized below.

Table **11-2**

| Item | Mercury | Apollo |
|---|---|---|
| Heat-Shield Weight | 315 lb | 1300 lb |
| Heat-Shield Weight/Vehicle Weight | 12 % | 13.7 % |
| Vehicle Weight | 2600 lb | 9500 lb |
| Total Heat Load | 210,000 Btu | 7,760,000 Btu |
| Heat-Shield Efficiency (Total Heat Load/Heat Shield Weight) | 666 Btu/lb | 5,970 Btu/lb |

### [11-4.3] Planetary Mission Heat Protection

The geometric shape of an entry vehicle is highly dependent upon whether the dominant thermal environment is convective or radiative. If the environment is primarily convective, total heating is reduced by the use of blunt shapes; if radiative heating is predominant, total heating is reduced by the use of fine shapes. Packaging of a manned payload in an entry vehicle also plays a role in determining the shape of the external heat-protection system, since total heat-shield weight is a function of the external surface area, and surface area can be minimized by employing blunt shapes.

A study was made of these major trade-offs (Ref. 18) by employing both a slender- and blunt-vehicle concept for achieving lift-to-drag ratios of 0.5 and 1.0. Side views of the configurations analyzed are shown in Fig. 11-44, along with the growth of configuration sizes according to requirements for 3-, 6-, and 10-man crews. The M-1 and M-2 configurations, developed by NASA's Ames Research Center, characterize fine-body approaches to achieving lift-to-drag ratios of 0.5 and 1.0, respectively, while the Apollo and lenticular configurations are representative of blunt-body approaches to the same lift-to-drag ratios. It is of interest to note that the size of the lenticular and M-2 vehicles is relatively insensitive to changes in the size of the crew. This phenomenon is due to the minimum height required to pack-

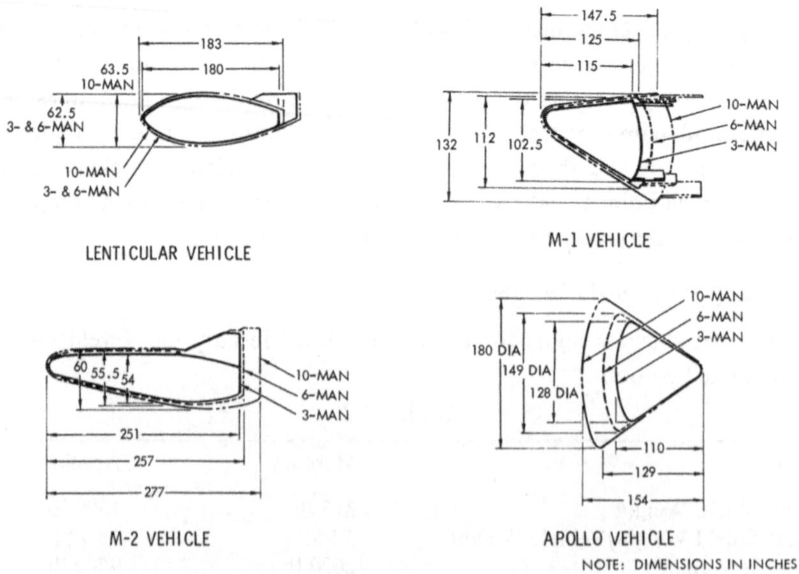


**Fig. 11-44.** Earth entry module configurations (Ref. 18).

| p (ATMOS) | T (R) | Δh SUBLIMATION (BTU/LB) |
|---|---|---|
| 0.01 | 5625 | 15,550 |
| 0.10 | 6110 | 14,800 |
| 1.00 | 6725 | 14,050 |
| 10.00 | 7500 | 13,000 |

| p (ATMOS) | T (R) | Δh SUBLIMATION (BTU/LB) |
|---|---|---|
| 0.001 | 5160 | 12,300 |
| 0.010 | 5400 | 11,300 |
| 0.100 | 5620 | 10,100 |
| 1.000 | 5920 | 9,250 |
| 10.000 | 6340 | 8,975 |

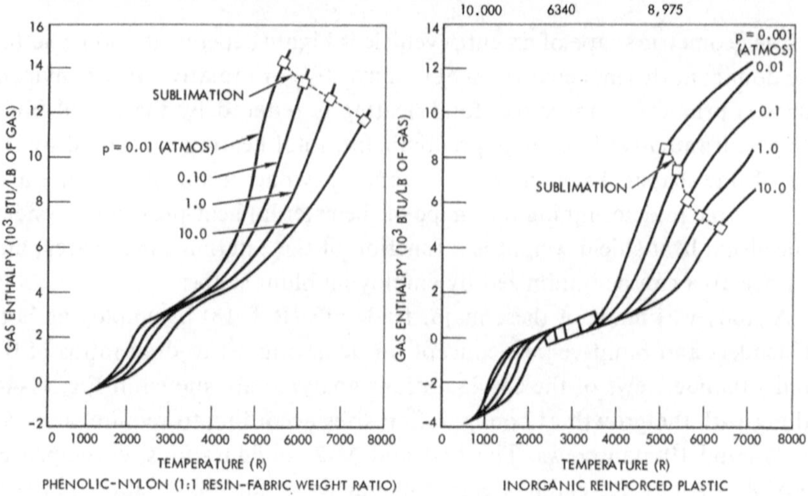


**Fig. 11-45.** Comparison of pyrolysis product enthalpies.

age just one man. When this critical dimension is reached in the lenticular and M-2 shapes, the vehicles are capable of carrying many men.

The convective and radiative stagnation-point heating rates and the heating distributions previously described for earth entry from planetary missions were employed in a charring-ablation computer program to obtain weights of heat-protection systems for earth entry of the above four configurations at velocities up to 75,000 feet per second. Two types of charring-ablation heat-protection systems were investigated: inorganic reinforced plastic and phenolic-nylon (50-percent phenolic and 50-percent nylon, by weight). Adequate experimental and theoretical studies of these types of charring ablators have been made to provide consistent analyses of their heat-protection capability at high speeds.

Because of the high peak heating rates experienced during the initial entry phase, which persist for approximately the first 100 seconds of the trajectory, currently available materials will either melt or sublime. During the later phases of the entry trajectory, surface erosion of charring-ablation systems will revert to the usual mode of chemical erosion due to oxidation. Both modes of surface-material ablation were accounted for in the investigation. However, at high entry velocities, sublimation is the primary mechanism for material ablation in the stagnation regions of the vehicles. The chemical equilibrium enthalpy data of the gaseous products of these materials are shown in Fig. 11-45 as a function of pressure and temperature, and demonstrate the relative efficiency of these two types of heat-protection systems. Also, the gas enthalpy and temperature where sublimation occurs are indicated as a function of pressure. Phenolic-nylon data indicate that about 14,000 Btu/lb are absorbed in subliming phenolic-nylon material at 1 atmo-

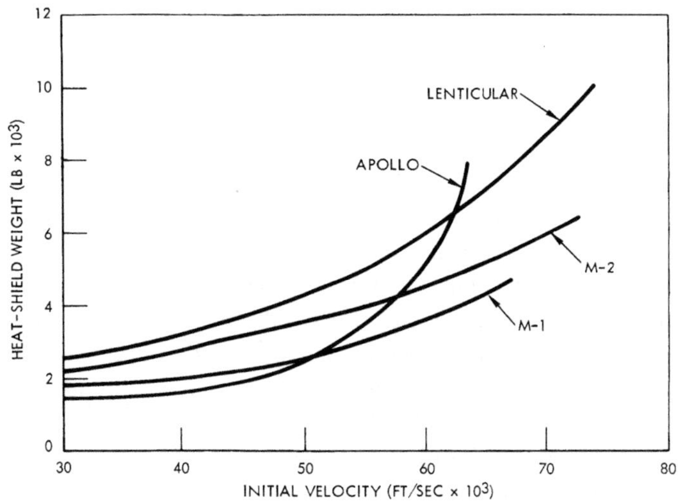


**Fig. 11-46.** Effect of configuration on heat-shield weight—6-man earth-entry vehicle (Ref. 18).

sphere of pressure, while about 9200 Btu/lb are absorbed by inorganic reinforced plastic.

A comparison of the weights of heat-protection systems for the 6-man vehicles is shown in Fig. 11-46 as a function of entry velocity for an inorganic reinforced plastic heat shield. This illustration indicates that the weight of the Apollo heat shield is lowest for velocities up to 50,000 feet per second, but increases rapidly thereafter, primarily because of the requirement to accommodate increasing radiative heat transfer encountered at the higher velocities. The M-1 configuration has somewhat higher heat-shield weights than does the Apollo shape up to 50,000 feet per second, where a crossover occurs because of the relatively lower radiative heating encountered by the finer M-1 vehicle. The lenticular and M-2 vehicles each have large heat-shield weights because of their large surface areas and the longer flight times associated with a lift-to-drag ratio of 1.0.

## [11-5] SUMMARY

Despite the relatively severe environment encountered by an entry vehicle returning from a lunar mission, the improvement in efficiency of heat-protection systems since the early Mercury missions allows such an entry to be made with only an insignificant increase in the ratio of heat shield-to-total-vehicle weight. Confidence in the reliability of these advanced charring-ablation heat-protection systems and in the predicted weight requirements has been gained from many ground tests in plasma facilities that simulate the mission thermal environment and from limited unmanned flight tests using probes. Unmanned flight tests of the full-scale Apollo command module have been made at orbital velocities to test the entire Apollo system. Soon, unmanned entry flight tests will be conducted at earth-escape velocity to qualify the Apollo command module for the manned lunar mission.

In the 1980 period, manned missions to Mars and Venus may be undertaken, The results of recent studies, some of which have been presented here, indicate that earth entry at high velocities appears feasible, even in light of the criticality of the entry-corridor depth and the severe design requirements placed on the heat-protection system. In addition, the use of Venus to alter the course of a vehicle returning from Mars missions would significantly reduce earth-entry velocities. Such an operational mode would enhance the feasibility of extending blunt-shaped bodies, characterized by the Apollo command module configuration, to planetary missions.

Some of the major points developed in this chapter are summarized in Fig. 11-47 through 11-50, which cover the major problems of flight perfor-

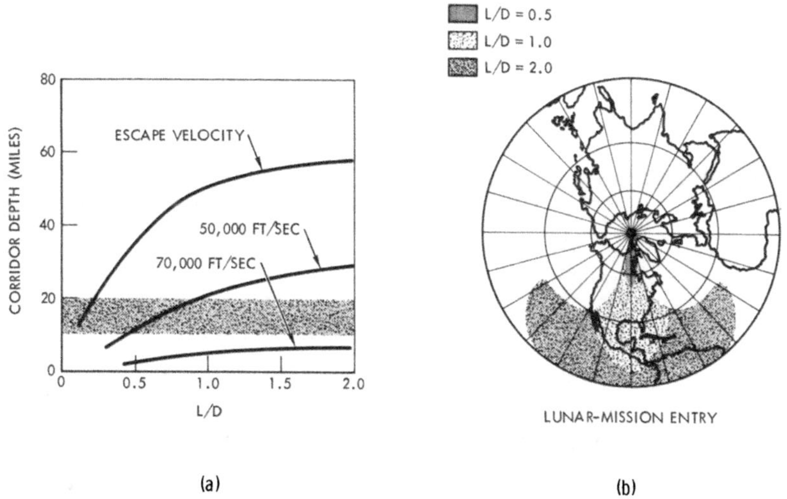


**Fig. 11-47.** Corridor depths and maneuvering envelopes.

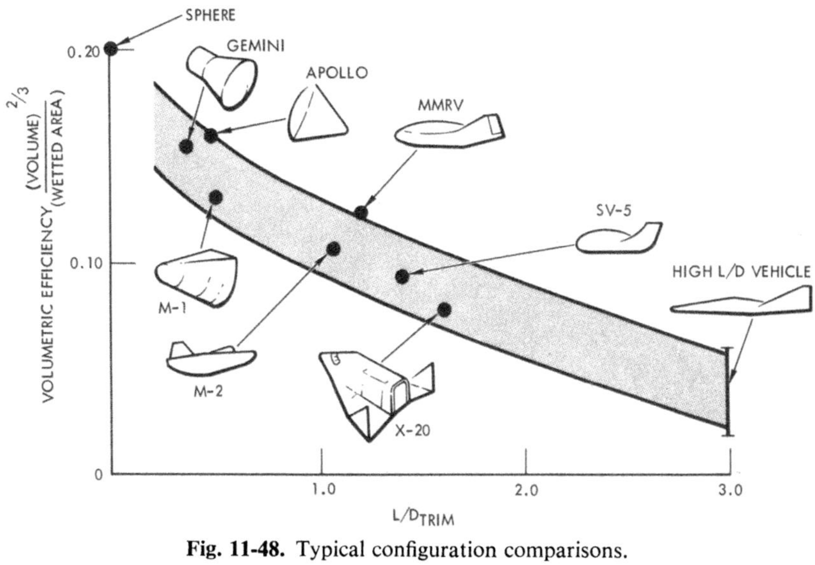


**Fig. 11-48.** Typical configuration comparisons.

mance and thermal environments associated with earth entry from lunar and planetary missions. Fig. 11-47a shows the depth of the entry corridor and the maneuvering envelopes as a function of lift-to-drag ratio. From this illustration, it can be concluded that, for velocities up to 50,000 feet per second, an $L/D$ of 0.5 is sufficient to allow entry into a 10-nautical-mile corridor—the accuracy limitations of existing guidance systems. Increasing the $L/D$ beyond 1.0 does not significantly increase the corridor depth. Envelopes of

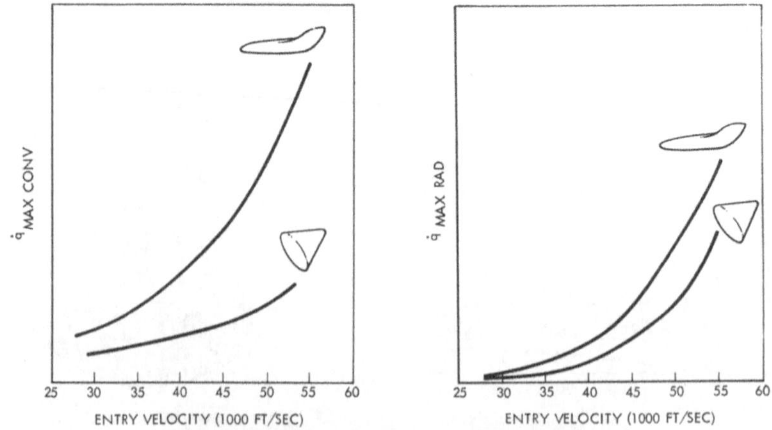


**Fig. 11-49.** Stagnation-point Heating trends.

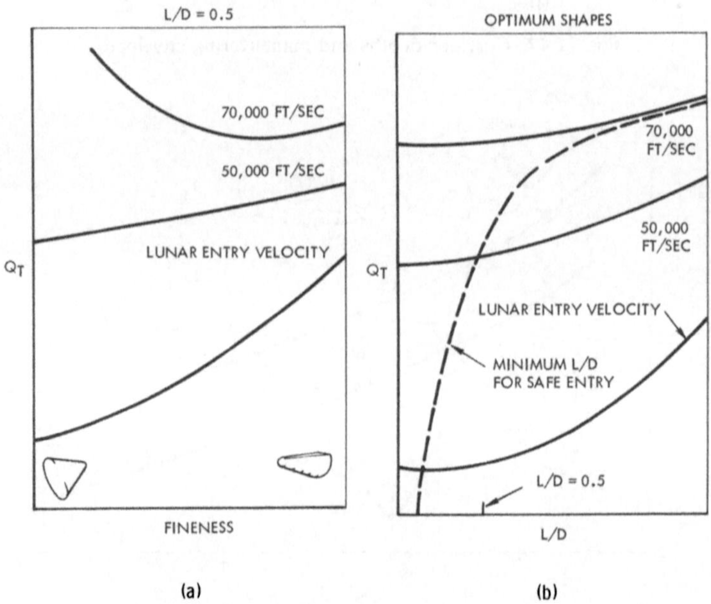


**Fig. 11-50.** Total heat load.

maneuvering capability are shown in Fig. 11-47b for entry from lunar missions. The envelopes for $L/D$'s of 1.0 and 2.0 are significantly larger than those for an $L/D$ of 0.5. However, the requirement for maneuvering following entry is adequately satisfied by vehicles having an $L/D$ of 0.5 because the geographical entry position can be adjusted with a small propellant expenditure prior to earth entry.

Fig. 11-48 illustrates the volumetric efficiency of configurations as a function of $L/D$. Since entry vehicles must become finer as $L/D$ increases, the surface area per unit of volume increases with increasing $L/D$, resulting in decreased volumetric efficiency. The weights of structural and heat-protection systems tend to increase with surface area; consequently, vehicle weight can be minimized by choosing the lowest $L/D$ consistent with mission requirements.

For entry velocities associated with lunar missions, it has been shown that radiative heating contributes a relatively small percentage of the total heat load for vehicles of practical interest, convective heating being the major component of the total load. On the other hand, for velocities characteristic of earth entry from planetary missions, radiative heating has been shown to be predominant. Stagnation-point heating trends for convective and radiative peak heating rates are shown in Fig. 11-49. Since convective heating is inversely proportional to the square root of the radius, blunt bodies minimize convective heating. Slender bodies, however, tend to minimize radiative heating, assuming a given altitude and velocity, because such heating is directly proportional to the radius. But factors other than geometrical shape should be considered in practical design. For example, slender bodies would tend to decelerate at a lower altitude because of the higher ballistic coefficient. As was shown in Subsection 11-2.1, the ballistic coefficient is an altitude-scaling parameter. Since the maximum deceleration occurs at lower altitude, and thus higher density, both convective and radiative peak heating rate per unit area would increase with the slender body (see table in Subsection 11-3.1).

Fig. 11-50 shows the trends in total heat load. In Fig. 11-50a, $L/D$ is fixed at 0.5, and the vehicle fineness is varied. In Fig. 11-50b, "optimum shapes" are shown, i.e., the degree of fineness for minimum total heating at different lift-to-drag ratios. Because of the rapidly increasing radiant heat transfer, finer shapes tend to reduce the total heat load at very high entry velocities. If an optimum shape (one that minimizes total heating) were selected as a function of $L/D$ and entry velocity, the trend in total heating would be that shown in Fig. 11-50b, which indicates that the lowest possible $L/D$ is preferable to reduce total heating. If, however, the entry-corridor requirement is imposed as a constraint on the minimum acceptable $L/D$ (as defined in Fig. 11-47a), an $L/D$ of at least 0.25 is required for lunar missions and an $L/D$ of at least 0.5 is required for entry velocities of 50,000 feet per second. Earth entry at velocities above 70,000 feet per second appears marginal, regardless of the $L/D$, because of diminishing corridor depth (see Fig. 11-47a).

It has been shown that heat-protection systems employing charring ablators can withstand the severe thermal environment associated with entry velocities as high as 55,000 feet per second. However, because of the significant influence of radiative heating on vehicle design, the current uncertainties in determining the gas-dynamic radiation at high velocities require that further theoretical and experimental studies be performed for this flight regime. Flight tests at hyperbolic velocities similar to those completed for the FIRE program will be required to obtain substantiating data. Although current theoretical estimates indicate that existing systems utilizing charring ablation for heat protection are adequate for earth entry from planetary missions, existing experimental facilities cannot adequately simulate this environment for testing heat-shield materials. Consequently, additional facilities must be developed and flight testing must be conducted at hyperbolic velocities to obtain data to substantiate the theoretical estimates.

Based upon available data on the radiative properties of air and its constituents, it appears that blunt bodies with heat-protection systems utilizing charring ablators can withstand the thermal environment of earth-entry velocities up to approximately 55,000 feet per second. Earth-entry velocities of 55,000 feet per second or less are associated with Venus missions and Mars missions employing Venus swingby (see Subsection 11-1.4). For earth-entry velocities greater than 55,000 feet per second, where radiative heating is predominant, slender bodies may offer some advantages. Although the stagnation-point radiative heating is greater for finer shapes, the area affected by radiative heating is smaller. However, transition to turbulent flow occurs earlier for slender bodies than for blunt bodies, resulting in higher heat-transfer rates and reduced ablative heat-protection efficiency. In any case, regardless of the body shape, the 10-mile corridor characteristic of present guidance accuracy must be decreased to accommodate entry velocities greater than approximately 70,000 feet per second.

## SYMBOLS

| | |
|---|---|
| $A$ | = reference area |
| $C_D$ | = drag coefficient |
| $C_H$ | = convective heat-transfer coefficient |
| $C_L$ | = lift coefficient |
| $C_p$ | = specific heat of material |
| $\bar{C}_p$ | = average specific heat of material between $T_{melt}$ and $T_{initial}$ |
| $D$ | = drag |
| $e$ | = 2.718.... |
| $f$ | = function |
| $f'$ | is defined by $V_x = xf'$ |
| $g$ | = gravity |
| $h$ | = altitude or gas enthalpy |
| $h_{eff}$ | = effective heat of ablation |

$I$ = radiation intensity
$j$ = integer
$k$ = density ratio across a normal shock
$K$ = thermal conductivity of material
$L$ = lift
$m$ = mass
$\dot{m}$ = mass rate of material lost
$p$ = pressure
$\dot{q}$ = heating rate
$\bar{q}$ = dimensionless radiative heat-transfer rate at stagnation point
$\dot{q}_{rr}$ = reradiated heat-transfer rate
$Q_T$ = total heat load
$r$ = $R_E + h$
$R$ = gas constant
$R_C$ = radius of curvature
$R_E$ = earth radius
$R_N$ = nose radius
$R_S$ = shock-wave radius of curvature
$s$ = flight-path distance
$\dot{s}$ = recession rate
$S$ = distance along surface
$t$ = time
$T$ = temperature in degrees Rankine (R) or Fahrenheit (F)
$T_C$ = cabin temperature
$V$ = velocity
$V_m$ = oxidizing species diffusion velocity
$V_n$ = surface gaseous products diffusion velocity
$V_x$ = velocity component in $x$ direction
$V_y$ = velocity component in $y$ direction
$V_C$ = circular-orbit velocity
$x$ = coordinate along shock-wave front
$X$ = distance from leading edge of flat plate
$y$ = coordinate locally normal to body ($y=0$ is located at the shock wave)
$\bar{y}$ = dimensionless coordinate ($y/\Delta$)
$Z$ = compressibility factor
$\beta$ = scale height
$\gamma$ = flight-path angle
$\xi$ = $\theta + \gamma$
$\Delta$ = shock-detachment distance
$\Delta h$ = change in enthalpy

$\Delta_0$ = shock-detachment distance for adiabatic flow
$\Delta t$ = time of flight
$\varepsilon$ = surface emissivity
$\eta$ = $\left[\frac{(\ln I)}{(\ln h)}\right]_{p=\text{constant}}$
$\theta$ = range angle
$\varrho$ = density
$\bar{\varrho}$ = dimensionless density ($\varrho/\varrho_i$)
$\sigma$ = Stefan-Boltzmann constant
$\sigma_{TU}$ = ultimate tensile strength
$\tau$ = thickness
$\phi$ = roll angle
$\psi$ = ratio of convective heat-transfer rate with injection to the convective heat-transfer rate without injection

## Subscripts and Superscripts

$c$ = convective
$c_1, c_2, c_3$ = constants
$EN$ = initial entry condition
$FP$ = flat plate
$g$ = gas
$i$ = initial conditions
$0$ = conditions at stagnation point
$p$ = gaseous combustion products
$R$ = recovery conditions
$RAD$ = radiative
$s$ = solid
$SL$ = sea level
$W$ = wall or surface conditions
$*$ = reference condition
$\infty$ = free-stream conditions
$\delta$ = conditions at edge of boundary layer

## REFERENCES


1. Sohn, R. L.: Venus swingby mode for manned mars missions, *Journal of Spacecraft and Rockets*, **1**: No. 5, 565–567 (September-October 1964).
2. Allen, H. J. and A. J. Eggers, Jr.: A Study of Motion and Aerodynamic Heating of Ballistic Missiles Entering the Earth's Atmosphere at High Supersonic Speeds, NASA, Report 1381 (1958).
3. Chapman, D. R.: An Analysis of the Corridor and Guidance Requirements for Supercircular Entry into Planetary Atmospheres, NASA, TR-R-55 (1960).
4. ———: An Approximate Analytical Method for Studying Entry into Planetary Atmospheres, NACA, TN 4276 (May 1958).
5. Loh, W. H. T.: A second-order theory of entry mechanics into a planetary atmosphere, *Journal of Aerospace Sciences*, (October 1962), pp. 1210–1221.
6. Probstein, R. F.: Shock wave and flow field development in hypersonic reentry, *American Rocket Society Journal*, Vol. **31**, (February 1961), pp. 185–194.
7. Hoshizaki, H.: Heat transfer in planetary atmospheres at supersatellite speeds, *ARS Journal* (October 1962).
8. Hanley, G. M.: Hypervelocity laminar convective flat-plate heating, *ARS Journal*, (November 1962), pp. 1740–1742.
9. Eckert, E. R. G.: Survey on Heat Transfer at High Speeds, U.S. Air Force, WADC, TR 54-70 (April 1954).
10. Yoshikawa, K. K. and D. R. Chapman: Radiative Heat Transfer and absorption Behind a Hypersonic Normal Shock Wave, NASA, TN D-1424 (1962).
11. Hanley, G. M. and K. D. Korkan: Inviscid, nonadiabatic flow in the stagnation region of blunt bodies, *Proceedings of the XVth International Astronautical Congress*, Warsaw, Poland (September 1964).
12. ———, ———: Approximate, inviscid, nonadiabatic stagnation region flow field solution, *AIAA Journal*, **3**: 8, (August 1965), pp. 1537–1538.
13. Kivel, B. and K. Bailey: Tables of Radiation from High-Temperature Air, Avco, Research Report 21 (December 1957).
14. Korkan, K. D. and G. M. Hanley: Apollo command module aerothermodynamic characteristics at hyperbolic earth-entry velocities, AIAA paper, No. 65-491 (January 1965).
15. Gravalos, F. G., I. H. Edelfelt, and H. W. Emmoms: The transonic flow about a blunt body of revolution for gases in chemical equilibrium. Paper presented at the 9th Annual Congress of International Astronautical Federation, Amsterdam (August 1958).
16. Offenhartz, E.: Reentry/ablative systems, Short course on Design Factors for Materials Application in Spacecraft, given at the University of California at Los Angeles (May 11, 1965).
17. Kratsch, K. M., L. F. Hearne, and H. R. McChesney: Thermal performance of heat-shield composites during planetary entry. Paper presented at the AIAA-NASA National Meeting, Palo Alto, California (September 1963).
18. Hanley, G. M., K. D. Korkan, and S. Turner: Manned Mars Landing and Return Mission Study, First Quarterly Report, North American Aviation, Inc., Space and Information Systems Division SID 63-1129 (August 1963).

# 12

# *Atmospheric Braking Entry and Associated Technologies*

K. A. EHRICKE, Assistant Division Director
*Astrionics, Autonetics Division,*
*North American Aviation, Inc.*

## [12-1] INTRODUCTION

The use of planetary atmospheres for planetocentric maneuvers offers itself in the case of the planets Venus, Earth and Mars. In all three cases, the atmospheres are comparatively fluffy and thin, if compared to atmospheres of the jovian planets, and the entry velocities are usually within the range of technical feasibility. The highest entry velocities are encountered upon return from Mars along a mono-elliptic transfer orbit. Maximum velocities are in the range of 70,000 to 75,000 ft/sec. It is possible, however, to reduce the speed at least as far down as 50,000 to 60,000 ft/sec, either by a properly arranged mission profile or by a braking maneuver or by Venus fly-by on return. The braking maneuver is most economically applied at perihelion transit rather than at Earth approach. If the velocity is reduced to the level of about 60,000 ft/sec by mission profile adjustment, direct atmospheric entry along an unretarded approach hyperbola, is technically, as well as possibly space medically, the most difficult case of atmospheric braking; but also the most rewarding case in terms of effect on the reduction in orbital departure weight of the interplanetary vehicle. One factor which in all cases makes it easier to brake at Earth than in the atmospheres of Venus or Mars is the fact that the vehicle weight is down to the level of the entry model, whereas a much larger mass must be designed for the thermal and load stress accompanying atmospheric entry in those cases where the atmospheres of Mars and Venus are involved. The resulting rather heavy weight of the drag brake system and associated heat shield is one of the reasons why atmospheric braking pays off comparatively less in the case of these two planets. Even so, by proper shaping of the mission profile and proper design of the drag brake, worthwhile reductions in orbital departure weight can be achieved. These reductions can be translated into dollars.

The subsequent material discusses the atmospheric braking at Venus and Mars, the associated requirements and illustrates on a number of examples the potential cost savings resulting from such a maneuver.

## [12-2] THE ATMOSPHERES OF VENUS AND MARS

The overall atmospheric models of both planets, though still quite uncertain, are better understood owing to the successful missions of Mariner II to Venus in 1962 and Mariner IV to Mars in 1965. For the purposes of atmospheric braking on planetary missions, however, only the outer regions of the Venus atmosphere are of interest, whereas the thin atmosphere of Mars requires closer approach to the planet's surface.

### [12-2.1] Venus Atmosphere

The results gained from the Mariner II mission (Ref. 12-1) indicate the upper level of the Venus cloud cover is at about 90 km altitude. Temperatures in the $-30$ to $-70\,^{\circ}$F range were observed in the atmospheric layer above the clouds. Not enough $CO_2$ was observed to be present above the clouds to effect appreciable absorption in the 10-micron region. With negligible water and $CO_2$ content, it appears that the Venus atmosphere above 90 km (about 300,000 ft) should not be very much different from that of Earth. The principal constituent must be nitrogen. De Vaucouleurs (Ref. 12-2) determined, from the occultation of Regulus by Venus, a pressure of $2.10^{-6}$ atm and a temperature of about $290\,^{\circ}$K at a height of about 95 km. This compares with a pressure of about $2.1 \cdot 10^{-6}$ atm and $200\,^{\circ}$K at 95 km in the Earth's atmosphere. The pressures are practically identical. The difference in temperatures is larger; but, considering the conclusions from the Mariner II experiments, the more likely temperature of the Venus atmosphere at this altitude is of the order of 230 to $250\,^{\circ}$K. For these reasons, it is concluded on the basis of available evidence that the atmospheric conditions for atmospheric braking near Venus are comparable to those in the upper atmosphere of Earth.

### [12-2.2] Mars Atmosphere

Absence of barriers such as the cloud covers of Venus and the weaker gravitational force of Mars do not only cause the atmosphere to be more rarified, but probably also better mixed than that of Venus or Earth. Uncertainties in surface conditions, therefore, are more intensely reflected in the evaluation of conditions at higher altitudes than in the case of Venus.

The density and pressure gradients in the case of Mars are only about one third to one fourth of the terrestrial values. For a Mars atmosphere of 85 mb surface pressure, this results in a pressure-density equivalence with the terrestrial atmosphere in the 25–30 km altitude region and in higher Martian pressure and density at higher altitude. A surface pressure of 85 mb (0.085 atm) was long considered a representative value for the Mars atmosphere. The occultation experiment carried out by Mariner IV, however, showed that the surface pressure is only of the order of 5 to 15 mb. A pressure of 10 mb (0.01 atm) exists in the Earth atmosphere at about 33 km (100,000 ft). If the Martian surface pressure were 10 mb, this would displace the equilibrium altitude of pressure-density with that of Earth to about 45 km (150,000 ft).

### [12-2.3] Conclusions

Fig. 12-1 compares the atmospheric pressures versus altitude for the three planets Venus, Earth and Mars. In view of the somewhat weaker gravitational

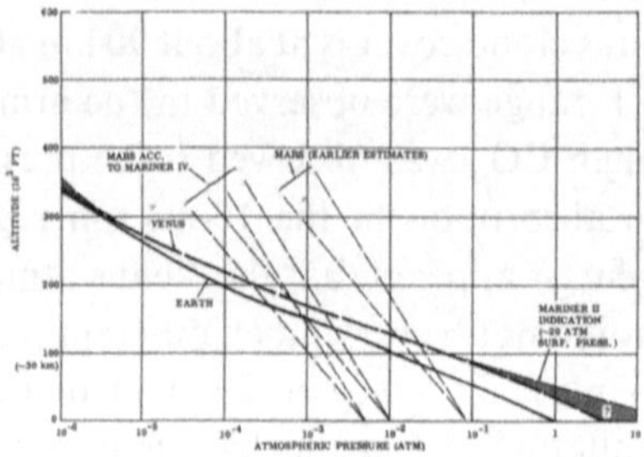


**Fig. 12-1.** Comparison of atmospheric pressures (Venus, Earth, Mars).

field of Venus and the greater proximity to the sun, one might expect a decrease in the pressure and density gradient above the cloud layer, as indicated in the upper left corner of the Venus curve. On the lower right of the curve, there is also an uncertainty left. The conclusions drawn from the Mariner II experiment (Ref. 12-1) indicate a surface pressure of about 20 atm. The uncertainty at the lower end of the Venus pressure curve, however, is of lesser importance than the one at the upper end for purposes of atmospheric braking, rather than landing on Venus. The dense cloud-cover region of the atmosphere seems to imply a pressure-density gradient not too much different from that of the terrestrial atmosphere. If this trend were to continue above the cloud layer (i.e. above about 90 km) it could imply the necessity for the HISV to dive into the outer cloud layers, namely, to about 70 km (250,000 ft) altitude. But, while this possibility has only moderate probability, it could almost certainly be avoided if the atmospheric gradient above the cloud layer would decrease only to the level of 75–50 percent of that of the terrestrial atmosphere. The atmospheric constituents should be primarily nitrogen in molecular, dissociated and ionized form, to which some minute amounts of argon, oxygen, carbon and hydrogen are added. Such atmosphere is, technologically speaking, considerably more "friendly" than the expected lower atmosphere of Venus. Atmospheric braking of an HISV should, therefore, be a comparatively less problematic ask, from the environmental standpoint, than landing on he planet's surface.

The enormous reduction of the Martian surface pressure, as compared to earlier estimates, augments the technological problems. This, however, affects primarily the task of soft-landing on the surface and was, in part, the cause for the change in Voyager launch vehicle from Saturn IB/Centaur to Saturn V. However, so far as atmospheric braking for capture of an HISV is concerned, the consequences are less severe. Even at a surface pressure of 5 mb ($5 \cdot 10^{-3}$ atm), the pressures in the 150,000 to 200,000 ft altitude region should be similar and the Martian atmosphere should feature higher pressures above roughly 200,000 ft. Since the Mars approach velocities of the HISV are in the range of 22,000 to 27,000 ft/sec for manned missions, it is unlikely that a surface approach to less than 150,000 ft is required. Thus, the atmospheric braking maneuver would still occur at altitudes at which the atmospheric pressure is higher, equal or only slightly lower than that of the terrestrial atmosphere. This favorable circumstance is due to the lower pressure-density gradient of the Martian atmosphere which, in turn, is caused by the presumed greater vertical dynamic and more thorough mixing process of the atmospheric gases. It must, therefore, be assumed that the upper atmospheric layers of Mars, those which would be traversed by the HISV, are more enriched with CO and $CO_2$ than on Earth or Venus. This fact, and the longer deceleration path through the "fluffy" atmosphere indicate a potentially more severe heating problem than for braking at Venus. This problem is alleviated only by the lower arrival velocities involved at Mars than at Venus.

## [12-3] ATMOSPHERIC BRAKING (AB)

The "standard" atmospheric braking (AB) process for the purpose of planetary capture consists of two primary maneuvers: (1) atmospheric passage, and (2) periapsis-raising maneuver at the apoapsis of the resulting non-circular capture orbit. This process, which is shown in Fig. 12-2, assumes precise atmospheric passage, so that the apoapsis of the capture orbit coincides with the desired altitude of the circular terminal capture orbit. If the HISV lost more velocity than intended during atmospheric passage, the

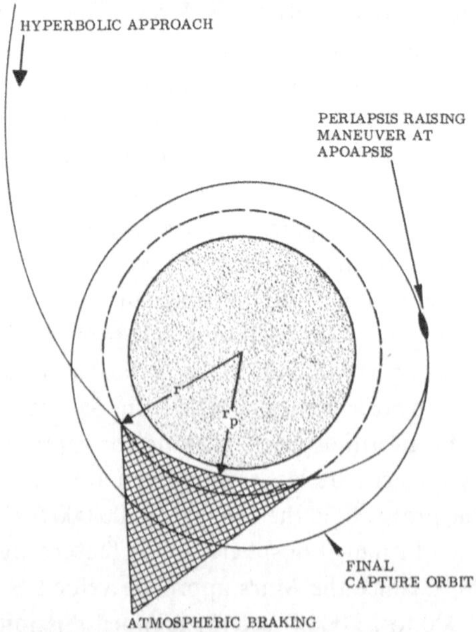


**Fig. 12-2.** Atmospheric braking to circular orbit (Schematic).

subsequent periapsis-raising maneuver must raise the apsis to the desired circular orbit altitude, followed by a second powered maneuver at that point to establish circularity. Atmospheric braking to a circular capture orbit is referred to subsequently by the abbreviation ABC.

Atmospheric braking to an elliptic orbit (ABE) is shown in Fig. 12-3. This process reduces the demands on the atmospheric braking proper, thereby reducing heating period and alleviating heat loads.

## [12-3.1] Venus

Fig. 12-4 shows the variation of hyperbolic excess velocities for a typical fast round-trip mission (about one year) to Venus. If atmospheric braking is to be applied, it is desirable to keep the HISV from being unduly large. This means that $v^*_{\infty 3}$, the hyperbolic excess velocity which is a measure of the magnitude of the Venus departure maneuver, should be as small as possible.

Also, $v^*_{\infty 4}$, which is a measure of the atmospheric entry speed at Earth return, should be small enough to permit entry without preceding retro-maneuver, to keep the terminal mass small. These requirements mean that departure should be as early as possible in a given window, i.e., before minimum $v^*_{\infty 1}$, thereby raising the magnitude of $v^*_{\infty 2}$, that is, the only maneuver which requires no propellant mass (or very little) if atmospheric

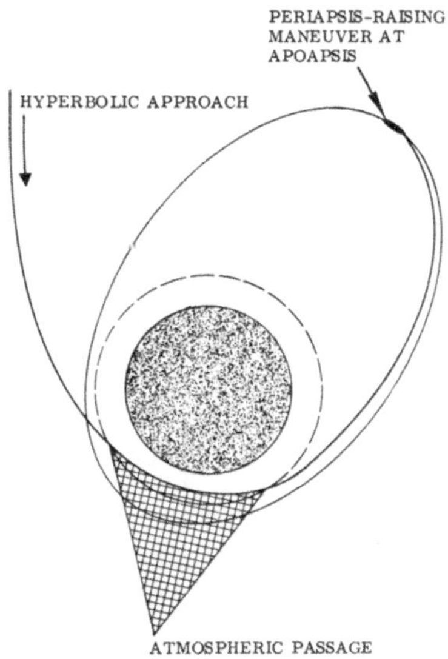


**Fig. 12-3.** Atmospheric braking to elliptic orbit (Schematic).

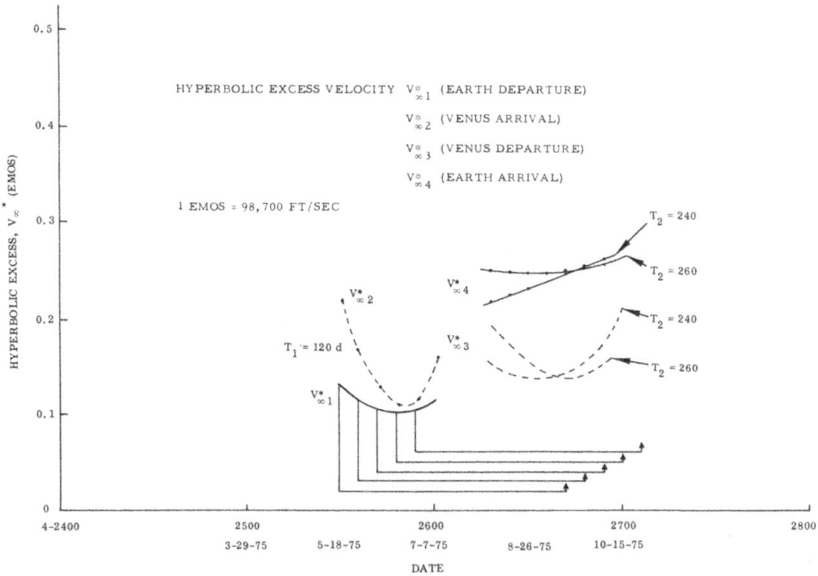


**Fig. 12-4.** Earth-Venus round-trip mission 1975: hyperbolic excess velocities.

braking is applied. This causes $v^*_{\infty 2}$ to be larger than if the mission is optimized for minimum Earth orbital departure weight (ODW) with powered maneuvers on all transfer terminals. The values of $v^*_{\infty 2}$ for any 120 to 150 day transfer corridor to Venus lie between 0.15 and 0.3, mostly between 0.15 and 0.25 EMOS. Figure 12-5 shows that the latter range corresponds to entry

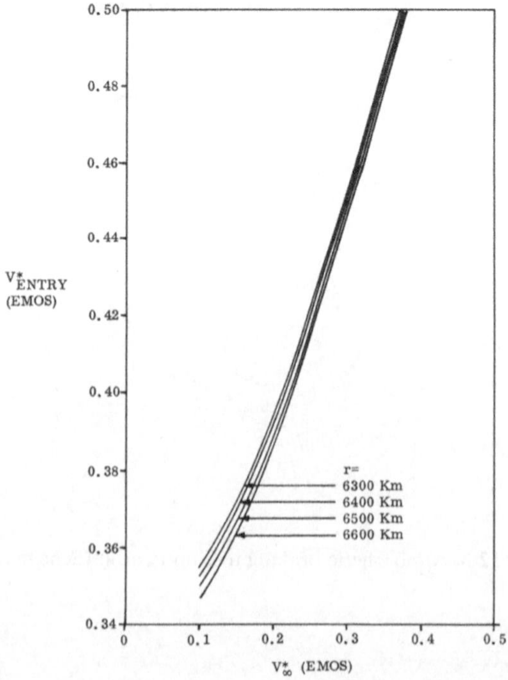


**Fig. 12-5.** Relation between hyperbolic excess velocity and entry velocity for Venus.

between 0.22 and 0.26 EMOS, corresponding to about 43,000 and 45,000 ft/sec entry velocity at Earth return. The circular velocity of Venus is given by the relation

$$v_c = 22{,}000/\sqrt{r^*}\ (\text{ft/sec})$$

where $r^*$ is the distance from the planet's center in units of the planet's radius. At $r^* = 1.1$ the circular velocity is 21,000 ft/sec. At the above Venus

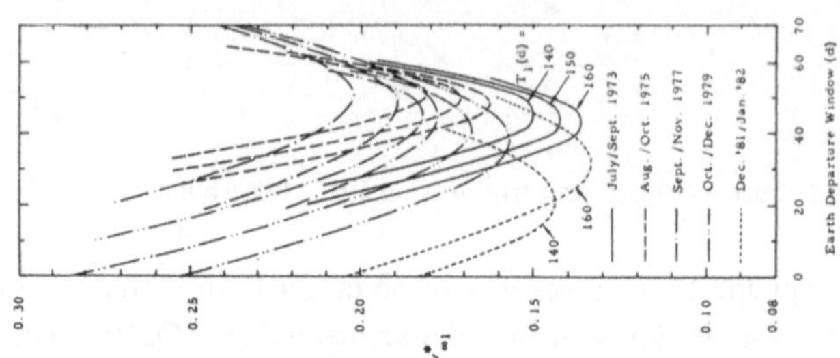


**Fig. 12-6.** Hyperbolic excess velocity at earth departure for earth departure windows 1973 through 1982. (The three curves of each group represent transfer periods of 140, 150 and 160 days in the sequence shown for the 1973 curves).

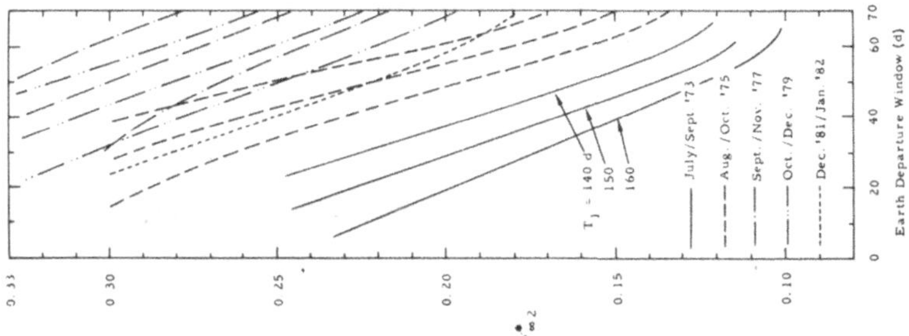


**Fig. 12-7.** Hyperbolic excess velocities at mars approach for earth departure windows 1973 through 1982. (The three curves in each group represent transfer periods of 140, 150 and 160 days in the sequence shown for the 1973 curves).

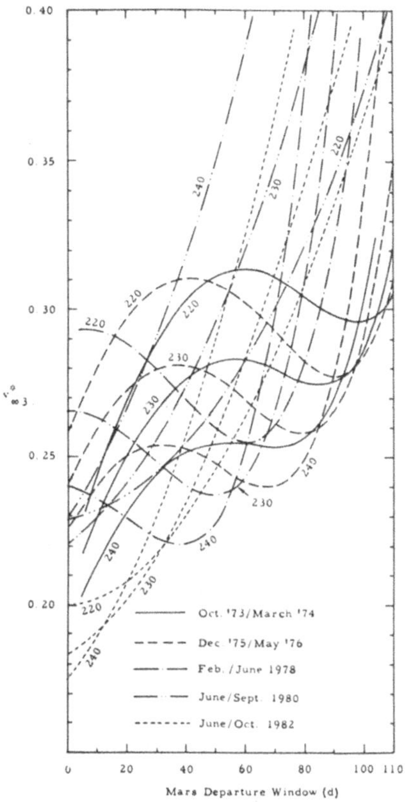


**Fig. 12-8.** Hyperbolic excess velocity at mars departure for mars departure windows 1973 through 1982. (Numbers on curves designate return transfer time $T_2$ in days).

arrival velocities, the velocity reduction which must be achieved for capture to a near-circular velocity is, therefore, of the order of 15,000 to 20,000 ft/sec.

### [12-3.2] Mars

Figs. 12-6 through 12-9 show the hyperbolic velocities for fast Mars round-trip missions. Since conditions, especially for Earth return, vary considerably over a complete cycle of mission opportunities, a total of five opportunities is shown, covering the more difficult opportunities. Again, as in the case of Venus, early departure at a given transfer window, results in lowering the Mars departure impulse and the atmospheric entry velocity at Earth return. It is seen that the hyperbolic excess velocity $v^*_{\infty 2}$ lies mostly between 0.2 and 0.3. Fig. 12-10 shows that this corresponds to about 26,000 to 34,000 ft/sec.

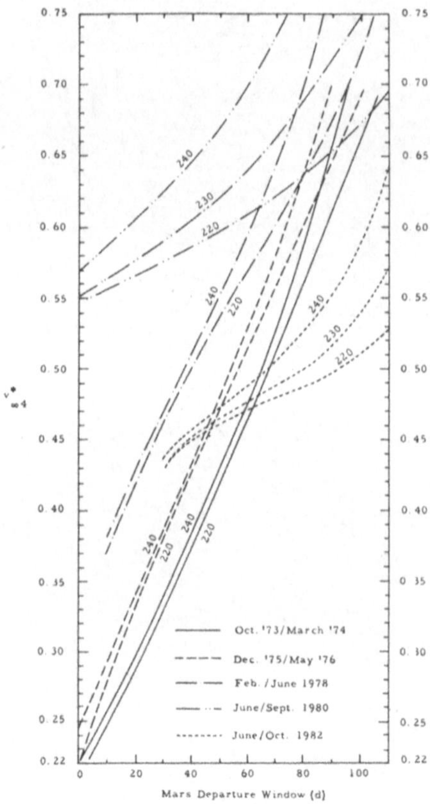


**Fig. 12-9.** Hyperbolic excess velocities at earth approach for mars departure windows 1973 through 1982. (Numbers on curves designate return transfer time $T_2$ in days).

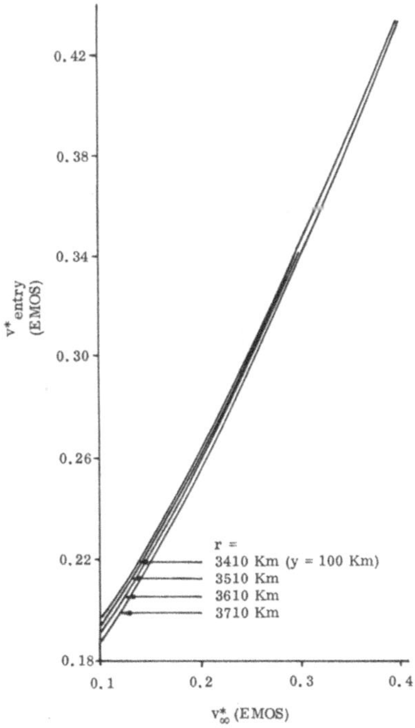


**Fig. 12-10.** Correlation between hyperbolic excess velocity (v*) to atmospheric entry velocity ($v^*_E$) at DIFFERENT altitudes for planet Mars.

The circular velocity at Mars is given by the relation

$$v_c = 11{,}800/\sqrt{r^*}\ \text{(ft/sec)}$$

At $r^* = 1.3$ the circular velocity is 10,300 ft/sec. Thus, the atmospheric braking maneuver must reduce the velocity near Mars by some 15,000 to 22,000 ft/sec.

### [12-3.3] Conclusions

It is interesting to note that the amount of velocity reduction is approximately the same, but the entry velocity in the case of Venus is approximately 10,000 ft/sec higher than for Mars.

The Venus departure maneuver is generally lower, namely, of the order of 14,000 to 17,000 ft/sec, compared to 18,000 to 24,000 ft/sec for Mars (Fig. 12-11). The unretarded atmospheric entry velocity at Earth return is significantly lower when returning from Venus compared to return from Mars (Fig. 12-11). For these reasons, the terminal vehicle returning into the Earth's atmosphere is lighter and the planet departure stage smaller (for a dual reason, namely, lighter Earth entry vehicle and smaller planet departure

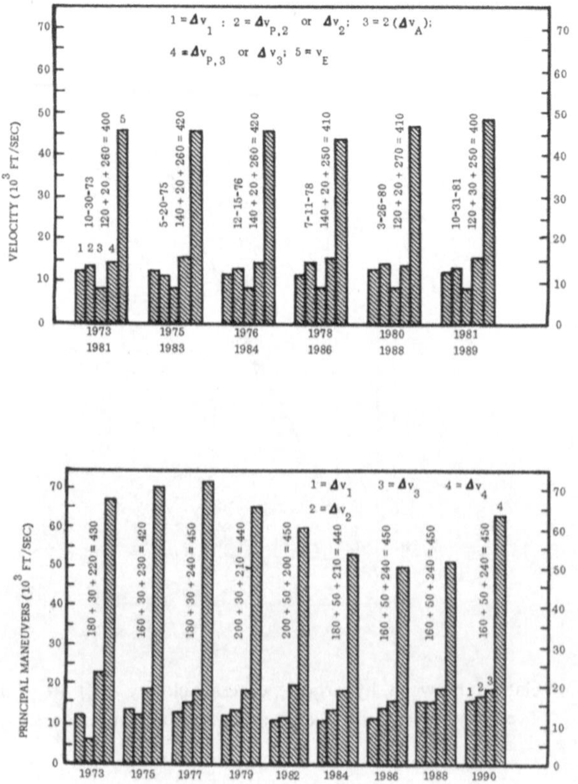


**Fig. 12-11.** Mono-elliptic round-trip velocity profiles for Venus and Mars missions. Velocity profiles are based on initial mass Minimization. Powered maneuvers represent earth departure, target planet arrival and departure. Fourth bar represents earth entry velocity along unretarded hyperbola on return from target planet.

maneuver) in the case of Venus than of Mars. Therefore, the size and mass of the HISV upon arrival at the target planet is smaller in the case of Venus. A lighter and smaller vehicle reduces the problems of thermal protection and aerodynamic drag braking. From these viewpoints, then, atmospheric braking at Venus appears to pose comparatively fewer problems than braking at Mars.

From the viewpoint of absolute velocities, Mars offers more favorable conditions. However, these might be balanced by more severe radiative heat transfer conditions if the stronger mixing process in the Martian atmosphere does, indeed, lead to a higher CO and $CO_2$ content in the brake region of Mars than of Venus.

## [12-4] DRAG BRAKE SYSTEM (DBS)

Fig. 12-11 shows maneuver velocities based on mission timing which minimizes the Earth orbital departure weight (ODW) of the HISV on the assumption that both capture and escape at the target planet are accomplished by means of powered maneuvers. As noted in the preceding section, the capture maneuver tends to be smaller and the escape maneuver larger than when the mission is optimized (for minimum ODW) for atmospheric braking. In the latter case, greater advantage is taken of the fact that the DBS weight varies far less with the delta-$v$ required for capture than would the propellant weight in the case of a powered maneuver. Therefore, the conditions shown in Fig. 12-11 are comparatively less advantageous to the application of atmospheric braking. If atmospheric braking can be shown to result in lower ODW under these conditions, it can be assumed to be even more advantageous for mission profiles optimized for the application of aerodynamic braking.

### [12-4.1] Propellant Fractions

For Mars, low capture maneuver requirements lie between 11,000 and 13,000 ft/sec for establishment of circular orbits at $r^* = 1.3$. If the capture maneuver is to be carried out propulsively, the propulsion system could be either chemical, assuming a specific impulse $I_{sp} = 450$ sec, or nuclear (solid core reactor) in which case $I_{sp} = 800$ sec can be postulated. In the more likely case of nuclear propulsion, taking 12,000 ft/sec for the capture maneuver, the resulting mass ratio is 1.6, corresponding to a propellant weight fraction of 0.375. In the case of chemical propulsion, the mass ratio is 2.29 with a propellant weight fraction of 0.564. Mass ratio and propellant weight fraction

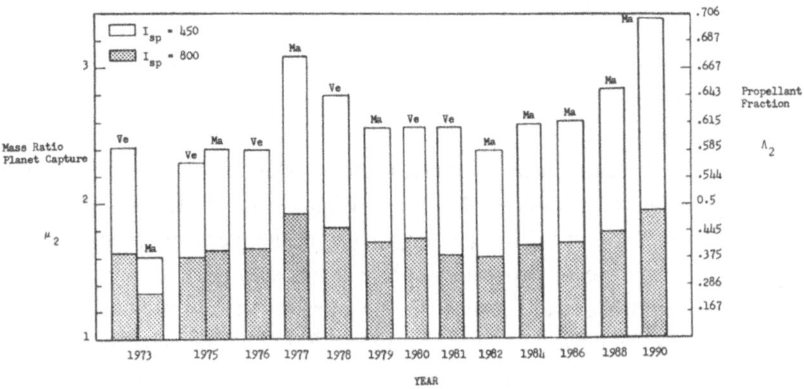


**Fig. 12-12.** Variation of mass ratio for mars round-trip mission: Mars capture (M-2) (Circular orbit; $r^* = 1.3$).

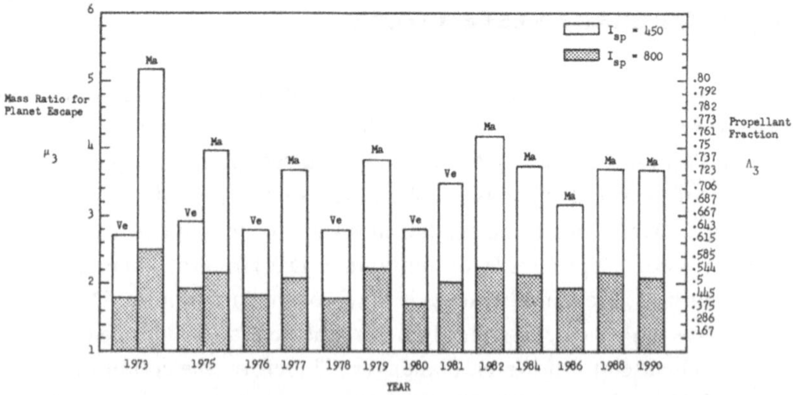


**Fig. 12-13.** Variation of mass ratio for mars round-trip mission: Mars departure (M-3) (Circular orbit; $r^* = 1.3$).

for Mars capture missions in the period of 1975 through 1990 (corresponding to those in Fig. 12-11) are shown in Fig. 12-12 for the capture maneuver and in Fig. 12-13 for the departure maneuver with $I_{sp} = 800$ and 450 sec, respectively*. It is seen that the propellant required for the capture maneuver (no losses included) constitutes between 20 and 50 percent of the weight of the arriving HISV for nuclear propulsion, and between 58 and 70 percent for chemical propulsion (which is one of the reasons why it is unlikely that chemical propulsion will be used for a fast mono-elliptic Mars round-trip capture mission). For the departure maneuver, the propellant fractions range from 47 to 56 percent (nuclear) and from 69 to 76 percent (chemical).

For Venus, Fig. 12-12 shows that the capture maneuver propellant fraction varies far less than for Mars, but is generally higher, lying between 39 and 44 percent (nuclear) and 58 and 64 percent (chemical). The departure maneuver propellant fraction, on the other hand, is smaller for Venus, ranging from 42 to 51 percent (nuclear) and from 61 to 71 percent (chemical), as seen in Fig. 12-13. For elliptic capture at Venus (which reduces the overall mission velocity requirement for round-trip missions with powered maneuvers only**) at an apsidal ratio of $n = r_A/r_P = 8$ and $r_P^* = 1.1$, the capture velocity requirement is reduced significantly, namely, to 5000–6000 ft/sec. Using 6000 ft/sec and chemical propulsion ($I_{sp} = 450$ sec) the mass ratio is 1.273 and the propellant weight fraction 0.214. The departure maneuver, on the other hand, is increased, because in order to achieve a net saving in mission velocity, departure should occur at or near the periapsis of the elliptic capture orbit. Unless the capture period is quite short (1 to 3 days), this is likely to require a rotation of the ellipse's major axis by a certain angle. This rotation ma-

* The departure maneuver assumes coplanar and cotangential departure from the capture orbit.

neuver, which must precede the escape maneuver proper, can be as large as 8500 ft/sec for an $n=8$, $r_p^*=1.1$ elliptic orbit about Venus. While this still results in a saving compared to powered capture and escape involving a circular orbit, it represents a shift in the wrong direction from the standpoint of atmospheric braking, since it reduces the velocity change accomplished by braking and increases the departure maneuver for which propellants are required.

### [12-4.2] Drag Brake System (DBS)

When atmospheric braking is substituted for a propulsive capture maneuver, the weight of the DBS replaces the weight of the propulsion system. The latter consists of propellant, tanks and thrust system. The thrust system must be accounted for in full, if a different thrust system is used for the departure maneuver. If the same thrust system is used in both maneuvers, only propellant and tanks required for the capture maneuver should be considered. The drag brake system consists of the heat shield (ablation material, insulation and structure) for the HISV body and of the drag brake proper (ablation material, high-temperature insulation, structure and actuators). The HISV body consists of the propulsive weight required for the periapsis-raising maneuver following atmospheric passage, of the planet departure stage, of the propulsive weight required for Earth approach maneuvers and of the life support section, i.e., the sum of all mission modulus plus the Earth entry module for terminal dive into the Earth's atmosphere.

The problem, therefore, boils down to reducing the DBS weight fraction as far as possible below the propellant weight fraction required for the capture maneuver. It is apparent from the above data that this is not difficult, if chemical propulsion is used. For more advanced propulsion systems, however, the possible gains are smaller and eventually become sufficiently marginal to lie in the uncertainty region of determining the DBS weight in view of the many gaps in engineering data required to determine DBS design and weight.

An important factor determining the maximum heat load is the peak acceleration. The higher this peak value, the deeper the HISV must dive into the atmosphere and the more intense will be the maximum heat load. Also, the mechanical stresses, hence the structural weight of the HISV, will increase with increasing peak deceleration. When propulsive forces are used for the capture maneuver, peak thrust acceleration can be kept well under $2g_{Ea}$*. Therefore, the deceleration load associated with atmospheric braking should be kept as low as possible. For a given drag brake configuration, the desired

* The same is not true for Mars, because of its weaker gravitational field.

** $g_{Ea}$ = Earth-$g$

velocity reduction is attained at a given low peak deceleration more readily in a fluffy atmosphere, i.e., in the Mars atmosphere rather than the Venus or Earth atmosphere. For a given atmosphere, the required peak deceleration for a given velocity reduction is lowered by lowering the $W/C_DA$ (drag parameter) value, i.e., by enlarging the drag brake area $A$ for a given HISV weight $W$ and configuration drag coefficient $C_D$. Unfortunately, the weight of the drag brake tends to rise rapidly with its size, unless special designs are considered. Following several trajectory computations, a peak deceleration of $5g_{\text{Ea}}$ was tentatively selected.

Regarding the configuration of HISV and DBS, three basic alternatives exist: (1) High-drag body shape of the HISV; (2) Low-drag body shape of the HISV with a rigid drag body design which flares out prior to atmospheric entry; (3) Low-drag body shape of the HISV with a flexible drag body design which is unfolded prior to atmospheric entry.

In the first alternative HISV and DBS are completely integrated. Whether or not this results in a light heat shield design depends on the heat transfer rates experienced and on the propellants used. In any case, alternative (1) has the following disadvantages: (a) The configuration of the HISV is preempted and, therefore, difficult to standardize for Venus and Mars missions. Also, accomodation of changes due to propulsion system improvements or changes in crew size or type and shape of destination payload becomes a more complex, hence expensive task; (b) Flexibility to account for uncertainties in knowledge of the respective planetary atmospheres or to accomodate differences in atmospheric entry velocities (which can be fairly large, especially for Mars, as was discussed in Section 2) is greatly reduced. Therefore, primary attention is given to alternatives (2) and (3), although this does not mean that possibilities should be disregarded which are offered by shaping the HISV so as to provide a certain amount of drag, as much as is feasible without unduly emphasizing the above disadvantages. In the subsequent portion of the discussion, atmospheric braking at Mars as the possibly more difficult problem (for reasons outlined at the end of Section 3) is considered.

On alternative (2), preliminary stress analyses were carried out to size the drag brake configuration. The calculations are based on a uniform drag load on the frontal area, sufficient to produce $5g_{\text{Ea}}$ deceleration with an ultimate load factor of 1.4 times this deceleration. The reference computations assumed a vehicle gross weight (i.e. including drag brake) of $10^6$ lb (Earth). Constant drag brake dimensions were assumed. Therefore, the weight of ablation material and high temperature insulation stays constant. The drag brake structural weight required for lower vehicle gross weights was proportioned in a rational method. The weights were based on aluminum material at room temperature with a specific allowance for ablation and high temperature insulation material. For estimates of altitude and velocity in the Mars atmosphere, the NASA Mars Model Atmosphere No. 2A* was used.

The results are shown in Figs. 12-14 and 12-15 for two diameters (156′

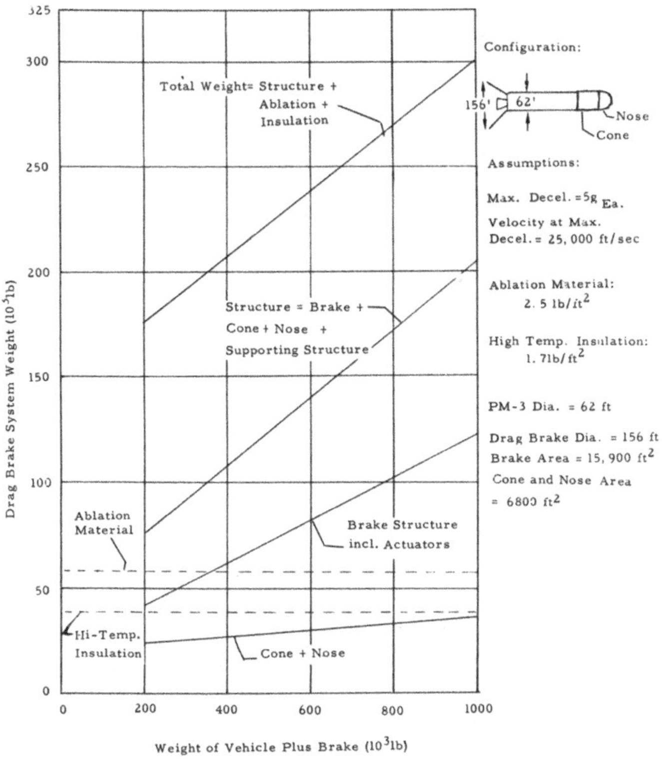


**Fig. 12-14.** Estimated weights for drag brake system (156′ dia) for mars atmospheric braking.

and 130′) of the open drag brake for a velocity of 25,000 ft/sec at the $5g$-deceleration point.

Altitude and associated velocity at which the $5g$-condition is reached are shown in Fig. 12-16 together with the structural weight of the drag brake, as function of the diameter of the open brake. The altitude was determined on the assumption of NASA Atmosphere Model 2A and $C_D = 1.0$. The variation of maximum dynamic pressure is plotted in Fig. 12-17 as function of diameter of the open brake. These weight calculations result in the DBS mass fractions presented in Fig. 12-18 for two diameters. Figure 12-18 shows

* 25 mb surface pressure.

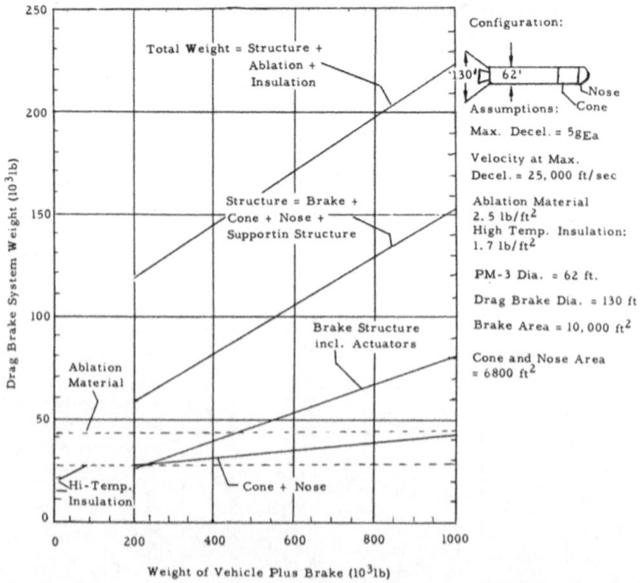


**Fig. 12-15.** Estimated weights for drag brake system (130′ dia) for mars atmospheric braking.

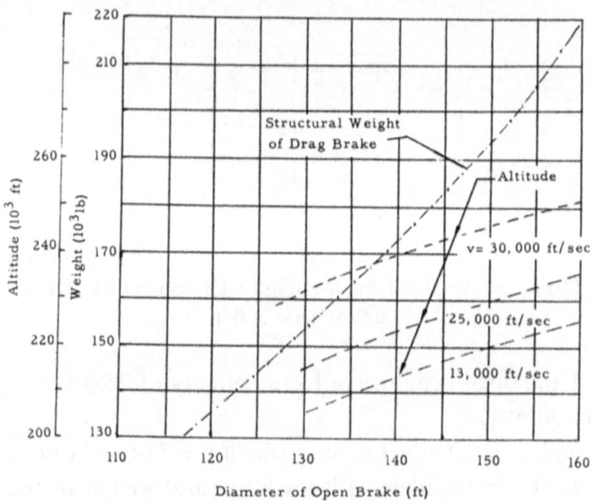


**Fig. 12-16.** Structural weight of drag brake for a vehicle gross weight of $10^6$ b at mars atmosphere entry and altitude at which $5\,g_a$; deceleration is reached for 3 velocities (NASA model mars atmosphere No. 2A (Mean)).

that, in view of the large mass of the HISV, very large drag brakes are required to obtain low dynamic pressure, hence, low convective heat transfer. Such dimensions, however, lead to unacceptably high DBS mass fractions

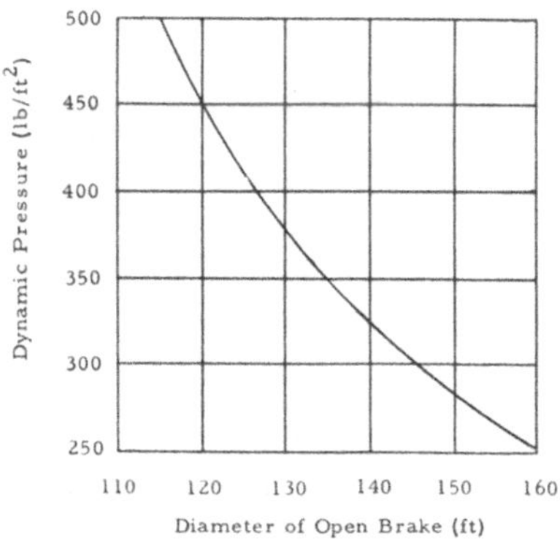


**Fig. 12-17.** Dynamic pressure vs. diameter of open brake at point of attaining 5 $g_{Ea}$ deceleration.

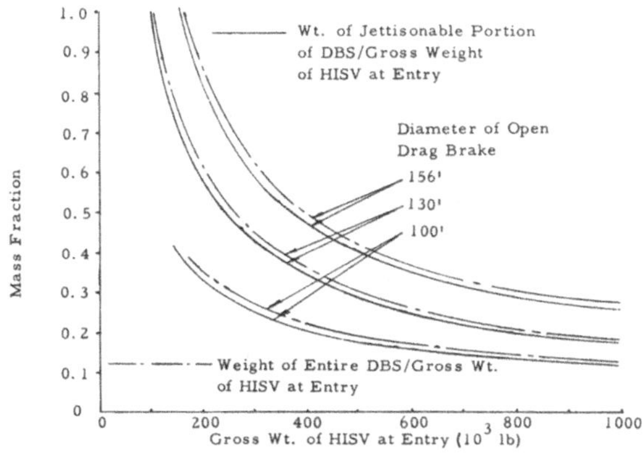


**Fig. 12-18.** Variation of mass fraction of drag brake system.

(Fig. 12-18). The high percentage of nitrogen in the Martian atmosphere and, hence, between shock wave and body surface tends to raise the temperature behind the shock (Ref. 12-3). Addition of $CO_2$ leads to the formation of the strongly radiating molecule CN. Continued addition of $CO_2$, however, will reduce the stagnation temperature (Ref. 12-3). The peak radiative heat transfer rate is, therefore, a function of a particular mixture ratio. Radiative heat transfer data (Ref. 12-4) for a mixture of 7.5 percent $CO_2$ and the remainder nitrogen, was correlated by AVCO by the equation

$$q_R = 2\delta\,(v/10^4)^{7.7}\,(\varrho/10^{-6})\;(\text{Btu/ft}^2\,\text{sec}) \tag{12-1}$$

where $\delta$ (ft) is the shock detachment distance. The peak value occurs when (Ref. 12-3)

$$v = v_E e^{-\frac{1}{7.7}} = 0.88 v_E \,(\text{ft/sec}) \tag{12–3}$$

and

$$\varrho = -\frac{2}{7.7}\,\frac{m \sin\gamma_E}{C_D A}\,(\text{slugs/ft}^3) \tag{12–3}$$

The peak radiation, therefore, is, with $\varrho = (\varrho/\varrho_{\text{Ea}})\,\varrho_{\text{Ea}}$ and $\varrho_{\text{Ea}} = 2.5 \cdot 10^{-3}$ slugs/ft$^3$,

$$q_{R,\,\max} = -\,475\delta\,\frac{m\beta}{C_D A \varrho_{\text{Ea}}}\,\sin\gamma_E \left(\frac{v_E}{10^4}\right)^{7.7} \tag{12–4}$$

where subscript E refers to entry conditions (300,000 ft altitude), $\beta$ is the inverse atmospheric scale height ($1/\beta = 50{,}000$ ft), $m$ = vehicle gross mass, $\gamma$ = flight path angle (positive with increasing altitude), $\varrho$ = local density in the Martian atmosphere, $\varrho_{\text{Ea}}$ = surface density in Earth atmosphere. The integrated radiative heating is then given by

$$Q_R = -\,\frac{3.12\, q_{R,\,\max}}{\beta v_E \sin\gamma_E} = -\,1480\,\frac{m}{C_D A \varrho_{\text{Ea}}}\,\frac{1}{v_E}\left(\frac{v_E}{10^4}\right)^{7.7} \tag{12–5}$$

Comparing Eqs. (12–4) and (12–5) one sees that $Q_R$, as does $q_{R,\,\max}$, increases with the vehicle gross mass and, of course, strongly with the entry velocity; but that $Q_R$ is independent of $\beta$ and $\gamma_E$. This conclusion depends on the validity of the assumption that the radiative heat coefficient varies with the first power of the density.

Using a correlation of Detra (Ref. 12-5) for maximum radiative heating in equilibrium air

$$q_{R,\,\max} = 80\delta \left(\frac{v_E}{10^4}\right)^{8.5} \left(\frac{-\,\beta m \sin\gamma_E}{\varrho_{\text{Ea}} C_D A}\right)^{1.6} \tag{12–6}$$

$$Q_R = -\,\frac{2.9\, q_{R,\,\max}}{\beta v_E \sin\gamma_E} \tag{12–7}$$

and assuming $-(m/C_D A)\sin\gamma_E = 1$, $v_E = 24{,}000$ ft/sec and $\beta_{\text{Ma}} = \beta_{\text{Ea}}$, AVCO determined the ratio of peak radiative heating for Mars and Earth to be

$$\frac{q_{R,\,\max,\,\text{Ma}}}{q_{R,\,\max,\,\text{Ea}}} = \frac{475}{80}\left(\frac{v_E}{10^4}\right)^{-0.8}\left(\frac{\beta}{\varrho_{\text{Ea}}}\right)^{-0.6} = 31 \tag{12–8}$$

$$\frac{Q_{R,\,\text{Ma}}}{Q_{R,\,\text{Ea}}} = 33.5 \tag{12–9}$$

The greater severity of radiative heating for Mars compared to Earth is apparent from these figures.

Pressure loads during entry and peak pressure loads are given by the normal shock approximation

$$P_S = p + v^2\varrho\left(1 - \frac{1}{2}\frac{\varrho}{\varrho_s}\right) \tag{12–10}$$

where the subscript $s$ refers to conditions behind the shock wave. The peak ambient pressure in the case of atmospheric braking, in distinction from entry and descent, is given by the maximum permissible deceleration which, in turn, determines $\gamma_E$. Fig. 12-19 shows a typical high-drag flight path

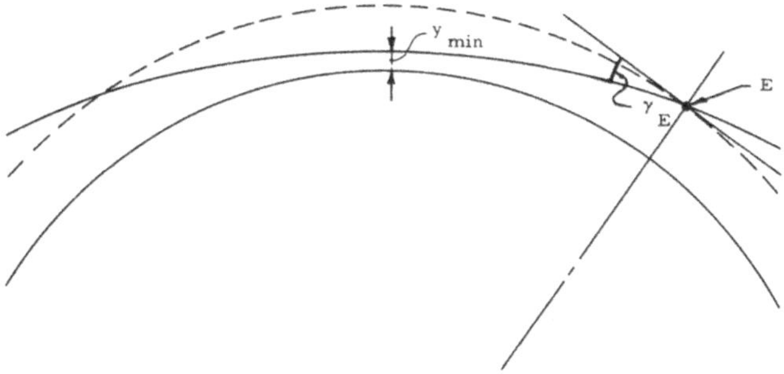


**Fig. 12-19.** High-drag flight path for atmospheric braking.

through the atmosphere, in distinction from a glider flight path which more closely follows the surface curvature due to operation of the glider at negative and positive angles of attack. AVCO has derived a simplified equation for the peak stagnation pressure.

$$P_S \simeq 1.93\frac{m}{C_D A}g_{Ea}\frac{\dot{v}_{max}}{g_{Ea}} \tag{12–11}$$

where $g_{Ea}$ is the Earth's gravitational acceleration and $\dot{v}_{max}$ the peak deceleration (ft/sec$^2$). The peak stagnation pressure behind the shock wave increases with the vehicle gross mass and decreases with the cross sectional area. Therefore, for equal peak values of $P_S$, a 500,000 lb vehicle needs half the drag brake area or 70 percent of the diameter, provided $C_D$ is the same for both vehicles. Referring to Fig. 12-18, it follows that if a 130 ft diameter drag brake is used for a vehicle of $10^6$ lb$_{Ea}$ gross weight, a 500,000 lb vehicle would obtain the same peak stagnation pressure with a 92 ft brake. For a 300,000 lb vehicle a diameter of $130/\sqrt{3} = 75$ ft would be needed.

The same considerations apply to Eq. (12–4). If $\delta$ is the same (i.e. the general configuration and the angle of attack of the drag brake the same) and for the same entry angles and entry velocity, the maximum radiative heat transfer rate is the same for equal values of $m/A$ or $mg_{Ea}/A$.

Fig. 12-20 shows the variation of the overall weight as well as of the

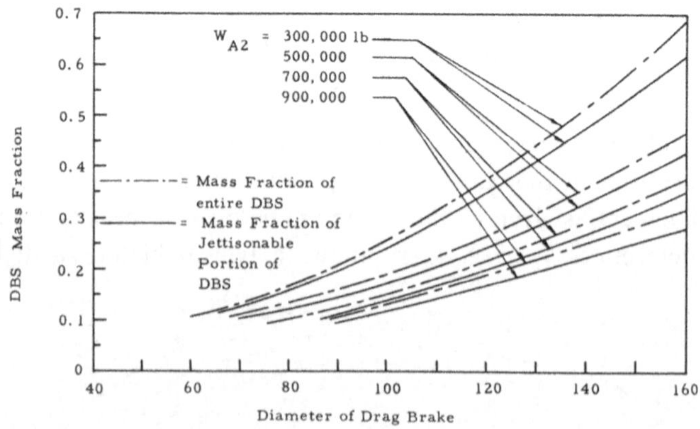


**Fig. 12-20.** Variation of DBS mass fraction vs. diameter for several vehicle gross weights $W_{A2}$.

jettisonable mass fraction of the DBS, as defined in Fig. 12-18 versus diameter for vehicle gross weight of 300, 500, 700 and $900 \cdot 10^3$ lb. Assuming 130 ft diameter for a 900,000 lb vehicle, equal peak radiation heating rates and peak stagnation pressures should be obtained with a 114 ft drag brake on a 700,000 lb vehicle, a 97 ft brake on a 500,000 lb vehicle and a 75 ft brake on a 300,000 lb vehicle. Fig. 12-20 shows that the approximate mass fractions for these configurations are 0.27, 0.2, 0.175 and 0.14. This trend clearly favors smaller vehicle masses if atmospheric braking is contemplated. Comparison with the propellant mass ratios mentioned above also shows that a greater gain is realized if atmospheric braking is used to the fullest extent, i.e., to reduce velocity to near-circular, rather than to a more or less highly elliptic capture orbit, because the propellant mass fraction decreases faster with decreasing velocity reduction than the DBS mass fraction.

Referring to Fig. 12-13, the mass ratio for the Mars capture maneuver in 1982 at $I_{sp}=800$ sec is 1.62. Hence, the propellant mass fraction is 0.383. Thus, in terms of $W_{A2}$, the payload gain, corresponding to the above DBS mass fractions is 0.113, 0.183, 0.208 and 0.243 or, in lb (Earth), 102,000, 128,000, 105,000 and 73,000 lb; or a corresponding reduction in orbital departure weight (ODW). In order to compute the reduction in ODW one must know the non-propellant weight lost between Earth departure and Mars arrival. The biggest item in this weight group is the net inert weight of the Earth departure stage (PM-1). Assuming a representative figure of 0.2 $W_{A2}$ for the non-propellant weight lost, it follows for $W_{A2}=900$, 700,

500 and 300 $k$ and for a mass ratio of 1.59 for $I_{sp}=800$ for the 1-2-82 reference mission, the following ODW values are obtained: $1.72 \cdot 10^6$, $1.34 \cdot 10^6$, 955,000 and 573,000 lb; whereas, with atmospheric braking, for $W_{A2}=$ 798,000, 572,000, 395,000 and 227,000 lb, the ODW values are: $1.52 \cdot 10^6$, $1.09 \cdot 10^6$, 755,000 and 433,000 lb. The ODW ratios for the vehicles with drag brake compared to vehicles using retro propulsion are therefore: 0.885, 0.815, 0.79 and 0.757; or savings between 12 and 24 percent in ODW.

It is realized, however, that in computing the weight of the drag brake system, not all weight penalties may have been considered. For instance, no additional weight for the heat shielding of the cylindrical section of the model vehicle was included. The thermo-meteoroid shield surrounding the tanks is adequate for space conditions but not for entry conditions. Furthermore, no allowance has been made for the high load to which the structure of the entire vehicle is subjected during the braking process.

Finally, provisions have to be made for a small chemical propulsion system capable of raising the periapsis altitude and adjusting the capture orbit following atmospheric braking. Allowing for a trimming capability of 2000 ft/sec and $O_2/H_2$, the weight of propellant and hardware for this trim stage is approximately 9 percent (0.09) of $W_{A2}$. This substantially reduces the above calculated payload gain, particularly for 900,000 and 700,000 lb vehicle weight, showing again the greatest penalty for the largest vehicles.

Under these conditions, the payload gain is, respectively: 0.023, 0.093, 0.118 and 0.153, or: 19,000, 65,000, 59,000 and 46,000 lb. The corresponding ODW's are, respectively: $1.6 \cdot 10^6$, $1.22 \cdot 10^6$, 845,000 and 485,000 lb. The ODW ratios are now reduced to 0.975, 0.91, 0.885 and 0.85. It is unlikely that $W_{A2}$ is smaller than 500,000 lb. Thus, atmospheric braking can be expected to produce a reduction in ODW of approximately 10 to 15 percent.

This is far less than the savings attainable by means of high-speed Earth entry (60 to $70 \cdot 10^3$ ft/sec), or by means of perihelion braking or Venus powered fly-by. On the other hand, because the potential savings offered by atmospheric braking increase with decreasing $W_{A2}$, atmospheric braking becomes more attractive in connection with other weight saving measures such as those mentioned and, thereby savings of more than 15 percent through atmospheric braking at Mars alone may be attainable. In order to assess the weight saving more accurately, additional trade-off studies are necessary which systematically investigate the effectiveness of atmospheric braking with and without high-velocity Earth entry, perihelion braking or Venus powered fly-by; as well as for conditions which drive the ODW up, such as orbital re-capture at Earth, rather than atmospheric entry.

The above result of 10 to 15 percent ODW reduction applies to the Mars reference missions which are not optimized for maximum use of atmospheric braking. When such optimization is carried out, the reduction in ODW due to atmospheric braking approaches 20 to 25 percent.

Although the radiative heat transfer must be expected to be much higher for Mars entry than for Earth entry, the absolute entry conditions are nevertheless not as severe as those for Earth. One reason, of course, is that the entry velocities are smaller. For different entry velocities, but otherwise the same conditions, the ratio of the maximum radiation rates for Mars and Earth (Eqs. (12–4) and (12–6)) become

$$\frac{q_{R,\,\max,\,\mathrm{Ma}}}{q_{R,\,\max,\,\mathrm{Ea}}} = \frac{475}{80}\,\frac{\delta_{\mathrm{Ma}}}{\delta_{\mathrm{Ea}}}\,\frac{(v_{E,\,\mathrm{Ma}}/10^4)^{7.7}}{(v_{E,\,\mathrm{Ea}}/10^4)^{9.5}}\left(\frac{\beta}{\varrho_{\mathrm{Ea}}}\right)^{-0.6} \tag{12–12}$$

This relation shows that the ratio $v_{E,\mathrm{Ma}}/v_{E,\mathrm{Ea}} < 1$ is further reduced by the fact that $v_{E,\mathrm{Ea}}$ is raised to a higher power. Thus, while the strong radiation component renders sharp-nosed bodies with smaller surface areas preferable, the not very stringent heating conditions still make blunt drag bodies acceptable. This is important, because the large size of most of the vehicles which had to be considered in this study would not lend itself very well to a glider type configuration, or even a conical configuration without increasing shear forces and convective heat transfer on more sensitive parts of the vehicle, such as the propellant tanks.

Studies conducted by AVCO have shown three groups of materials to be suitable for low $m/C_DA$ vehicles entering Mars at velocities in the neighborhood of 24,000 ft/sec: (a) low temperature ablators, such as Teflon or Avcoat 5019, (b) low conductivity, high temperature radiators such as epoxy resins (Avcoat II and 19), and (c) heat sinks (beryllium).

In view of the size of the vehicles and drag bodies, serious manufacturing problems are involved. Teflon is very difficult to manufacture in large sections. Other plastics require long-lead time molding dies, equipped for high pressures. Avcoat 19 is available to the vacuum bag process of fabrication, but again the large sizes involved pose great difficulties. Avcoat II is an epoxy resin with the requisite rheological properties necessary for spray applications.

What appears most advantageous is a design which minimizes the heat input into the HISV proper and which provides a drag brake which consists of individual sections or facets, rather than consisting of one large unbroken surface. These sections or facets can be manufactured more conveniently. Produced in quantity, they become elements of the drag brake. They can be separated from each other by a certain distance. This would, of course, be unacceptable in a heat shield design which represents the outer shell of a body (integrated heat shield design). The drag brake is separate and leaks have only one possible slight disadvantageous effect, namely, that of reducing the base pressure, hence, the base drag of the brake. This effect is small, however, if the slots between the facets are not too wide, because of their choking due to supersonic flow conditions. In supersonic flow, surfaces with porosities as high as 50 percent behave like solid surfaces.

On the basis of this design philosophy, it is necessary to convert the life support section from the radial or cross shape, shown for pure space application, to conical shapes. The cone extends from the life support section to the tank section covering the spine section. The tank section is cylindrical. At the end of the tank section is located the drag brake which opens up prior to entry, forming a large drag flare.

The conical section can be covered with Apollo Material which is of low density (0.56 gm/cc) and has a very low heat conductivity ($k = 0.07$). Both characteristics are distinct advantages in view of the large surface areas involved and the period of atmospheric flight which is about 250 seconds. Along the sides of the tank section where the heat flux rates are lower, Avcoat 5016 can be used. Its characteristics are given in Ref. 12-5.

For the drag brake, the following partly competitive or conflicting requirements must be considered.

1) Large size at low weight desirable, because low $m/C_D A$ is generally desirable. It minimizes heat loads on the HISV life support and tank sections. It lengthens the path of atmospheric flight, hence reduces maximum deceleration load. It permits selection of relatively high periapsis altitudes and thereby reduces propellant requirements for the subsequent orbit trimming operation.

2) As insensitive as possible to radiation heat transfer so that the resulting penalties connected with a large size drag brake remain acceptable.

3) Not to consist of a single large surface or even a small number of large panels, but of smaller sections which are manufactured readily and economically and which permit ready repair of individual, damaged parts. Such characteristics would allow continued application of standardized methods of design, manufacturing and assembly, independent of the size, or growth in size with time, of the drag brakes.

4) In spite of its large size the brake must be packageable in such a manner that it can be transported into orbit in one piece within the confines of a Saturn V payload section.

5) It must be capable of area variation for drag control (either thermally or by deceleration loads) in not fully known atmospheric conditions, and in order to assure maximum practical insensitivity to approach errors.

6) In view of its size, the drag brake must rely on passive cooling methods, preferably radiation cooling because that is the simplest, most reliable method, and involves the smallest weight penalty.

7) Unfolding of drag brake should be simple and reliable, with minimum probability of "sticking" during unfolding due to long dormancy in space.

8) Must be resistant to space environment, because it is not practical from a weight standpoint to provide adequate cover for the brake. Insensitivity to meteoroid impact is especially important.

9) Drag brake should be applicable in the atmospheres of Venus, Mars and possibly Earth.

These requirements are best met by one of two drag brake designs.

DESIGN I: Rigid design, consisting of individual segments which, when folded, are oriented parallel and adjacent to the cylindrical tank section of the Earth departure module (PM-1) and which are swung out prior to entry. Each of the segments consists of an outer frame between which "ribbons" of steel mesh, covered with Teflon, are mounted in tension, like the strings in a tennis racket (Fig. 12-21).

DESIGN II: Flexible design, consisting of a cone-shaped mesh of woven steel bands, covered with Teflon to prevent critical overheating of the mesh. This design is shown in Figs. 12-22 through 12-28.

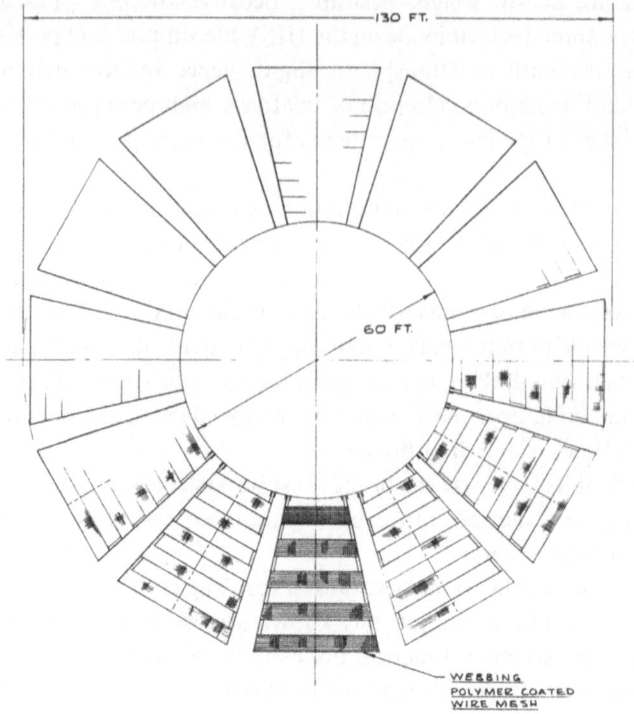


**Fig. 12-21.** Drag brake design I

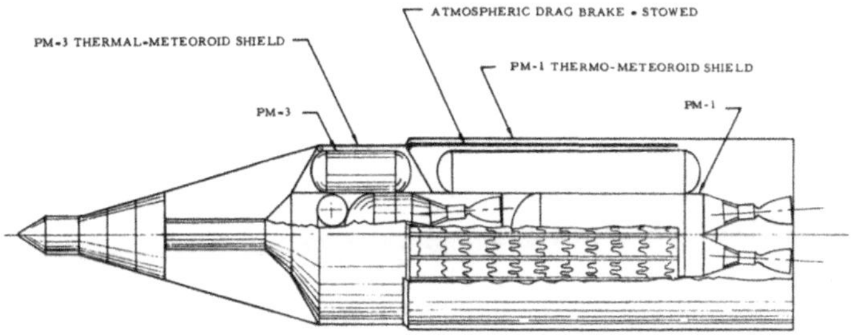


**Fig. 12-22.** Mars or Venus HISV designed for atmospheric drag brake capture.

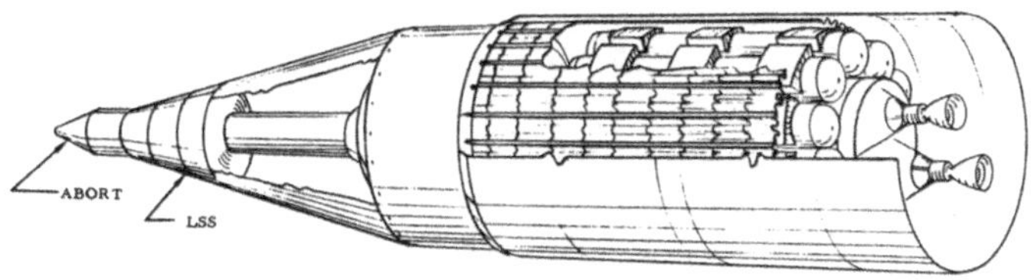


**Fig. 12-23.** Three-dimensional view of HISV with drag brake packaged around $H_2$-tanks of PM-1 (earth departure propulsion module).

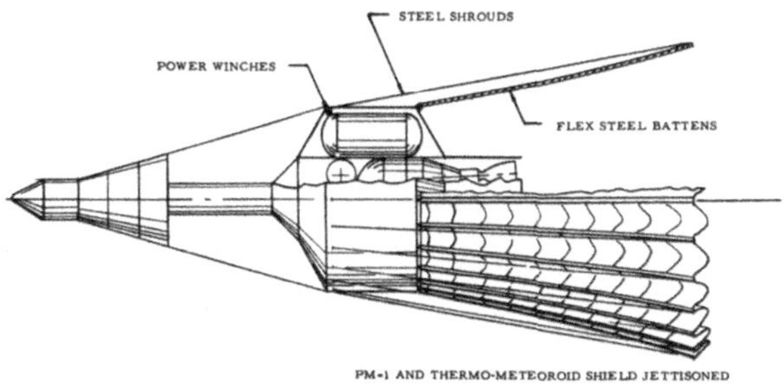


**Fig. 12-24.** Side view of HISV during earth-mars transfer.

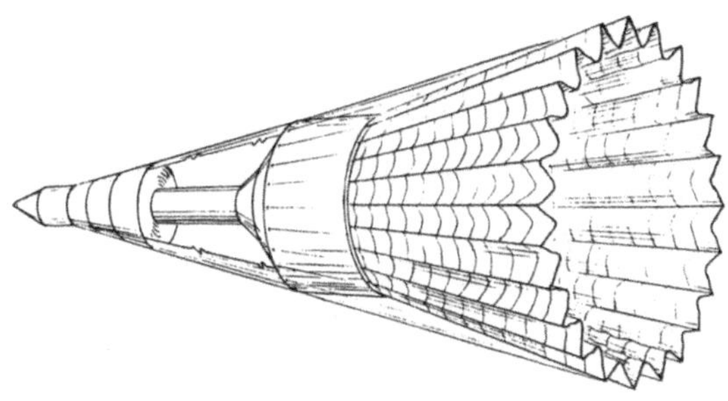

**Fig. 12-25.** Three-dimensional view of HISV during earth-mars transfer.

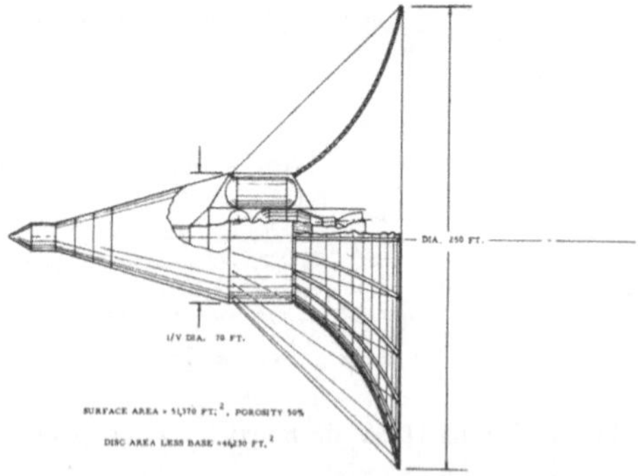


**Fig. 12-26.** Side view of HISV in atmospheric braking mode with drag brake fully released.

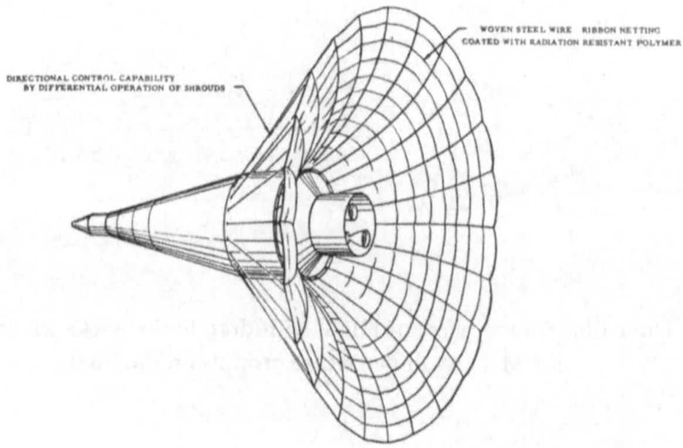


**Fig. 12-27.** Three-dimensional view of HISV in atmospheric braking mode.

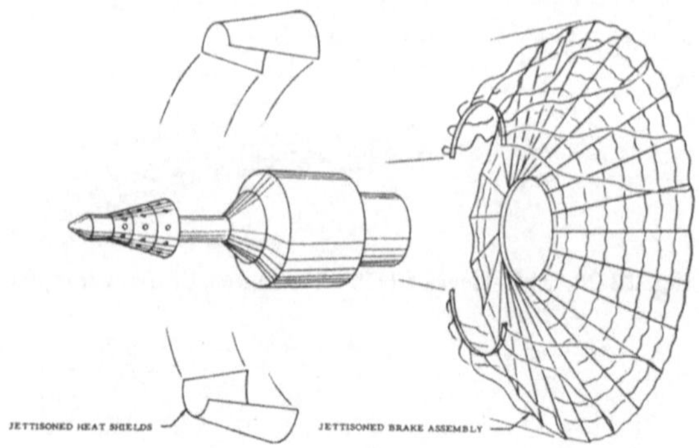


**Fig. 12-28.** HISV in post-brake mode, jettisoning drag brake and frontal fairings.

The flexible mesh brake (with about 50 percent porosity) offers particular promise in terms of low weight. While for Earth entry the most worthwhile area of weight reduction lies in an increase in heat of ablation at comparatively moderate temperatures, the greatest return in the case of planetary drag brakes of large size is offered by reducing the structural weight. This is done in the case of the flexible mesh drag brake where stiffening in operation is provided by dynamic pressure. The principal problem area in a drag device of this type is the possibility of flutter. Flutter can be discouraged by large angles of attack of the mesh ribbons and by varying the porosity in such a manner that the system is in tension. This means that the porosity should be reduced as one moves aft, generating higher area density in the outer region adjacent to the periphery of the drag brake; and lower area density in the inner region, adjacent to the aft end of the vehicle. The effectiveness of such arrangement requires further theoretical analysis and wind tunnel testing.

An advantage common to both designs is the capability of radiating from both the outer and the inner area which provides high radiation cooling effectiveness. The rigid design avoids sensitivity to flutter. The flexible design is potentially lighter and more easily stored for transport into orbit.

Because of the porosity of the design and because of the fact that the flexible drag brake offers ready radiation cooling in both directions normal to its surface, lower area weights should be attainable than with the rigid drag brake design, thereby further increasing the prospect that the ODW can be reduced significantly. A preliminary estimate, based on the flexible design, indicates a reduction in ODW by 25 percent for Mars and 32 percent for Venus. These figures apply to the reference missions shown before which are not optimized for atmospheric braking. If this is done, the savings in ODW climb to 30 to 36 percent, respectively.

Finally, it should be pointed out that the selection of a peak deceleration of $5g_{\mathrm{Ea}}$ is somewhat arbitrary, although not without a reason. Higher peak decelerations would reduce the area, hence the weight, of the drag brake proper, especially for Mars. For Venus, there exists a peak acceleration which if exceeded, should force the vehicle deep enough into the planet's atmosphere to encounter CO and $CO_2$. This would result in a rapid rise in radiation heating which makes large blunt surfaces far less suitable (i.e., of low weight) than slender, streamlined shapes. The use of large, highdrag surfaces causes large shock stand-off distances and, therefore, intense radiation heating. In the case of Mars, this too would tend to become a greater problem if higher peak decelerations and deeper entry into that atmosphere would be selected; but less so than for Venus.

Therefore, if blunt drag brakes are used as standard means of atmospheric braking at Venus as well as Mars (which would reduce development costs), the "brake strip" at Venus appears to be more definitely restricted to the outer atmosphere, above the cloud cover, hence to comparatively lower peak decelerations than in the case of Mars. This estimation, if verified, would allow the use of higher peak decelerations at Mars. The most suitable peak deceleration must be determined by a rather comprehensive and complex trade-off study, involving the entire manned space vehicle and the mission profile. A similar trade-off study should, of course, be carried out also for Venus. If the results show that the optimum peak acceleration for Mars is significantly higher than for Venus (i.e., by a factor of 1.5 to 2 or more) this fact may enable the ODW reduction due to atmospheric braking for Mars to become similar to those for Venus; both may be as high as 40 percent ODW reduction.

### [12-4.3] Conclusions

Preliminary configuration studies indicate that with chemical propulsion and nuclear graphite reactor, a high-drag configuration of the HISV appears less attractive for purposes of atmospheric braking than a combination of vehicle and drag brake (DB). The reasons for this conclusion are that the combination of vehicle and drag brake permits greater flexibility of design, better standardization, lower heat loads on the sensitive (cryogenic) portions of the HISV, better possibilities of standardizing the HISV design as well as the drag brake design for Venus and Mars and superior flexibility in accomodating a variety of HISV sizes (due to pay-load variations), of target planet arrival velocities (due to varying transfer conditions and mission profiles) and to accomodate uncertainties in atmospheric conditions.

The combination of vehicle and drag brake offers two alternatives, namely, a rigid and a flexible drag brake design. The rigid design appears to be heavier, the lightest version being the one designated as Design I ("tennis racket" design) and shown in Fig. 12-21. This design consists of individual segments. Each segment consists of a rigid outer frame between which ribbons of Teflon covered steel mesh are mounted in tension. The flexible design, referred to as Design II, consists of a cone-shaped mesh of woven steel bands covered with Teflon (Figs. 12-22 through 12-28). Both designs assume a certain amount of porosity, in fact anticipating 50 percent porosity. This and the fact that radiation cooling takes place on front and back sides reduces the design weight below that of a conventional heat shield design. The reduction may amount to 50 percent in the case of the flexible design, but that design has more unsolved problems at this time than the rigid Design I. These problems include the dynamics of unfolding the large flexible structure and the possibility of flutter.

The use of blunt forms such as the drag brake appears to be acceptable for the Mars atmosphere, but restricts the braking path at Venus to the outer atmospheric region, essentially above the cloud layer, where the CO and $CO_2$ content appears to be low, or at least to be not more influential than in the Mars atmospheric braking path.

It was found that, even under the most conservative assumptions, a reduction in (Earth) orbital departure weight (ODW) of the HISV of 10 to 15 percent is obtained, for departure weights of the order of 1.5 to $2.10^6$ lb. These conservative assumptions include drag brake systems of conventional rigid plate-type heat shield design (Figs. 12-14 and 12-15), and a mission profile not optimized for taking maximum advantage of atmospheric braking. If the mission profile is optimized in that manner the reduction in ODW amounts to 20–25 percent. If the Design I or Design II is used, and if the HISV-DB system is optimized for highest possible peak deceleration, rather than the $5g_{Ea}$ assumed in the preceding sample analysis, then, together with mission profile optimization, as ODW reduction by 40 percent appears attainable.

## [12-5] EVALUATION OF RESULTS

The only objective of atmospheric braking is reduction in orbital departure weight (ODW) of the HISV so that the logistic requirements, hence the logistic transportation cost, can be reduced and the entire manned planetary mission cost effectiveness be improved.

The significance of an ODW reduction by 10 to 40 percent depends on several factors. It depends primarily on the ratio of ODW to payload capability of the Earth launch vehicle (ELV) and on the Earth-to-orbit transportation cost. The latter is obvious. The former determines not only the number of ELV's required to deliver the ODW of one HISV into orbit; it also determines the procurement requirements, that is the sum of minimum number of ELV's (and their payloads) *plus* the number of redundant (backup) ELV's (and pay-loads) which must be provided as a function of success probability of orbital delivery and desired overall delivery probability. The larger the number of ELV's required for successful delivery (i.e., the larger the ratio of ODW to payload weight delivery capability of the ELV), the larger is the number of redundancies which must be provided for a given individual delivery probability and a given desired overall delivery probability.

A secondary factor is orbital assembly which consists of the mating of HISV modules in orbit and of orbital fueling. It is readily seen that the success probability of orbital mating adds to the number of ELV redundancies as well as HISV module redundancies, especially if one assumes that failure to mate disqualifies for mission use *both* modules involved. The success probability of orbital fueling adds to the redundancy of ELV's and of orbital tankers.

It is the effect of ODW reduction on the procurement cost of ELV's HISV modules and orbital tankers which is of primary cost significance. Compared to this, the savings in HISV production cost proper is insignificant, because most of the ODW reduction is reduction in propellant weight. In fact the HISV production cost may increase, because the cost per pound of the drag brake system is likely to be at least 10 to 100 times as high as that of the propellant. This cost increase, however, is readily overcompensated the reduction in Earth-to-orbit transportation cost, so that an overall net mission cost reduction can be expected.

To evaluate the effect of a reduction in the orbital departure weight of the HISV, a hydrogen-fueled, solid core nuclear powered HISV was assumed of the following ODW: 1, 1.5, 2, $3 \cdot 10^6$ lb. Three Earth launch vehicles (ELV) were assumed: (1) Saturn V (220,000 lb payload weight; 100,000 $ft^3$ payload section volume; direct transportation cost to orbit 350\$/lb), (2) Saturn V Modified (chemical) (330,000 lb; 150,000 $ft^3$; 290\$/lb), (3) Post-Saturn (chemical; reusable) (1,000,000 lb; 1,000,000 $ft^3$; 100\$/lb).

Except for (3), the HISV has to be delivered piecemeal (by modules) into orbit, where the modules are checked out, mated and the final mission readiness test and final countdown for orbital launch are performed. The modules are laid out to utilize to a maximum degree the available volume of the ELV payload section, so as to maximize their size and minimize the number of matings required. Therefore, the payload weight capability of the ELV is often exceeded if the module is fueled to capacity. It is assumed that the modules are fueled to the degree permissible by the payload weight constraints of the respective ELV. For the remaining fuel and for make-up fuel (assumed to be 30 percent of total) orbital tankers are needed, carried into orbit by the respective ELV, which fuel the modules in orbit following mating of the modules. It is assumed that the success probability of the individual orbital delivery is $P_D = 0.95$. The success probability with which the delivery of all modules in each particular case has to be accomplished is $P_D^* = 0.91$. For the success probability of orbital mating of the modules and of orbital fueling, 1.0 (i.e., 100 percent) was assumed for the sake of simplicity. Even so, the procurement cost is strongly influenced by whether or not the individual modules, making up the HISV, are interchangeable. Therefore, two cases were considered, one with interchangeable and the other with non-interchangeable modules to illustrate the difference.

The step-by-step data computation of the procurement is shown in Table 12-1 which, in that form, is self-explanatory. The result is plotted in Fig. 12-29. For the solid curve (1), the average gradient is seen to be about

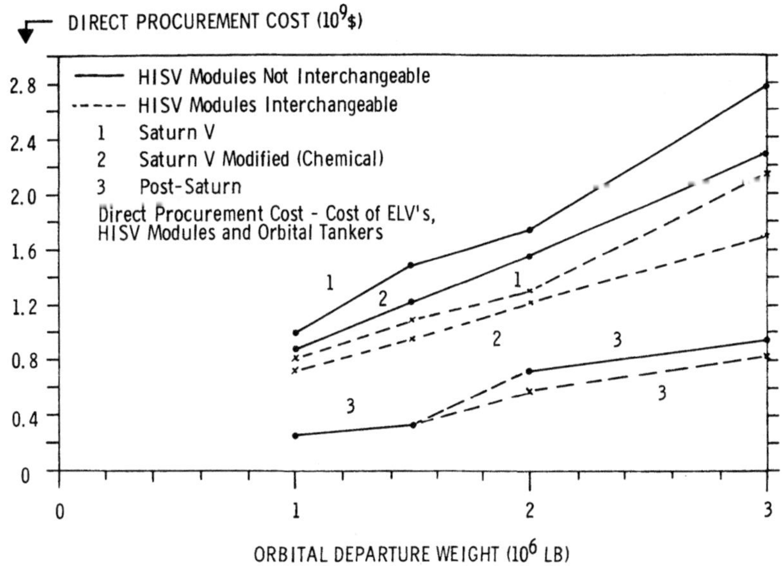


**Fig. 12-29.** Variation of direct procurement cost with orbital departure weight of heliocentric interorbital vehicle.

\$900M per million lbs ODW of the HISV in the 1 to 3M lbs ODW regime. For the solid curve (2), it is about \$700M and for the solid curve (3) about \$300M. For the broken curves (modules interchangeable), the corresponding values are about \$700M, \$500M and \$250M per million lbs ODW of the HISV in the 1 to 3M lbs ODW regime. It is seen, therefore, that in cases (1) and (2) (ELV limitations) a reduction in ODW by 15 to 25 percent pays off, let alone a reduction by 30 to 40 percent. The amount depends, of course, on the particular ODW value. In the lower ODW region of the dashed curve (2) and for all curves (3), however, the payoff becomes small, and in the lower ODW region of (3) it becomes insignificant.

Table **12-1.** Computation of Procurement Cost

| | | | | | |
|---|---|---|---|---|---|
| 1. | Orbital Departure Weight (ODW) of HISV ($10^3$ lb) | 1000 | 1500 | 2000 | 3000 |
| 2. | Initial Payload Weight ($10^3$ lb) (=0.08 (1)) | 80 | 120 | 160 | 240 |
| 3. | Weight of Fueled Propulsion Modules ($10^3$ lb) (=(1)—(2)) | 920 | 1380 | 1840 | 2760 |
| 4. | Volume of Modules ($10^3$ ft$^3$) @ 3 lb/ft$^3$ average density) | 307 | 460 | 614 | 920 |
| 5. | Hardware Weight of Modules ($10^3$ lb) (=0.12 (3)) | 110 | 158 | 220 | 335 |

Table **12-1** (*Continued*)

| | | | | | |
|---|---|---|---|---|---|
| 6. | Number of Modules is ELV Payload Section has Volume of | | | | |
| | (a) 100,000 $ft^3$ | 3 | 5 | 6 | 9 |
| | (b) 150,000 $ft^3$ | 2 | 3 | 4 | 6 |
| | (c) $\geqslant$ 1,000,000 $ft^3$ | 1 | 1 | 1 | 1 |
| | (= (4)/(Payload Section Volume) (round figures)) | | | | |
| 7. | No. of ELV's for HISV Payload if ELV Payload Capacity is | | | | |
| | (a) 220,000 lbs | 1 | 1 | 1 | 1 |
| | (b) 330,000 lbs | 1 | 1 | 1 | 1 |
| | (c) $\leqslant$ 1,000,000 lbs | 0 | 0 | 0 | 0 |
| 8. | Number of ELV's as Module Carriers | | As (6) | | |
| 9. | Minimum Number of ELV's for HISV Payload and Modules (= (7) + (8)) | | | | |
| | (a) | 4 | 6 | 7 | 10 |
| | (b) | 3 | 4 | 5 | 7 |
| | (c) | 1 | 1 | 1 | 1 |
| 10. | Hardware Weight per Module Delivery ($10^3$ lb) | | | | |
| | (a) | 34 | 33 | 37 | 37 |
| | (b) | 52 | 53 | 55 | 56 |
| | (c) | 190 | 278 | 380 | 575 |
| 11. | Propellant Weight per Module Delivery to Fill ELV up to its Payload Capacity ($10^3$ lb) (= ELV Payload Cap. —(10)) | | | | |
| | (a) | 180 | 180 | 180 | 180 |
| | (b) | 270 | 270 | 270 | 270 |
| | (c) | 810 | 720 | 620 | 420 |
| | (round figures) | | | | |
| 12. | Overall Propellant per ODW ($10^3$ lb) (= (1)—(2)—(5)) (round figures) | 810 | 1220 | 1620 | 2580 |
| 13. | Propellant Delivered with Modules (= (11)·(6)) | | | | |
| | (a) | 540 | 900 | 1080 | 1620 |
| | (b) | 540 | 810 | 1080 | 1620 |
| | (c) | 810 | 720 | 620 | 420 |
| 14. | Remaining Propellant ($10^3$ lb) to be Delivered by Orbital Tanker (= (12)—(13)) | | | | |
| | (a) | 270 | 320 | 540 | 960 |
| | (b) | 270 | 410 | 540 | 960 |
| | (c) | 0 | 500 | 1000 | 2160 |

Table **12-1** (*Continued*)

| | | | | | |
|---|---|---|---|---|---|
| 15. | Propellant Loss ($10^3$ lb) Assumed to be 30 % of (12) (round figures) | 240 | 370 | 490 | 770 |
| 16. | Overall Propellant Delivered by Orbital Tankers ($10^3$ lb) (—(14)+(15)) | | | | |
| | (a) | 510 | 690 | 1030 | 1730 |
| | (b) | 510 | 780 | 1030 | 1730 |
| | (c) | 240 | 870 | 1490 | 2930 |
| 17. | Number of ELV's for Carrying Orbital Tankers (=(16)/(Tanker Capacity)) | | | | |
| | (a) Tanker Capacity = 200K | 3 | 4 | 5 | 9 |
| | (b) Tanker Capacity = 300K | 2 | 3 | 4 | 6 |
| | (c) Tanker Capacity = 900K (round figures) | 1 | 1 | 2 | 3 |
| 18. | Module Carrier Procurement Requirements (based on (9)) (Case 1: Modules not Interchangeable) ($P_D = 0.95$; $p_D^* = 0.91$) | | | | |
| | (a—1) | 7 | 11 | 13 | 19 |
| | (b—1) | 5 | 7 | 9 | 13 |
| | (c—1) | 1 | 1 | 2 | 2 |
| 19. | As (18): Case 2: Modules Interchangeable. This means ELV must be deduced sub (9) in all cases where the number sub (7) is 1 instead of 0. | | | | |
| | (a—2) | 4+1 | 6+1 | 7+1 | 11+1 |
| | (b—2) | 3+1 | 4+1 | 5+1 | 7+1 |
| | (c—2) | 1 | 1 | 1 | 1 |
| | The "+1" means that the ELV identified sub (7) is added back. | | | | |
| 20. | Tanker Carrier Procurement Requirements Based on (17) ($P_D = 0.95$; $P_D^* = 0.91$) Tankers are Interchangeable. | | | | |
| | (a—3) | 4 | 5 | 6 | 11 |
| | (b—3) | 3 | 4 | 5 | 7 |
| | (c—3) | 1 | 1 | 3 | 4 |
| 21. | Total Procurement of ELV's Case (1): (18)+(20) | | | | |
| | (a) | 11 | 16 | 19 | 30 |
| | (b) | 8 | 11 | 14 | 20 |
| | (c) | 2 | 2 | 5 | 6 |
| 22. | Same as (21) | | | | |

Table **12-1** (*Continued*)

| | | | | | |
|---|---|---|---|---|---|
| | Case (2): (19) + (20) | | | | |
| | (a) | 9 | 12 | 14 | 23 |
| | (b) | 7 | 9 | 11 | 15 |
| | (c) | 2 | 2 | 4 | 5 |
| 23. | ELV Cost ($10^6$ $) | | | | |
| | Case (1) | | | | |
| | (a) 220K @ 350 $/lb = 77M/ELV | 850 | 1240 | 1460 | 2300 |
| | (b) 330K @ 390 $/lb = 90M/ELV | 730 | 990 | 1260 | 1800 |
| | (c) 1000K @ 100 $/lb = 150M/ELV | 200 | 200 | 500 | 600 |
| 24. | ELV Cost ($10^6$ $) | | | | |
| | Case (2) | | | | |
| | (a) as (23) | 700 | 920 | 1080 | 1780 |
| | (b) as (23) | 630 | 810 | 990 | 1360 |
| | (c) as (23) | 200 | 200 | 400 | 500 |
| 25. | Total Module Weight to be Procured; Case 1 ($10^3$ lb) (= (10)·[(18)—1]*) | | | | |
| | (a) | 200 | 330 | 450 | 630 |
| | (b) | 210 | 330 | 440 | 670 |
| | (c) | 190 | 278 | 760 | 1150 |
| | * Except for (c) | | | | |
| 26. | As (25); Case 2 ($10^3$ lb) (= (10)·[(19)—1]*) | | | | |
| | (a) | 136 | 198 | 259 | 405 |
| | (b) | 156 | 211 | 275 | 390 |
| | (c) | 190 | 278 | 380 | 575 |
| | * Except for (c) | | | | |
| 27. | Add Weight of Initial Payload from (2) to (25)* ($10^3$ lb) | | | | |
| | (a) | 280 | 450 | 610 | 870 |
| | (b) | 290 | 450 | 600 | 910 |
| | (c) | 190 | 278 | 380 | 575 |
| | * Except for (c) | | | | |
| 28. | Add Weight of Initial Payload from (2) to (26)* ($10^3$ lb) | | | | |
| | (a) | 216 | 318 | 419 | 645 |
| | (b) | 236 | 231 | 435 | 630 |
| | (c) | 190 | 278 | 380 | 575 |
| | * Except for (c) | | | | |
| 29. | Procurement Cost of HISV Hardware Case 1 ($10^6$ $) (= (27)·500 $/lb)) | | | | |
| | (a) | 140 | 225 | 305 | 435 |
| | (b) | 145 | 225 | 300 | 455 |
| | (c) | 95 | 139 | 190 | 288 |
| 30. | Procurement Cost of HISV Hardware Case 2 ($10^6$ $) (= (28)·500 $/lb)) | | | | |

Table **12-1** (*Continued*)

| | | | | | |
|---|---|---|---|---|---|
| | (a) | 108 | 159 | 210 | 323 |
| | (b) | 118 | 116 | 218 | 315 |
| | (c) | 95 | 139 | 190 | 288 |
| 31. | Procurement Cost of Orbital Tankers (assumed expendable) ($10^6$ \$) | | | | |
| | (a) 20K lb @ 200 \$/lb = \$ 4M/Tanker | 12 | 16 | 20 | 36 |
| | (b) 30K lb @ 180 \$/lb = \$ 5.4M/Tanker | 11 | 16 | 22 | 32 |
| | (c) 100K lb @ 100 \$/lb = \$ 10M/Tanker | 10 | 10 | 20 | 30 |
| | (=(17)·Cost of Tanker) (round figures) | | | | |
| 32. | Summary of Direct Cost Items Case 1 (=(23)+(29)+(31)) | | | | |
| | (a) | 1002 | 1481 | 1785 | 2771 |
| | (b) | 876 | 1231 | 1582 | 2287 |
| | (c) | 305 | 349 | 710 | 918 |
| 33. | Summary of Direct Cost Items Case 2 (=(24)+(30)+(31)) | | | | |
| | (a) | 820 | 1095 | 1310 | 2139 |
| | (b) | 759 | 932 | 1230 | 1307 |
| | (c) | 305 | 349 | 610 | 818 |

## List of Abbreviations

AB Atmospheric braking
ABC Atmospheric braking to a circular orbit
ABE Atmospheric braking to an elliptic orbit
DBS Drag brake system
HISV Heliocentric interorbital space vehicle
ODW Orbital departure weight

## List of Symbols

$A$ Projected area (relative to flow direction)
$C_D$ Drag coefficient
$I_{sp}$ Specific impulse
$n$ Ratio of apsidal distances, $r_A/r_P$
$q_R$ Radiative heat transfer rate
$r^*$ Radial distance from planet, measured in units of planet's radius
$r_A$ Apoapsis distance
$r_P$ Periapsisdistance
$v_c$ Local circular velocity about planet

| | |
|---|---|
| $v_E$ | Entry velocity into planetary atmosphere |
| $v_*^\infty$ | Hyperbolic excess velocity, measured in units of Earth mean orbital speed (subscript 1 = Earth departure; 2 = target planet arrival; 3 = target planet departure; 4 = Earth arrival) |
| $W$ | Local weight |
| $\gamma$ | Flight path angle |
| $\delta$ | Shock detachment distance |
| $\varrho$ | Atmospheric density |

## List of Subscripts

| | |
|---|---|
| $E$ | Entry conditions |
| Ea | Earth |
| Ma | Mars |
| Ve | Venus |
| $R$ | Radiative |

## REFERENCES

1. Jet Propulsion Laboratory: *Mariner, Mission to Venus*, McGraw-Hill Book Co., New York, 1963, p. 105–106.
2. de Vaucouleurs, G.: *Proceedings of Lunar and Planetary Exploration Colloquium*, II: 1, Sept. 1959, p. 42.
3. Levine, P.: *Entry of Space Vehicles into Planetary Atmospheres, Part I: Ballistic Entry*, Rep. AVCO-RAD-TM-62-20, AVCO Corp., April 1962.
4. Hidalgo R. W. Detra: *Generalized Heat Transfer Formulae and Graphs*, AVCO Everett Rep. RR72, AVCO Corp., March 1960.
5. Anon: *Heat Shield Materials Property Handbook*, AVCO/RAD Rep. No. 61-1951 (*Confidential*).

13

# *Entry Propulsion and Lower Systems Technology*

R. W. BUSSARD,
*Corporate Chief Scientist*
*Electro-Optical Systems, Inc., a Xerox Company*

## [13-1] REQUIREMENTS AND CONSTRAINTS

In this section we examine some of the bounding conditions on the problem of finding optimum choices of power and/or propulsion systems for planetary entry and atmospheric re-entry vehicles. To do so we first examine the classes of vehicles which are likely to be of interest in the near and far future of man's efforts to explore solar system space, and second review the classes and types of missions currently planned and rationally projected for this time span. Finally, the matrix of these vehicles and mission needs defines—in a broad way—the general requirements and constraints which should be met by future power and propulsion systems for such vehicles.

### [13-1.1] Classes of Re-entry and Planetary Entry Vehicles

We are all familiar with current planning for unmanned lander capsules for near-term Mars exploration purposes. These are simply the current extension of the original re-entry vehicle work carried out in the middle 1950's for ICBM RV warhead systems. Of course RV devices are much older than this; the earth's surface has been struck for millions of years by nature's own form of ablative entry vehicles—meteorites.

This is the image conjured up in our minds by the word "re-entry" vehicle; a hot object plummeting to earth in a cloud of ionized air. But today the word has taken on new meaning and greater scope with research programs aimed at lifting-body, low-speed RV systems (e.g., the NASA M-series of RV lifting-bodies) and the considerable body of study arising from detailed consideration of recoverable launch vehicles, recovery of large boosters, development of aerospace planes, etc. It is clear that the needs of each type of

RV or entry vehicle will be different, and it is useful to categorize these by type and functional behavior.

For simplicity let us take the first subdivision between *unmanned* and *manned* systems as indicated in Fig. 13-1. Then we readily distinguish a

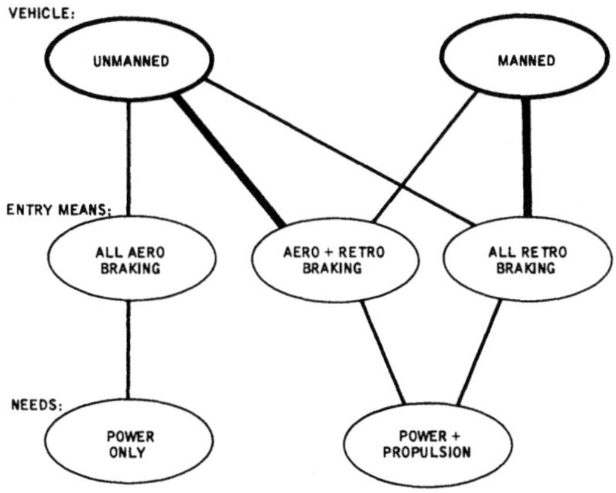


**Fig. 13-1.** Vehicle and mission classes for landers and entry vehicles.

useful further division between the two modes of entry/landing of *all-aerodynamic* braking and *all-retro-rocket* braking, plus a third mode which combines *both aero and retro* mechanisms. Aerodynamic braking in its many forms has been discussed earlier and will not be reviewed further here. Note that unmanned systems will most likely involve all-aero or aero-plus-retro, but are unlikely to involve full retro in planetary landing situations in which atmosphere is present. The absence of atmosphere on the Moon leads to the all-retro Surveyor, but aerodynamic effects will be important on all atmospherically shrouded bodies. In manned vehicles, however, it is highly unlikely that all-aero systems will be used, for the results of error and lack of flexibility are all-too-easily catastrophic in such a situation.

Lander and entry vehicles categorized in this way obviously span a large size range and a considerable time schedule in the future. The first unmanned planetary landers may well weigh the order of a few tens of pounds and land on Mars or Venus before 1970. The Voyager spacecraft lander will weigh a few thousand pounds, but is unlikely to be employed operationally before 1975 under the current program. The Apollo/LEM vehicle will weigh a few tens of thousands of pounds, and is the first (US) manned lander. First manned planetary landers are not likely before the middle 1980's and will also probably weigh the order of a few tens of thousands of pounds in early versions. Growth in size beyond this by much more than a factor of 10 or so seems unlikely until an era of mass transport methods is upon us—climbing up planetary gravitational hills is just too costly to justify a large vehicle unless there are repetitive, frequent, large payloads. Of course, many special cases exist. Certainly landing on Vespa or Eros (asteroid belt) does not require a large retro-rocket system as does landing on Ganymede. However, most of the manned vehicles will most likely travel to and land upon objects whose gravity fields are the same order as (generally less than) ours on earth, thus descent-phase propulsion requirements will tend to be of similar level for most systems. Hard landers of small size designed for launch from Jupiter probes will also be different than hard landers of large size and high drag for Martian surface landing. Each specific case deserves its own examination. Here we can only survey the general classes of such systems.

### [13-1.2] Classes of Missions

Mission types range from landers on one-to-fractional-$g_0$ surfaces on bodies with and without atmospheres, to landers on near-zero-$g_0$ bodies of interest throughout the solar system. Before assessing propulsion characteristics for these types it is useful to divide the flight operation into its several phases of: (1) Planetary capture; (2) Orbit shrinkage and de-orbiting; and (3) Descent (to surface).

The planetary capture phase is a subject of concern to the transport vehicle system which must carry the payload, manned or unmanned, from earth or earth-orbit to the target planet or other body. We do not concern ourselves here with this, but rather start from a presumed state of capture in an orbit about the target planet. If this orbit is nearly un-bound (i.e., the vehicle is only weakly captured) then it becomes important to assess optimum ways of delivering maximum payload from high orbit to low orbit, from whence it is de-boosted to the descent phase. In general, it is always true that a high-$I_{sp}$, low thrust system for orbit transfer or orbit shrinkage will yield payload gains over a low-$I_{sp}$, high-thrust transfer system if the $I_{sp}$ of the former system is above about 1000 sec or so. Thus, low-thrust systems involving isotope, electrical, or (small) fission reactor propulsion devices may prove attractive for such application. Likewise if the 'descent'' phase is that of matching heliocentric orbit parameters with a small and near-zero-$g_0$ body (e.g., asteroids, etc.) a low-thrust, high-$I_{sp}$ system may prove advantageous. On the other hand, low thrust requires precision guidance or command control of spacecraft motion, thus may be more suited to use with manned vehicles with the human on-board observer than to unmanned systems which require monitoring from extreme earth-to-target distances. In the latter case the operational predictability of the orbit mechanics of high-thrust short duration thrustor use may favor these for unmanned missions. Each case requires its own individual assessment to ascertain the optimum subsystem selection. Finally in the descent phase, the all-aerodynamic vehicle clearly does not require prime propulsion, but some thrust function may be needed for attitude control during re-entry or entry. The U.S.'s Mercury, Gemini, and Apollo programs all contemplate aerodynamic re-entry, following the trail blazed by meteorites and ICBM RV's. But, in fact, these also did involve parachute braking (a form of "retro-propulsion") and surface entry into the sea. This method of recovery has proven to be workable, but it is hardly practical, economically, and it is obviously impossible or very undesirable for manned vehicle operations in landing on distant planets at this stage of space flight development. In contrast it may be the simplest and most reliable method for landing unmanned vehicles on distant bodies, as the degree of precision control needed with powered descent trajectories is much much greater than for a falling body to impact safely.

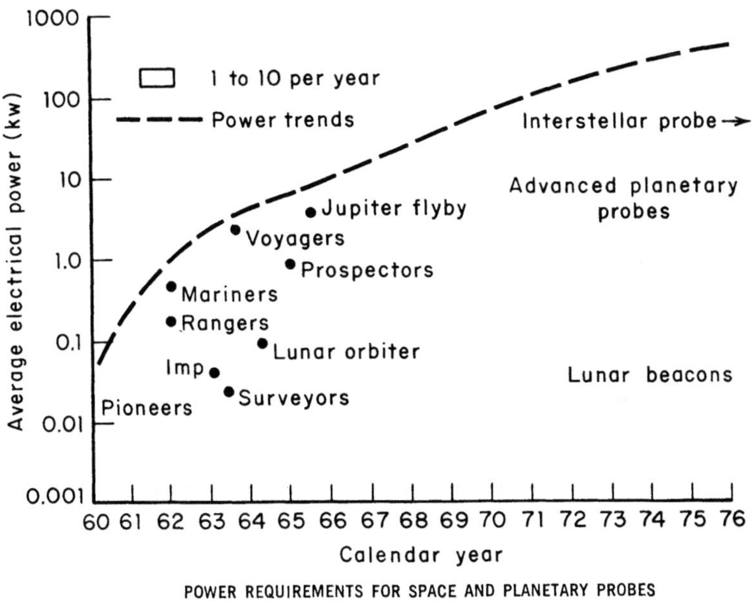


**Fig. 13-2a.** Projected power needs. 1960.

## [13-1.3] General Needs and Bounding Conditions on Power and Propulsion Systems

In attempting to assess power needs for space vehicles of the future it is immediately evident that projection much beyond current programs is a highly uncertain and unreliable guide. Programs come and go and "requirements" fall prey to the need to launch with whatever is available at the moment. In this regard it is instructive to review some past projections for then-future space activities. The series of Figs. 13-2 shown are drawn from these past projections for power requirements. The series of Figs. 13-3 show past assessments as to the optimum areas of application of various types of energy conversion systems to meet these requirements.

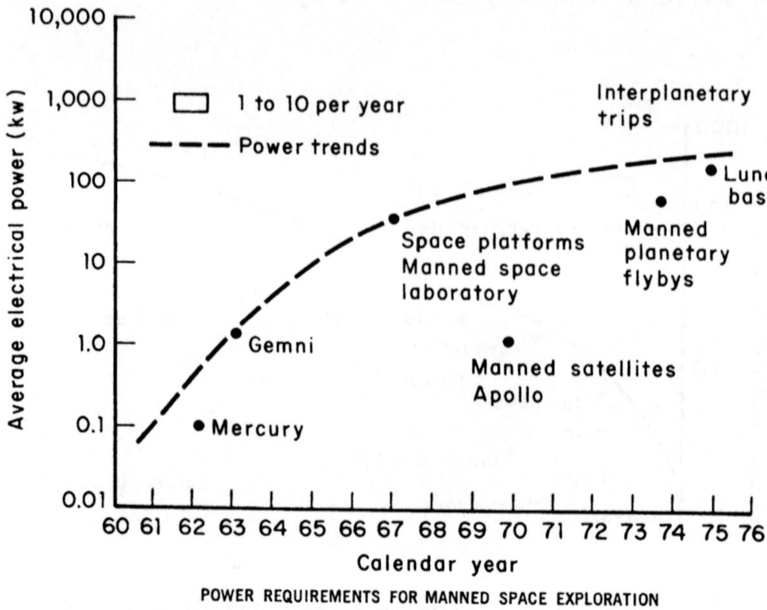


**Fig. 13-2b.** Projected power needs. 1960.
(Ref: W. R. Corliss, "Survey of Space Power Requirements", Prog. in Astro. & Aero., Vol. 11, 1962.)

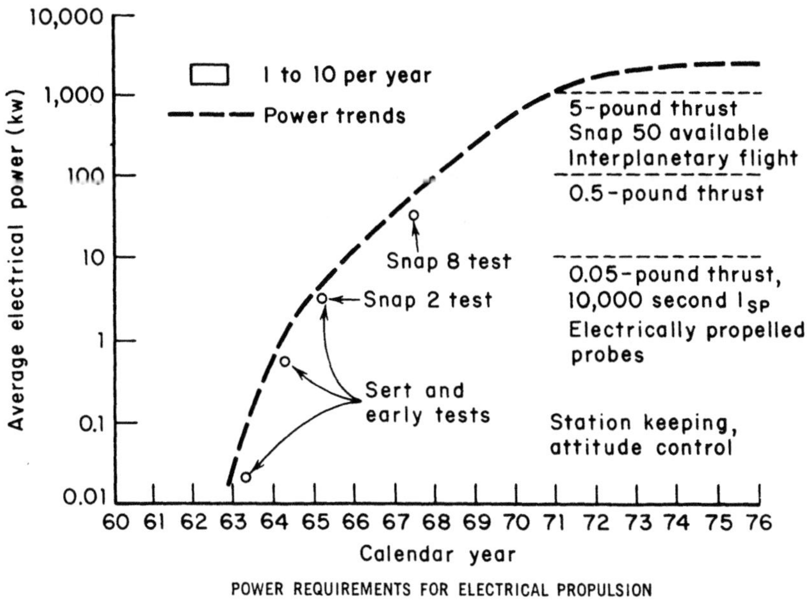


**Fig. 13-2c.** Projected power needs. 1962.
(Ref: W. R. Corliss, "Survey of Space Power Requirements", Prog. in Astro. & Aero., Vol. 11, 1962.)

Examination of Figs. 13-2 shows that in this field, the long-range guessing has usually been far too optimistic (in contrast to the generally-held belief to the contrary) and Figs. 13-3 reflect this optimism in terms of projected availability of advanced power systems. Attempting to counter these trends toward enthusiastic optimism without swinging to a negative outlook, Fig. 13-4 has been constructed as a "best-guess-ca. 1966" at the real needs of the future and the ways in which they may best be met.

Of course it is not possible to show all of the complex and inter-related phenomena which affect the attainability and/or application of each specific type of system; the diversity and detailed character of the various contending systems is so great as to render this impossible. All we can do here is cite general results and guidelines chosen for further discussion. Examining RV and planetary entry vehicle needs from this generalized view, a range of "needs" has been constructed, as shown in Table 13-1. Note how the "advanced-technology-future" has receded further into the calendar-future in this synthesized 1966 "best-guess." This is in part a result, not of pessimism, but of continued development and advancement in the older and more

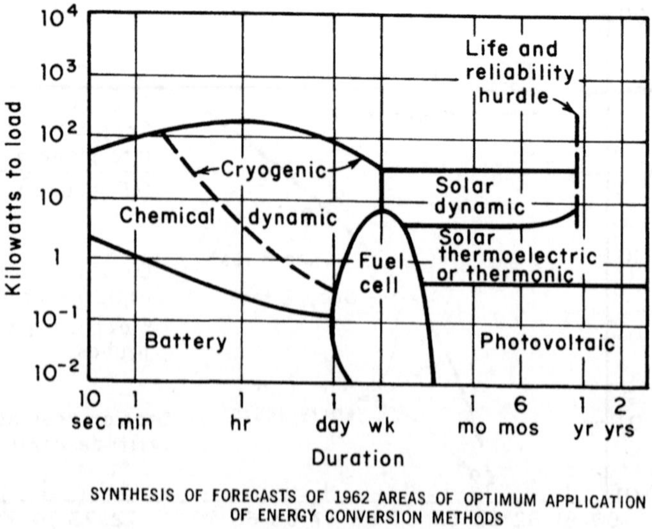


**Fig. 13-3a.** Projected technology application. 1960.
(Ref: C. Kelly, "Flight Vehicle Power Forecast", Prog. in Astro. & Aero., Vol. 4, 1960.)

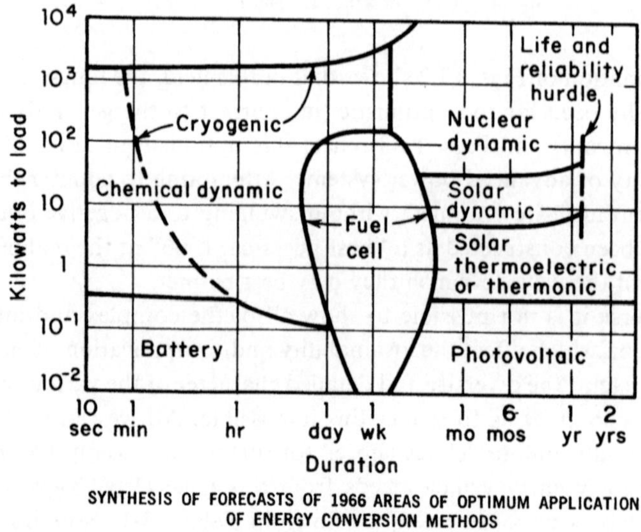


**Fig. 13-3b.** Projected technology application. 1963.
(Ref: G. C. Szego and Ernst M. Cohn, "Fuel Cells for Aerospace Application", Prog. in Astro. & Aero., May 1965.)

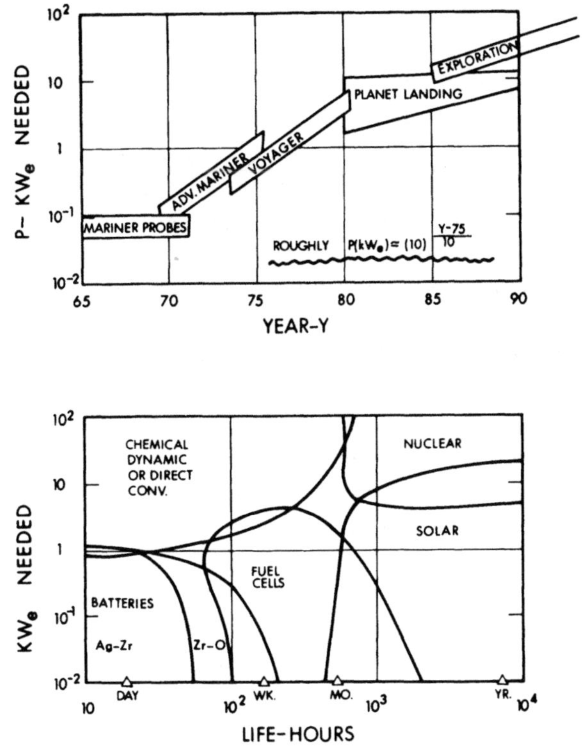


**Fig. 13-4.** 1966 "Best Guess" of RV/PE power needs and optimum technologies to satisfy these.

"standard" ways of doing business in space power. For example, current national programs are now underway which will lead to solar/photovoltaic power systems in the range of 25 to 40 lb./$kw_e$ within 2 to 3 years, as compared with over 110 lb./$kw_e$ as used in recent space flights (e.g., Mariner IV).

Table **13-1.** Estimated Range of Power Needs for RV and PE Vehicles

| Class | Weight | Power | Lifetime | Initial era | Remarks |
|---|---|---|---|---|---|
| *Unmanned* | 0–100 lb. | < 100 watts | Few days | 1965/72 | First landed probes |
| | 100–1000 lb. | 0.1–1 kw | Few days | 1969/75 | Mariner class |
| | 1000–10,000 lb. | 0.3–3 kw | Few weeks | 1973/80 | Voyager class |
| *Manned* | 10,000–50,000 lb. | ⩽ 10 kw | Few months | 1980/90 | First planetary landings |
| | 50,000–250,000 lb. | 10–100 kw | Few years | 1985/2000 | Major manned exploration |

Apparently it is important to be aware of the growth factor in known technologies, as we peer dimly into the future.

As regards propulsion; requirements here are in many ways easier to assess than for power. Either our vehicles will be aero-braked (no propulsion) or will use some form of descent thrusting. These two remaining classes involve either: (a) Retro-rocket de-orbiting with parachute or body-drag descent followed by impact or by thrusting for soft-landing; or (b) Retro-rocket descent all the way to landing. Which course is followed depends entirely on the mission and on the atmospheric features of the target planet. Hard-landing systems need at most only de-orbiting retro-rockets. The vehicle force accelerations (and hence thrust levels) required depend again upon the mission. Low-thrust, high-$I_{sp}$ can be used if time and guidance permit; if not, high-thrust low-$I_{sp}$ systems must be used. Typically, high-thrust solid or liquid rockets in the several hundred pound thrust range may find application on hard-landed vehicles up to several thousand pounds-weight. Larger vehicles (e.g. manned systems) would require proportionately larger engines.

Soft-landing systems must have thrust capability exceeding local vehicle weight. Since man will likely land only on bodies of one $g_0$ or less (by 1990) landing engine thrust levels will generally not need to exceed vehicle earth weight. Typically, Mars landers may require propulsion system thrust of 0.4 to 0.5 earth weight, while Venus landers must be roughly twice as high. Landings on small bodies—e.g. Deimos, Phobos, Vespa, etc.—will almost certainly be done with still smaller engine systems. In general a maximum thrust of no more than twice local weight is probably desirable. In addition, thrust throttling is essential for all soft-landing manned systems, simply to allow flexibility and control of the complete trajectory and thus to permit over-ride and/or compensation for all of the total systems errors which lead to trajectory misplacement as well as to give some freedom of choice of landing site to the spacecraft crew. Throttling range needed may be as high as 10:1, but future systems may find this degree of flexibility unnecessarily great as trajectory guidance systems and equipment improve in quality and reliability. Table 13-2 summarizes the general features which likely candidate propulsion systems should possess for the class of RV and PE vehicles of interest here.

Although not summarized in Table 13-2, it is obvious that operational features of each type of propulsion system will color and in some cases set or limit its use for various missions. For example, large nuclear engines used for throttlable landing engines on 250,000 lb manned landers of the 1990 era present the grave difficulty of residual radiation on-board after landing, rendering dis-embarkation impossible for humans; ergo, such engines can

Table **13-2.** Qualitative Needs of Propulsion Systems for RV and PE Vehicles

| Type of Landing | Type of Entry Braking | Maximum Thrust Requirement | Duration | Remarks |
|---|---|---|---|---|
| Hard | All-aero | None | None | ICBM RV mode |
| Hard | De-orbit retro plus aero plus impact landing | High-thrust or low-thrust for de-orbit. Fixed thrust levels acceptable with variable firing time capability. | Short or long, depending on propulsion system choice. | Choice depends on other mission features, time, distance, guidance, etc. |
| Soft | Retro de-orbit plus retro soft landing. | 1.2 to 2 times local weight. Engines must be throttable as well as capable of operation over variable times. | Variable but typically should have capability for considerable excess over total $\Delta V$ requirement of lander vehicle system. | Flexibility, reliability, and safety margin are of principal concern. |

not be used for this function without complex ground equipment, which makes their use unlikely. Similarly, $F_2$-oxidized engines create difficult problems for landings on planets with noticeable atmospheric water content for quantities of aqueous hydrofluoric acid would be formed around the landing site. Further detailed discussion of these features and characteristics is given in the next section, for both power and propulsion systems.

## [13-2] POWER AND PROPULSION SYSTEMS AND SYSTEM TYPES

### [13-2.1] Power

In this section we survey the range of candidate power systems for RV and PE vehicles, in a broad sense, in order to provide a background for evaluation of optimal choices for real system use in the future.

#### [13-2.1.1] Nuclear Energy Sources

There are only two available today: Fission reactors and radioisotopes. Fission reactors have been developed largely around zirconium hydride moderator systems, as exemplified by the SNAP-10A power system. AI has spent many years in refining $Z_rH_x$, sodium or NaK-cooled, reactor technology to the point that such reactors could be built for thermal power levels up to several megawatts with electrical power output after conversion of up to several hundred kilowatts. Table 13-3 indicates the specific power of such reactor systems. We note in passing that the principal difficulty in projecting such systems is a lack of program continuity.

Table **13-3.** Typical Specific Power for Hydride SNAP Reactor Power Plants

| Reactor | Power (elec.) | $\alpha$ lb/kwe. | Remarks |
|---|---|---|---|
| SN-10A | 0.5 $kw_e$ | 1300 | Orbital flight, early 1965. |
| SN-2 | 2–3 $kw_e$ | 300–400 | No program. |
| SN-8 | 30–60 $kw_e$ | 100–200 | Cancelled. |

Referring to the previous section, since power needs are lacking, little justification has appeared for program extension and space-fission-reactor power programs today are nearly all reduced to the "technology-development" mode (except possibly for Romashka in the USSR). This is certainly the case with SNAP-50, which was a fast reactor using uranium and zirconium carbide core structures with lithium coolant, columbium (niobium) tubing, etc., developed from origins stemming back to the late 1940's or early 1950's in Aircraft Nuclear Propulsion activities. This program, aimed at power output of 500–1500 $kw_e$ at $\alpha \approx 10$ to 20 lb/$kw_e$, suffered from an inability to project exact $\alpha$ numbers on a defensible time scale. Each year of work led to higher $\alpha$ values, less performance, and less interest, and for what purpose? The need for power above 100 $kw_e$ is solely for electric propulsion or laser radiation weapons, and neither use appeared compelling for program continuation. Thus, it is likely that the best we can count on for the near and even the relatively far future is the ca. 1000 lb./$kw_e$ of early hydride SNAP systems.

Conversion systems are another matter. Many, many types of conversion systems have been studied and carried to fruition in development programs. These include Hg-working-fluid Rankine cycle turbomachine systems, magneto hydrodynamic (MHD) systems, thermo-electric converters, and thermionic diodes. Late-comers include Brayton cycle systems and electro-hydrodynamics (EHD or EGD). Most practical experience to date has been accumulated with thermoelectrics and Rankine cycle equipment. The SNAP-10A flight test used Ge–Si converters developed by RCA with AI, which operated successfully in-orbit for over 1 1/2 months. A typical turbomachinery program is that connected with SNAP-2, not now a *power systems* program. Here the TRW Corporation has been developing a 2 $kw_e$ rated rotating unit combining the turbine and alternator on one shaft and seal system. Extensive experience with this unit (ca. 40,000 hr. test life—cumulative) indicates that such equipment could be viewed as on the verge of reliable and reproducible fabrication. Alkali metal systems have also been studied extensively, by GE, for Rankine cycle operation at higher temperatures than feasible with Hg. Higher temperature means smaller radiators and, in principle, lower radiator and system weight. A full resolution of the capabilities and performance qualities of each type of system will not be reached for perhaps a decade or more. Reactor development will proceed on an even longer time scale.

Table **13-4.** Radioisotope Space Power Generators*

| Designation | Watts | Life (yr.) | Wt. (lb.) | Application | Fuel | Status |
|---|---|---|---|---|---|---|
| Snap 3 | 2.7 | 5 | 5 | 4A and 4B | Plutonium 238 | Launched June and November, 1961. Unit on 4A still powering transmitter signals. Converter on 4B shorted out June, 1962. |
| Snap 9A | 25 | 5 | 27 | Navigational Satellite (Navy) | Plutonium-238 | First unit launched Sept. 18, 1963, second in December, 1963. Generators operating satisfactorily; third unit burned up in April, 1964. |
| Snap 11 | 25 | $\frac{1}{3}$ | 33 | Surveyor (NASA) (intact re-entry) | Curium-242 | Electrically heated generator delivered to JPL May, 1965. First fueled unit scheduled mid-1966. Prototype flight, electrically heated generators to be delivered to Jet Propulsion Laboratory and Manned Spacecraft Center in mid-1966. A fueling demonstration will be made. Use of Snap 11 on Surveyor missions dependent upon operation of solar panels on the lunar surface. |

Table 13-4 *(continued)*.

| | | | | | | |
|---|---|---|---|---|---|---|
| Snap 13 | 12 | $\frac{1}{3}$ | 4 | Thermionic demonstration device | Curium-242 | Program still active. Martin Div. of Martin Marietta Corp. is the prime. |
| Snap 17 | 30 | 3-5 | 30 | Communication Satellite | Strontium-90 | Two generator design and component test contracts (Phase 1) terminated November, 1964. (Thermoelectric modules continue under test.) |
| Snap 19 | 30 | 5 | 30 | Nimbus-B | Plutonium-238 | Electrically heated generators on test at Goddard Space Flight Center since February, 1965. Flight date 1967. Phase 3 flight hardware development initiated September, 1965. |
| Snap 25 | 75 | 5 | 35 | Earth orbital mission | Plutonium-238 | Canceled. System developments and advanced technology programs under way or planned adequately cover critical areas of this effort. No firm mission requirements exist for this power system. |
| Snap 27 | 50 | 5 | 25 | Apollo Lunar Surface Experiment Packages (intact re-entry) | Plutonium-238 | Two Phase 1 design studies completed. General Electric selected to continue program. Phase 2 begun August, 1967. Prototype fueled units to be delivered in April, 1967. Flight hardware scheduled for delivery beginning in July, 1967. |
| Undesignated | 6-10 | 1+ | 3 | Lightweight demonstration device | Plutonium-238 | Under test at General Atomic Div. of General Dynamics Corp. |

* After M. L. Yaffee, "AEC Doubles Space Radioisotope Efforts", Aviation Week & Space Technology, January 10, 1966.

*Isotope sources.* Many isotopes have been proposed for space power use. Table 13-4 gives a brief summary of current and past AEC programs for such devices and Table 13-5 shows the range of isotope fuels usually considered. Interest in isotope power sources is growing as indicated by the NASA Center activity in this field. As lander vehicles come closer to reality it will be necessary to carry out extensive high power isotope developments to keep pace with the requirements previously delineated. Certain isotopes lend themselves more readily to one class of mission than another; e.g. short-lived alpha emitters may be greatly preferable to long-lived gamma or beta emitters for near-earth manned orbital power systems, etc. The general status of technology is as indicated in the table, however, some classified fuel forms exist which offer considerably greater performance capabilities.

Unfortunately, the operating temperature of many isotopes is too low to be used with most current thermionic convertors hence radioisotope sources have tended to be used with thermoelectric convertors (e.g., PbTe, GeSi). This situation could change in the future as thermionics operate at lower temperatures with good efficiency and isotope sources are driven to hotter operation. Low temperature thermionics could also revive out-of-pile reactor/thermionic systems and thus ease the pain of development of such systems.

MHD convertors have gained popularity in recent years and are currently under intensive research study. These devices simply use an ionized gas—a plasma—in motion to replace the moving element in a turbomachine generator (or in the linear equivalent thereof). Inherent disadvantages of high temperature operation, wall cooling, corrosion, etc. tend toward low efficiency. These devices will likely not be practically competitive for at least one, and possibly two, decades.

### [13-2.1.2] Solar Energy Sources

The sun is a gigantic thermonuclear reactor, thus all solar systems are really "nuclear" in character. The solar flux at one A.U. (earth's distance) is very closely 1 $kw_{th}/m^2$. It falls as $1/r^2$ with increasing distance from the sun, thus is not likely to compete with other sources for missions to Saturn and beyond.

Solar energy can be used with any form of conversion machinery. Vigorous programs have been carried out on solar-collector thermionic and turbomachine systems. EOS, working for JPL, has been quite successful in developing very light and efficient solar-thermionic units up to 5 and 10 feet in mirror diameter. Other programs (ASTEC) have contemplated 50 foot diameter unfolding mirror systems with Rankine cycle conversion equipment. None of these programs to date have seen interest from the community of

**Table 13-5.** Isotope Decay Characteristics for Typical Bare, Encapsulated Radioisotope Sources

| Parent Isotope | Half Life | Energy* per Disintegration (mev/disint) | Fuel Form | Power Density (kw/in³) | Gamma Dose at 1 meter (mr/hr-watt) | Neutron Dose at 1 meter (mr/hr-watt) | Total Dose at 1 meter (mr/hr-watt) |
|---|---|---|---|---|---|---|---|
| Po-210 | 138.4 day | 5.305 | metal | 19.9 | 0.110 | 0.010* | 0.120 |
| | | | $PoO_2$ | 18.4 | 0.110 | 0.047 | 0.157 |
| | | | monopolonide*** | 3.61 | 0.110 | 0.030** | 0.130 |
| Cm-242 | 162.5 day | 6.10 | metal | 25.85 | 0.0181 | 0.161 | 0.179 |
| Cm-244 | 19.2 yr | 5.78 | metal | 0.560 | 0.115 | 4.04 | 4.155 |
| Pu-238 | 86.4 yr | 4.485 | metal | 0.176 | 0.00626 | 0.0057 | 0.01196 |
| Pm-147 | 2.6 yr | 0.0649 | metal | 0.0256 | $0.374 \times 10^{-4}$ | — | $0.374 \times 10^{-4}$ |
| | | | $Pm_2O_3$ | 0.0293 | $0.374 \times 10^{-4}$ | — | $0.374 \times 10^{-4}$ |
| Sr-90 | 29.3 yr | 1.159 | $SrTiO_3$ | 0.0172 | 182.6 | — | 182.6 |
| Ru-106 | 1.004 yr | 1.532 | | | 735.3 | | 735.3 |
| Ce-144 | 284 day | 1.346 | $CeO_2$ | 0.404 | 1550 | — | 1550 |

* The energy per disintegration is the total energy liberated by the parent isotope and its daughter product (assuming secular equilibrium) per disintegration of the parent isotope.

** The neutron field is generated by an alpha-*n* reaction with oxygen, a common impurity in polonium.

*** The specific chemical compound of this fuel form is classified.

users. All solar-concentrator systems require relatively precise guidance and pointing and this has not been deemed desirable to date in space ventures. Future ground power requirements after landing may lead to a need for such systems, however.

By far the most versatile and useful of all solar convertors to date is the doped silicon chip. This semi-conductor, the solar cell, has been used for prime power on all but a few of US spacecraft. In the beginning of the Space Age photovoltaic systems were relatively heavy, inefficient, and cumbersome; typically at several hundred lb/$kw_e$ with 4–5% efficiency. By 1964–65 these figures had changed to roughly 115 lb/$kw_e$ with 10–11% efficiency for operation of oriented (i.e. sun-pointed) solar arrays. Future development of rigid structure systems promises 25lb/$kw_e$ at 10% efficiency and thin film cells with plastic flexible substrates may yield even greater performance in the farther future. Of course solar cells used in a non-oriented way will have considerably greater specific weight than for oriented systems. Typically a random (covered sphere, tumbling or spinning spacecraft) array will weight 2–2.5 times as much for each $kw_e$ out as an oriented array, but system operational considerations are often overriding in such cases.

With solar concentrator systems the effective equivalent solar-energy-produced electrical power will vary almost as $1/r^2$, the solar flux variation. However, with photovoltaic systems the rate of decrease is less rapid than this in the out-bound (from the Sun) direction, because the solar cells operate at higher efficiency as they become colder at greater distances from the Sun. The variation is thought to be as $1/r^{1.7}$. Conversely, moving inward, cell efficiency drops progressively more rapidly toward zero with the result that a solar cell system reaches maximum performance about half-way between the orbits of Venus and Earth. Inward from this point photovoltaic systems fail rapidly and are surpassed competitively by solar-collector systems (which also fail still closer-in because of defocussing) and solar-thermoelectric systems which only begin to work properly at less than 0.5 A.U. In assessing the probable future it is important to remember that continuing advances in these "old stand bys" will make the going much harder for new and novel "advanced concepts".

### [13-2.1.3] Chemical Energy Sources

Chemical oxidation or combustion releases heat energy which, again, can be used to drive any of the thermally-activated convertors discussed previously: MHD, turbomachines, thermionic or thermoelectric convertors. Many of you have no doubt seen the ingenious kerosene-lamp-flame-heated thermoelectric convertors distributed by the Soviet Union in Arab countries to power small radio sets. The only unique class of devices here are the short-time high-power APU's utilized in re-entry vehicles. These have been developed around turbomachine and MHD concepts to the point where 50 watt hr. is obtainable per lb. of unit for 15–20 minutes of operation. Future growth promises to yield over 100 (watt hr.)/lb. for such periods—a performance level which far exceeds battery capabilities.

Batteries were first to power orbiting spacecraft, long ago, in 1957. Today batteries are still with us and will likely remain so for a long time to come, at least for secondary or back-up power, line floating, fusing, shadow-time power in earth orbits, etc. For short-lived, low-power requirements batteries are superior to all other systems. Ordinary batteries for space use today are Ni-Cd, but Ag-Zn, and $ZnO_2$ systems offer growth potential for the future. Still better, in prospect, are fuel cells operating on high energy fuels. These latter are nothing more than a form of battery with external charge carrier and acceptor supplies. Conventional batteries store their fuels internally following charging, fuel cells do not. All such systems are nothing more than devices which convert electron-bound energy exchange differences directly to electrical energy in the attached circuitry. They achieve this by causing the oxidation (or "combustion") reaction between the battery constituents to occur on the surface of one electrode which is binding one of the constituents by surface-work-function-like electrical forces.

Fuel cells can use nearly any combustible combination. A particularly energetic one is $O_2/H_2$, both cryogenic liquids, which holds considerable promise for the future. Once burned, the combustion products may be dissociated into their original constituents later, by long-term auxiliary power, to be re-used in the fuel cell during a time of high power demand. Thus, such a fuel-cell-plus-fuel system is rechargable, as are conventional batteries.

## [132.2] Propulsion

### [13-2.2.1] Electric Thrustor Systems

These comprise arcs, plasmas, charged colloidal particles, and ion engines of various types. Work has progressed in the past decade on *all* types and today ion engines are operating well past 8000 hours in bench tests, with 10,000–20,000 hour life capabilities in real prospect. All such devices are inherently high-$I_{sp}$ low-thrust in nature. Table 13-6 indicates the general range of performance expected.

Thrust/weight ratio for such engines ranges from $10^{-6}$ to $10^{-4}$ $g_0$ depending on mission and power supply choices. Thrust levels are measured in micro- and milli-pounds.

Clearly these devices are never going to be useful for the descent or landing modes of RV or PE vehicles, but they may well find use in orbit shrinkage

Table **13-6.** Typical Electric Engine Performance Ranges

| Type | $I_{sp}$ | Thermal Efficiency | Remarks |
|---|---|---|---|
| Arcs | 800–1600 sec. | $\leqslant$ 50 % | Electrical heating |
| Plasmas | 1500–5000 sec. | $\approx$ 70 % | Magnetic acceleration |
| Ions | 2000–10,000 sec. | 70–80 % | Bombardment or Contact ionization. |
| Colloids | 600–2000 sec. | 80–95 % | Cold system |

and/or deorbit boosting. If planetary capture is made to occur into a high or highly elliptical orbit, then considerable $\Delta V$ must be expended to achieve a low enough orbit that atmospheric drag can aid the de-orbit phase. In the case of the Earth this $\Delta V$ is the order of 10,000 ft/sec. If such a condition characterizes the mission profile, and sufficient time is available, payload gains of 40% or so may be made by use of electric propulsion for orbit shrinkage as compared with a competing candidate high-thrust low-$I_{sp}$ system (e.g. LOX/RP). If payload in Earth orbit costs \$300/lb (low today) and 1/30 is deliverable to a Martian capture orbit, even a 20% saving in orbiting vehicle payload delivered to low orbit is worth about \$100 million. Fortunately our space program R and D efforts have already given us much of the technology needed to achieve such savings.

One problem with all such systems is that of integration with vehicle design, for electric propulsion systems have their own unique requirements which must be met. Attention to interface and integration questions is as essential here as in all other parts of the system.

### [13-2.2.2] Chemical Rocket Engines

The history of rocketry to date lies in chemical rocket engine development. Exotic systems were conceived long ago and are still under study, while much refinement has taken place in standard types of liquid and solid rockets. Today we see a spectrum of high-performance solid propellants, "storable" liquid mono- and bi-propellants, cryogenic liquids, and super-propellants using unstable compounds. Similarly, rocket engines have evolved to the point where we now see radiatively-cooled, regeneratively-cooled, and ablatively-cooled engines using pressure or pump feeds, with and without engine throttling. Not all combinations of specification parameters are equally developed, equally tractable, or equally desirable for specific missions.

In assessing the relative worth of rocket engine systems it is helpful to have a larger view of their range of performance. Fig. 13-5 indicates this, showing typical engine (maximum) thrust-to-weight ratio as a function of thrust. From the figure we note that storable propellants hold the key at

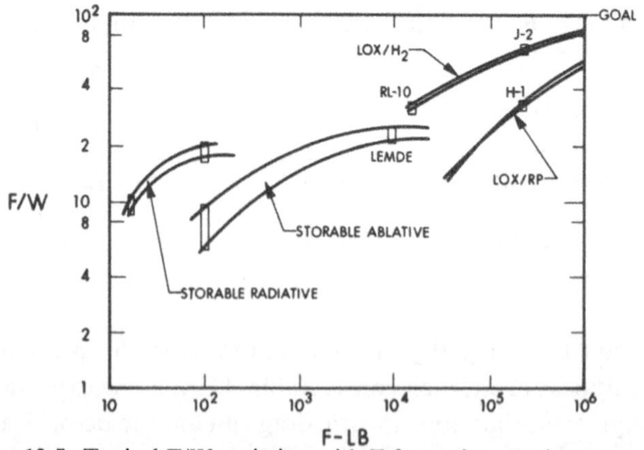


**Fig. 13-5.** Typical F/W variation with F for various engine types.

present to the thrust range below 10,000 lb., but cryogenic liquids offer most promise for higher thrust levels.

Propellant combinations of interest include those indicated in Table 13-7, which cites some of the features of each class. Note that little $I_{sp}$ growth is projected for the next decade or so. Hydrazine-fueled storable propellants are relatively highly developed and reliable, without truly noxious exhaust products. Similarly, $O_2/H_2$ cryogenic systems exhibit relatively sedate en-

Table **13-7.** Some Typical Chemical Rocket Propellants

| Propellant | $I_{sp}$ | Era | Remarks | |
|---|---|---|---|---|
| Mono-propellant $N_2H_4$ | 235 sec | 1965 | Current | storable |
| | 242 sec | 1968–9 | Improvement | storable |
| | 255 sec | 1972 | With additives | storable |
| $N_2O_4/N_2H_4$ | 300–310 sec | > 1965 | Little growth forseen | storable |
| $N_2O_4/N_2H_2(CH_3)_2$ | 310–320 sec | > 1965 | Little growth forseen | storable |
| $O_2/H_2$ | 420 sec | 1965 | At 300 psia today | cryogenic |
| | 440 sec | 1967 | At 3000 psia | cryogenic |
| $O_2/H_2/Be$ | 480 sec | 1967 | Very hazardous system | cryogenic |
| | 550 sec? | 1990? | No way known at present | cryogenic |
| $F_2/H_2/Be$ | 440–460 sec | 1967 | Hazardous system | cryogenic |

vironmental behavior. As indicated in the figure, engines of both types exist and will continue their development and test history as the future unfolds, thus leading to even greater reliability for future missions. Fluorine-oxidized systems are also under study and development, but exhibit the unfortunate exhaust products of HF, hydrofluoric acid when in an aqueous solution. This circumstance renders their use questionable if near, on, or from ground installations of value or which are subject to human occupation.

### [13-2.2.3] Nuclear Thrust Units

The fission rocket reactor is well along in development. $I_{sp}$ over 700 sec is on hand and 800 sec is likely as a reliable figure in the near-future. Thrust levels will rise from 60,000 lb. to 250,000 lb. in the next 5–10 years, and will also fall to 5000–10,000 lb. as smaller and lighter reactors are developed. Thrust/weight rating will most probably be in the range from 4 to 10 in this time-frame. For application to RV and PE vehicles, fission rockets are far worse than any chemical system, for residual radiation trapped in the reactor core and surrounding structure will always be so great as to prevent man from descending from a landed space craft without the aid of shielded, remotely operable exit walkways. This later day may come, but is unlikely before 1990.

Another class of nuclear thrustors is that derived from isotopes used as heat sources to heat hydrogen to $I_{sp}$ levels of 600–800 sec. As we have seen isotope power densities are always small, thus radioisotope thrustors will always be relatively small and of low-thrust capability. Typical thrust levels range from the order of a few tenths of a pound to several pounds and the thrust-to-weight ratio is the order of $10^{-2}$. This thrustor was conceived and developed by TRW Systems (STL) and dubbed Poodle, "a miniature Rover." Poodle thrustors fill an $I_{sp}$ and F/W ratio gap between high-thrust chemical rockets and low-thrust electrical propulsion. Analysis has shown that Poodles compete with each type at the ends of the mission spectrum of applicability. One such competive mission is that of the orbit shrinkage phase of planetary entry from a capture orbit. Here isotope thrustors could use boil-off from tanked $H_2$, if available, as the shrinkage and de-orbiting propellant, but even without this the gains in payload to be made by Poodle are the same order as from 40 lb/kw$_e$ electrical propulsion systems. The advantage of Poodles used this way is in their higher thrust-to-weight ratio, for only 4 or 5 days would be needed, while an electrical system would require 40 days or so of thrusting. The disadvantage is simply that isotopes are being used, with associated mission operational hazards and handling difficulties.

## [13-3] TECHNOLOGY AVAILABILITY AND THE MATCH WITH MISSION OPERATIONAL NEEDS

### [13-3.1] The Next Decade: 1965–1975

In prospect here we envision reliable systems using storable propellants, throttlable engines, continued expansion of photovoltaic and battery power systems, and use of high-thrust chemical engines for orbit shrinkage, de-orbiting, and planetary descent. Development of radioisotope power and electrical propulsion systems will continue. Across our vehicle/mission matrix we should expect to see:

1. Unmanned landers with small throttlable engines.
2. Manned landers with larger throttlable engines.
3. All using storable $N_2H_4$-based fuels and $N_2O_4$-based oxidizer systems.
4. Battery power on most lander vehicles.
5. Photovoltaic (PV) power on most bus/lander vehicle systems prior to de-orbiting and some small use on landers.
6. Some small use of radioisotope thermoelectric generators (RTEG) on landers.

### [13-3.2] The Farther Future: 1975–1990

As technology progresses we expect to see a fading of old concepts and growth and acceptance of new ones. In particular we should see the rise of cryogenic bi-propellant chemical rockets with throttling, as reliable hardware, the rapid growth of isotope power systems; the beginnings of fission reactor utilization, and the expansion of photovoltaic systems for use with manned systems. For our vehicle/mission matrix we should see:

1. Unmanned landers to the farther reaches of the solar system with both storable and cryogenic propellants.
2. Manned landers with cryogenic bi-propellants.
3. All engines throttlable.
4. RTEG power on most vehicles.
5. PV power to large sizes on most manned vehicles, both before and after landing.
6. Use of electrical propulsion low-thrust systems for orbit shrinkage and de-orbiting of unmanned systems, possible use of isotope power (perhaps derived from RTEG waste heat) for rapid de-orbit of manned landers.

Beyond 1990 the future is so vague that speculation on what *will be* is fruitless. It is easy to estimate what *could* be brought to life, but there is no known way to assess the future we will all jointly choose from the array of possible futures. The success of thermonuclear reactor development could revolutionize space power and propulsion—perhaps—and allow us, for the first time, to think seriously of mounting an unmanned interstellar probe using fusion propulsion. Time scale? This *might* prove possible by the mid-2000's. If so it must be only a century or so to interstellar manned explorations, and all that lies beyond. The problems we foresee for today's programs will appear miniscule under the eye of that future time, but that time will *not* be reached unless today's real problems are resolved. Planetary entry is thus a first (giant) step along a steep and lengthy highway to the stars.

## REFERENCES


1. Corliss, W. R.: Survey of Space Power Requirements, *Prog. in Astro.* and *Aero.*, Vol. **11**, 1962.
2. Finger, H. B. and F. Schulman: Power Requirements of the NASA Space Program, *Prog. in Astro.* and *Aero.*, Vol. **4**, 1960.
3. Kelly, C.: Flight Vehicle Power Forecasts, *Prog. in Astro.* and *Aero.*, vol. **4**, 1960.
4. Szego, G. C. and Ernst M. Cohn: Fuel Cells for Aerospace Application, *Prog. in Astro. and Aero.*, vol./?, May 1965.
5. Yaffee, M. L.: AEC Doubles Space Radioisotope Efforts, *Aviation Week and Space Technology*, vol./?, January 10, 1966.

14

# *Guidance and Control for Atmospheric Entry*

R. C. DUNCAN,
*Assistant Director for Systems*
*NASA Electronics Research Center*

## [14-1] INTRODUCTION

Previous chapters in this book provide the foundation of physics and technology required for the development of spacecraft systems for atmospheric entry. Hardware systems design involves assembling the technological blocks in such a way as to satisfy mission requirements while at the same time not violating critical constraints on size, weight, reliability, power, environmental provisions, and performance. It is clear that guidance and control are important elements in any recentry system, whether it be manned or unmanned. This chapter examines the concepts and hardware for guidance and control of the Mercury, Gemini, and Apollo spacecraft. The reentry technology for manned spacecraft systems has matured very greatly in the 1960's.

Manned planetary entry systems of the future will use concepts that have been developed and tested in the Apollo program. Apollo involves two major classes of entry problems: entry into a dense planetary atmosphere at supercircular velocities, and entry to a soft landing in a vacuum. The next major manned planetary exploration program, probably to Mars, will likely use these concepts. The Martian atmosphere may be sparse enough that Lunar Module concepts may be applicable for a manned landing. Until the recent Mariner photographs were obtained, it was assumed that Earth entry techniques would be used on Mars. Now, it is postulated that some aspects of lunar landing techniques may also be used.

Under any circumstances, it is of value to examine in some detail the reentry experience of Mercury and Gemini and the concepts and hardware applied in the Apollo program.

## [14-2] ENTRY TECHNIQUES

There are many different techniques which may be used to guide a spacecraft during the atmospheric entry maneuver. The selection of a particular technique is influenced by a number of considerations. The most important consideration which will influence entry guidance is the requirement for safety of the crew. The velocity and angle at which the spacecraft enters the atmosphere affects entry guidance. The entry guidance program must consider physical properties of the spacecraft and the amount of range control which is required. The entry guidance program is limited by the physical quantities which can be measured and by the size and speed of the onboard computer. There must be some technique for monitoring primary entry guidance in order to insure a safe entry. The primary guidance system and the entry monitoring system must be compatible with each other.

The maximum permissible acceleration in manned spacecraft is normally considered to be 10 $g$'s, although the maximum emergency acceleration may be as high as 20 $g$'s. Protection from aerodynamic heating is provided by the heat shield, but the entry guidance trajectory must be chosen to minimize heating problems.

The Mercury and Gemini spacecraft entered the earth's atmosphere at near circular velocities while the Apollo spacecraft returning from a lunar mission enters at near parabolic velocities. Spacecraft returning from interplanetary missions will enter at much higher velocities. The task of the entry

CLASSIFICATION OF AERODYNAMIC PROPERTIES OF ENTRY VEHICLES

| L/D | $W/C_DS$ | EXAMPLES |
|---|---|---|
| 0 | CONSTANT | MERCURY |
| 0 | VARIABLE | |
| CONSTANT | CONSTANT | GEMINI, APOLLO |
| VARIABLE | VARIABLE | |

**Fig. 14-1.**

guidance system becomes more difficult as the entry velocity increases. The entry angle must be kept within a value which will allow the spacecraft to be captured by the atmosphere and a value which will not result in excessive acceleration.

The aerodynamic properties of the entry vehicle may be divided into the general categories shown in Figure 14-1. The first group consists of vehicles such as the Mercury spacecraft which do not develop any lift and which have a constant ballistic number ($W/C_DS$). In this chapter the aerodynamic properties are considered to be variable only if the values of these quantities can be controlled by the guidance system. Although the aerodynamic properties are termed constant, they may vary as functions of Mach number and Reynolds number. The second group listed in Figure 14-1 are vehicles which do not develop lift, but whose drag properties can be varied. Such a vehicle would have a small amount of range control. The third group includes both the Gemini and Apollo spacecraft. The vehicles of this group develop lift, but their lift-to-drag ratio ($L/D$) and their ballistic number are constant. The trajectory which is followed by a spacecraft of this type can be controlled only by rolling the spacecraft. This changes the direction of any lift which is developed. Finally there is the last group of vehicles for which the ratio of lift-to-drag and the ballistic number can be varied.

The guidance parameters which can be measured must be considered when a guidance system is selected. The trajectory of the vehicle prior to the retrofire and entry maneuver can be determined by either earth-based tracking facilities or by measurements made onboard the spacecraft. Ground-based measurements cannot always be used for guidance during much of the entry maneuver, owing to loss of communication with the spacecraft due to the ion sheath. The inertial acceleration and attitude of the spacecraft may be measured onboard. Acceleration measurements are integrated to give the speed and position of the spacecraft during the entry. The guidance logic should use these measurements in a manner which will minimize the effects of errors in these quantities.

Any guidance logic which is considered must be compatible with the size and speed of any guidance computer which is available onboard. For obvious reasons, the simplest guidance logic which will give acceptable accuracy is the most desirable system.

## [14-3] MERCURY EXPERIENCE

The flight experience of manned spacecraft in this country has been in Projects Mercury, Gemini, and Apollo. A nonlifting shape was chosen for the Mercury spacecraft because of certain attractive operational aspects. Since the spacecraft would follow a ballistic path, there was a minimum requirement for autopilot, guidance and control equipment. Perhaps it would be more correct to say that the spacecraft was aimed at the target rather than guided to the target. In order to reach the target two maneuvers were performed. These maneuvers were the alignment of the spacecraft and the initiation of entry by the ignition of the retrograde rockets.

The location of the touchdown point was controlled by proper choice of the position of the retrofire maneuver, the total impulse of the retrograde rockets, and the attitude of the spacecraft during the retrofire maneuver. The retrograde propulsion of the Mercury spacecraft consisted of three solid propellent rockets. Each motor developed a maximum of 1,000 pounds and burned for 10 seconds. These motors were ignited at 5-second intervals.

It was attempted to maintain a constant attitude of the spacecraft during the retrofire maneuver. The desired yaw of the spacecraft was zero, and the roll would not affect the maneuver since the vehicle was symmetric. The desired pitch angle was approximately 34° as shown in Figure 14-2. Computer studies indicated that this pitch angle would minimize the effect of any errors in the pitch angle during retrofire.

The location of the touchdown point for a ballistic vehicle is directly related to errors in location (i.e. time) of the retrofire maneuver. In the typical orbit of a Mercury spacecraft, the touchdown point was shifted approximately 5 nautical miles by an error of one second in the beginning of the

NOMINAL RETROFIRE PITCH ANGLE

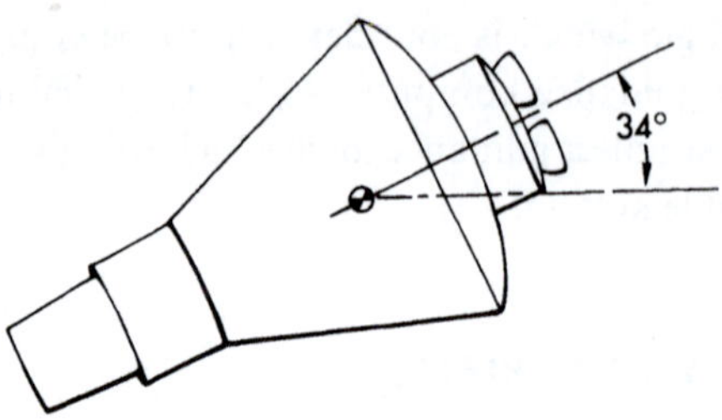


**Fig. 14-2.**

retrofire maneuver. In Project Mercury, the actual time for the beginning of the retrofire was determined from ground-based tracking, and the correct time was relayed to the spacecraft. The ignition of the retrograde rockets occurred on the integer second so that there was an inherent error of $\pm 2.5$ nautical miles.

Although a vehicle may be designed to develop no lift, this will occur only if the center of gravity is along the correct line. It is almost certain that the center of gravity will be displaced from this trim line and that the spacecraft will develop some lift. This lift can result in a large dispersion about the nominal touchdown point unless this factor is compensated for. The easiest way to reduce the effect of such lift is by rolling the vehicle about its longitudinal axis. The resulting effect on the geographic location of impact by establishing a roll rate is that the displacement from the ballistic path is averaged out. The results of computer studies for the Mercury spacecraft indicated that the lift resulting from a $\frac{1}{4}$-inch displacement of the center of gravity caused errors up to 4.5 nautical miles if the spacecraft were not rolled. If a roll rate of one degree per second was maintained the errors in the touchdown point were reduced to less than one-quarter nautical mile (see Figure 14-3).

A summary of the errors in the touchdown point of the four manned orbital Mercury missions is presented in Figure 14-4. The MA-6 spacecraft

USE OF ROLL TO REDUCE DISPERSION CAUSED BY CG OFFSET

| CG OFFSET INCH | ROLL RATE RPM | DISPERSION NM |
|---|---|---|
| 0.25 | 0 | 4.45 |
| 0.25 | 1 | 0.12 |
| 0.50 | 1 | 0.20 |

**Fig. 14-3.**

landed 40 nautical miles uprange, which was well within the primary recovery area. The primary recovery area was approximately 200 nautical miles long and 50 nautical miles wide. The MA-7 spacecraft landed 250 nautical miles downrange. Both the MA-8 and MA-9 spacecraft landed within sight of the primary recovery ship. It was planned to use the automatic stabilization and control system (ASCS) to control the attitude of the spacecraft during the retrofire maneuver on all flights. The planned mode of operation was used only on the MA-8 mission.

ERRORS IN LANDING LOCATION OF MANNED ORBITAL MERCURY MISSIONS

| MISSION | ASTRONAUT | LANDING ERROR, NM |
|---|---|---|
| MA-6 | GLENN | −40 |
| MA-7 | CARPENTER | +250 |
| MA-8 | SCHIRRA | −4 |
| MA-9 | COOPER | −1 |

**Fig. 14-4.**

The spacecraft was tracked by ground-based radar after the retrofire maneuver. The information obtained by this tracking was used to predict the landing point of the spacecraft. The predicted landing points of nine manned orbital missions and two unmanned suborbital missions are compared to the reported pickup points in Figure 14-5. The largest errors occurred on the MA-7 mission and were 6 minutes in latitude and 23 minutes in longitude.

During the final orbit of the MA-6 mission the proper time of retrofire was set into the clock and this time was checked. It was agreed that the retrofire maneuver would be under control of the automatic control system with the astronaut prepared to take over manually should a malfunction occur. The retrofire maneuver took place at precisely the right time over the California site. As a precautionary measure, the astronaut performed manual

## COMPARISON OF PREDICTED LANDING POINT AND REPORTED PICK-UP POINT OF SPACECRAFT

| MISSION | PREDICTED LANDING POINT BASED ON TRACKING AFTER RETROFIRE | REPORTED PICK-UP POINT OF SPACECRAFT |
|---|---|---|
| MA-6 | 21° 31.2' N<br>68° 52.9' W | 21° 25.6' N<br>68° 36.5' W |
| MA-7 | 19° 24' N<br>63° 52' W | 19° 30' N<br>64° 15' W |
| MA-8 | 32° 06' N<br>174° 31.8' W | 32° 05.5' N<br>174° 28.5' W |
| MA-9 | 27° 22' N<br>176° 29' W | 27° 22.6' N<br>176° 35.3' W |
| G II | 16° 26' N<br>49° 34' W | 16° 34' N<br>49° 46' W |
| G III | 22° 01' N<br>69° 52' W | 22° 26' N<br>70° 51' W |
| G IV | 27° 29' N<br>73° 25' W | 27° 44' N<br>74° 14' W |
| G V | 29° 43' N<br>68° 00' W | 29° 47' N<br>69° 45' W |

**Fig. 14-5a.**

## COMPARISON OF PREDICTED LANDING POINT AND REPORTED PICK-UP POINT OF SPACECRAFT (CONT)

| MISSION | PREDICTED LANDING POINT BASED ON TRACKING AFTER RETROFIRE | REPORTED PICK-UP POINT OF SPACECRAFT |
|---|---|---|
| G VI | 23° 35' N<br>67° 50' W | 23° 35' N<br>67° 50' W |
| G VII | 25° 23' N<br>70° 00' W | 25° 25' N*<br>70° 06' W |
| SA 201 | 8° 18' S<br>10° 49' W | 8° 28' S<br>11° 1' W |
| G VIII | 25° 15' N<br>136° 56' E | 25° 12' N<br>136° 05' E |
| G IX | 27° 52' N<br>75° 00' W | 27° 51' N<br>74° 56' W |
| G X | 26° 43' N<br>72° 00' W | 26° 45' N<br>71° 57' W |
| G XI | 24° 18' N<br>70° 00' W | 24° 15.4' N<br>70° 00' W |
| SA 202 | 17° 52' N<br>171° 52' E | 16° 5' N<br>168° 39' E |

*CIRCLE OF RADIUS 7 MILES WITH CENTER AT THIS COORDINATE

**Fig. 14-5b.**

control along with the automatic control during this maneuver. The attitudes during retrofire were held within about 3° of the nominal attitudes as a result of this procedure, but large amounts of fuel were expended.

After the firing of the retrorockets, the real-time computers on the ground predicted that the touchdown point would be approximately 40 nautical miles uprange of the intended touchdown point. The predictions were within a small distance of where the spacecraft and astronaut were finally recovered. As far as the ground was concerned, the entry into the earth's atmosphere was entirely normal. The ionization blackout occurred within a few seconds of the expected time. Although voice communications with the astronaut were lost for approximately 4 minutes and 20 seconds, the C-band radar units continued to track throughout this period and provided indication that all was well during the high heating period.

The touchdown point of the MA-7 spacecraft was approximately 250 nautical miles downrange. This error is believed to have been caused by a combination of an incorrect time of retrofire and by incorrect attitude of the spacecraft during retrofire. As in the other missions, it was intended to have the automatic control system maintain the spacecraft attitude during the retrofire.

At approximately 6 minutes prior to retrofire the pilot enabled the manual proportional control system as a backup to the automatic stabilization and control system as specified for an automatic retrofire maneuver. The pilot then engaged his automatic control system and almost immediately reported a discrepancy between the instruments and the external window references. In the 5 minutes prior to retrosequence, he attempted to analyze the automatic control system problem, and rechecked his manual control system in preparation for this event.

At 30 seconds prior to retrofire the pilot again checked his automatic stabilization and control system orientation mode upon ground request. While the pilot was making this check, the spacecraft attained an excessive pitchdown attitude. The pilot switched from automatic to the fly-by-wire mode and repositioned the spacecraft to the retrofire attitude using his view of the earth as a reference. During this maneuver a large error in the yaw of the spacecraft developed. The pilot cross-referenced the attitude gyro indications and external references for attitude information during the firing of the retrorockets.

The pilot backed up the automatic retrofire system by pushing the manual retrofire button when the event did not occur at the commanded time. Retrofire occurred between 3 and 4 seconds late. This delay would account for approximately 15 to 20 nautical miles of the total overshoot error.

Radar tracking data examined after the flight showed that the mean spacecraft pitch attitude during the retrofire maneuver was essentially correct. Some deviations were observed in the spacecraft's roll attitude during retrofire; however, roll errors of this magnitude have a negligible effect on touchdown point dispersion. Radar tracking data recorded an average yaw error of 27 degrees for the spacecraft during retrofire. This error in yaw was essentially corrected by the time the retrofire was terminated. It was concluded that the error in the touchdown point resulted primarily from the misalignment in the spacecraft yaw attitude.

During both the MA-8 and the MA-9 missions, the spacecraft landed within sight of the primary recovery ship, and the estimated errors in the touchdown points were less than 5 nautical miles. The MA-8 mission was the only flight on which the automatic stabilization and control system was used to control spacecraft's attitude during retrofire. During retrofire for the MA-9 mission, the pilot used manual control to supplement the automatic control system.

## [14-4] ENTRY GUIDANCE SCHEMES

An entry vehicle which is designed to develop lift to change the path of the vehicle will overcome many of the problems which are inherent with a ballistic vehicle. The guidance system of such a vehicle must be capable of utilizing this lift properly. Many of the characteristics of the entry guidance system will depend upon the amount of lift which can be generated and the manner in which the lift is generated. Both the Gemini and Apollo spacecraft are trimmed at a constant angle of attack. This results in a constant lift-drag ratio and a constant ballistic number. Therefore the only way that the path of the spacecraft can be controlled is by rolling the spacecraft about its longitudinal axis.

The possible entry guidance methods may be divided into two general classifications: (1) Guidance using predicted capabilities, and (2) Guidance using a nominal trajectory.

The choice of which type to use depends upon considerations such as the size and speed of the onboard computer and the range of entry conditions which the guidance system must be capable of handling. The dividing line between the two classifications is not well defined. It is possible that an entry guidance logic will use elements of both techniques.

In the method of guidance about a nominal trajectory, the state variables along the nominal path are precomputed and stored onboard the spacecraft. The variations in the measured variables from the stored values are used in the guidance logic either to control the spacecraft back to the nominal trajectory (path controller) or to establish a new trajectory to reach the destination (terminal controller). For this guidance logic, a desirable nominal trajectory must be selected. The desired nominal trajectory is selected prior to the entry by optimization procedures.

The method of guidance using predicted capabilities is capable of handling a wider variety of entry conditions than the guidance about a nominal trajectory. This guidance technique predicts the path which will enable the vehicle to reach the desired destination without violating heating and acceleration limits. The prediction of the future trajectory may be accomplished by a rapid forward integration of the equations of motion for the remainder of the flight, or by using approximate closed-form analytic solutions to the equations of motion. The main advantage of the fast prediction method is that it is able to handle any possible flight condition. The principal disadvantage of this method is the requirement for speed in the computer. The use of closed-form solutions reduces the required computational speed and flexibility of the guidance system.

## [14-5] GEMINI ENTRY GUIDANCE

Projects Gemini and Apollo furnish this country with experience in the entry guidance of lifting manned spacecraft. Two forms of entry guidance logic are used in the Gemini program. The same footprint capability of about 600 nautical miles downrange and 60 nautical miles across-range exists for each technique. The two techniques may be termed "rolling entry" and "fixed-bank entry."

In the rolling entry technique the steering logic is based on calculating the difference between the actual range-to-go and the predicted range based on a continuously rolling vehicle. During entry, the ratio of the downrange error to the crossrange error is used to compute the bank angle required to rotate the axis of symmetry of the footprint the amount required to pass through the desired touchdown point before the downrange error is reduced to zero. When this rotation is completed, the spacecraft flies at zero bank angle until the downrange error is zero. At this point a continuously rolling entry is initiated. Because downward lift is not used, the range calculation is biased slightly to predict a greater range than the expected value, thus preventing the target from moving outside the footprint. The continuous rolling is interrupted occasionally for correction by flying near zero bank angle.

The fixed-bank entry technique has certain advantages over rolling entry. The entry trajectory is flown at a fixed series of bank angles. The prediction is based on the range obtained at a fixed bank angle and the bank angle is adjusted to make the predicted range agree with the desired range to the touchdown point. Crossrange error is allowed to increase until it reaches a fixed percentage of the lateral range capable at that particular time. The bank angle is then reversed. Thus the entry is flown in a series of reversals of an essentially constant bank angle. Theoretically, any point in the footprint is available with only one bank angle reversal; but inaccuracies in the knowledge of the atmosphere, the aerodynamic characteristics of the spacecraft, and in performances of the guidance system will generally require additional maneuvering.

The fixed-bank-angle method has the advantage of being more compatible with the crew monitoring function than the rolling entry method. By viewing the horizon out the spacecraft windows, the astronauts can compare the maneuvers with those expected for the entry condition and take over control in the event of a malfunction. The manually controlled backup technique is similar to the automatic method, but uses a precomputed program of bank angles. The fixed-bank angle method will require less attitude control fuel than the rolling method. Figure 14-6 summarizes the results of reentry during the early flights in the Gemini program. The early flights of Project Gemini dramatically demonstrated the effect of the difference between the estimates of the aerodynamics obtained from wind tunnel tests and the aerodynamics of the actual spacecraft. The unmanned flight of GT-2 indicated that the $L/D$ ratio of the spacecraft was higher than the predicted value over most of the Mach number range. This led to the removal of 58 pounds of nonfunctional ballast for the first manned flight (GT-3) piloted by Grissom and Young.

ERRORS IN LANDING LOCATION OF MANNED GEMINI MISSIONS

| MISSION | ASTRONAUTS | LANDING ERROR, N MI |
|---|---|---|
| GEMINI II | (UNMANNED) | 18 |
| GEMINI III | GRISSOM YOUNG | 64 |
| GEMINI IV | MCDIVITT WHITE | 47 |
| GEMINI V | COOPER CONRAD | 97 |
| GEMINI VI | SCHIRRA STAFFORD | 7 |
| GEMINI VII | BORMAN LOVELL | 7 |
| GEMINI VIII | ARSTRONG SCOTT | 3 |
| GEMINI IX | STAFFORD CERNAN | 2.3 |
| GEMINI X | YOUNG COLLINS | 2.5 |
| GEMINI XI | CONRAD GORDON | 2.5 |

**Fig. 14-6.**

The rolling entry technique was used for the entry of GT-3. This required the crossrange error to be eliminated first and then the downrange error was to be eliminated. One purpose of this flight was to check out the onboard guidance system. For this reason the spacecraft was controlled manually to ground-computed commands and the commands generated by the onboard guidance were observed and compared to these ground commands. The crossrange error was eliminated first as planned. Although full lift was used for the remainder of the flight, the GT-3 spacecraft fell 64 nautical miles short of its intended target point. If the bank angles computed by the onboard guidance system had been used, the spacecraft would have come closer to the target but it would not have reached the target.

The GT-3 spacecraft fell short of the target chiefly because it developed a lower $L/D$ than had been estimated before the flight. The footprint of the Gemini spacecraft was reduced thereafter to about one-third of its originally estimated size because of the reduced $L/D$. The estimated size of the footprint for the remainder of the Gemini program was 200 nautical miles long and 34 nautical miles wide. The length of the footprint is defined as the distance from the zero lift point to the maximum lift point. For missions GT-4 and subsequent, the guidance logic was changed so that crossrange and downrange errors were eliminated simultaneously at the beginning of the entry when effective lift was the greatest.

The GT-4 spacecraft missed the target point by 47 nautical miles. The guidance computer failed prior to entry. The spacecraft was therefore flown in an open loop way on the basis of information furnished by groundbased computers prior to entry.

As in the case of the first two manned Gemini flights, GT-5 fell far short of its target point. In fact, it was the worst miss of the program. The spacecraft fell 97 nautical miles short of the target. Subsequent study disclosed that most of the miss was due to a ground error in failing to provide the onboard computer with the spacecraft's proper inertial coordinates at the time of retrofire. The information computed by the ground system ignored the fact that the Earth rotates approximately 361 degrees in 24 hours instead of 360 degrees. This error was present during the flights of GT-3 and GT-4, but the duration of these flights was not long enough for the effects of this error to become pronounced. By the end of the eight-day GT-5 mission this error was approximately 8 degrees. As a result of this error the onboard guidance system attempted to steer the spacecraft to a target which was approximately 480 nautical miles closer than the actual target point. The miss distance would have been greater except that the Gemini entry guidance logic prohibited the spacecraft from rolling over to develop negative lift. The entry guidance system called for a zero-lift entry which is the minimum lift possible under these circumstances. By the time the astronauts realized that the guidance system was giving erroneous commands, the range capability of the spacecraft was insufficient to reach the target point. During entry the guidance system was providing the correct commands to guide the spacecraft to the false target.

The flights of GT-6 through GT-12 proved that the problems previously experienced with the entry programs were solved. The actual landing point of GT-6 was within 7 nautical miles of the target point. The spacecraft was manually controlled to the bank angles commanded by the onboard computer. After retrofire and jettisoning of the retro-adapter section, the spacecraft was rolled to the full positive lift position. At 290,000 feet the onboard computer displayed bank angle commands on the flight director displays. The nominal bank was first 47 degrees left and then 47 degrees right. These bank angles were held to 80,000 feet. The drogue parachute was deployed at about 50,000 feet and the main chute was deployed at about 10,500 feet. The landing point of GT-7 was also within 7 nautical miles of the target point. The results for GT-8 through GT-11 are summarized in Figure 14-6.

## [14-6] APOLLO ENTRY

The obvious difference between the guidance problems of Project Gemini and Project Apollo is the difference between the entry velocities of the two missions. The Gemini spacecraft entered at earth orbital velocities while the Apollo spacecraft enters at near parabolic velocities. The much higher entry velocity of the Apollo spacecraft greatly increases the possibility that the acceleration limits will be exceeded. The range of a spacecraft entering at near parabolic velocities can be increased by allowing the spacecraft to skip out of the atmosphere. The time of flight may exceed the duration of the power supply or the life support system if the exit velocity is too high.

The Apollo entry guidance logic is a combination of the technique of using predicted capabilities and the technique of using a nominal trajectory. The guidance logic during the initial phases uses predicted capabilities while the guidance logic during the final phase uses a nominal trajectory. The logic is divided into portions which reflect the characteristics of the entry trajectory which the Apollo spacecraft will follow.

The typical Apollo entry trajectory can be divided into the general areas which are shown in Figure 14-7. At some time prior to entry, the entry vehicle is separated from the service module and aligned to the entry attitude; the IMU is aligned and the navigation system is updated for the last time. The initial bank angle depends upon whether the vehicle is entering at the top of the corridor or at the bottom. If the spacecraft is entering close to the top of the corridor, the bank angle will be such that the lift is directed downwards in the vertical plane. Otherwise, the lift is directed upwards in the vertical plane.

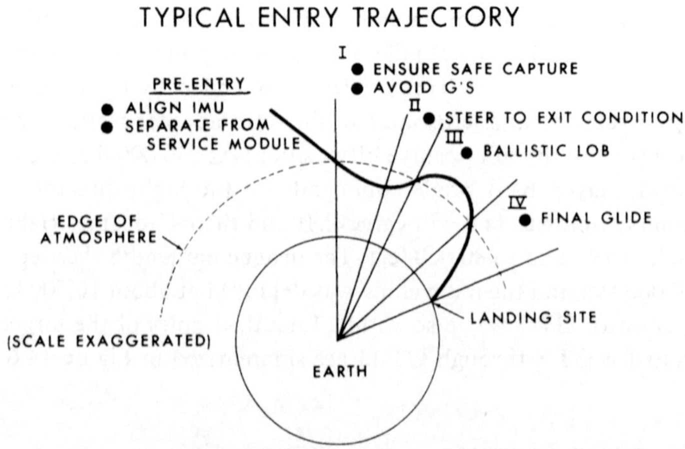


**Fig. 14-7.**

Upon encountering the sensible atmosphere, the spacecraft flies the following phases of the entry: (1) an initial phase during which a safe capture is ensured and excessive acceleration is avoided; (2) a second phase during which the vehicle is steered so that the final phase will be able to guide the spacecraft to the target; (3) a ballistic lob which may be bypassed if the range to the target is short, and (4) the final phase during which the spacecraft is steered to the target. This final phase usually includes the last six to eight hundred miles of the entry and is similar to the entry from a low orbit about the Earth.

The basic flow of the guidance logic is shown in Figure 14-8. Certain portions at the beginning and end of the logic flow are repeated each time the steering commands are computed. These portions are the NAVIGATION, TARGET DATA, and ROLL COMMAND sections. The LATERAL LOGIC section is also entered each time the steering command is computed except during the initial phase of entry and during the ballistic lob phase. In earth orbit the PHASE SELECTOR is set so that only the BALLISTIC PHASE and the FINAL PHASE are used.

The NAVIGATION section calculates the inertial position and velocity of the vehicle by using simple numerical integration techniques. The measured ac-

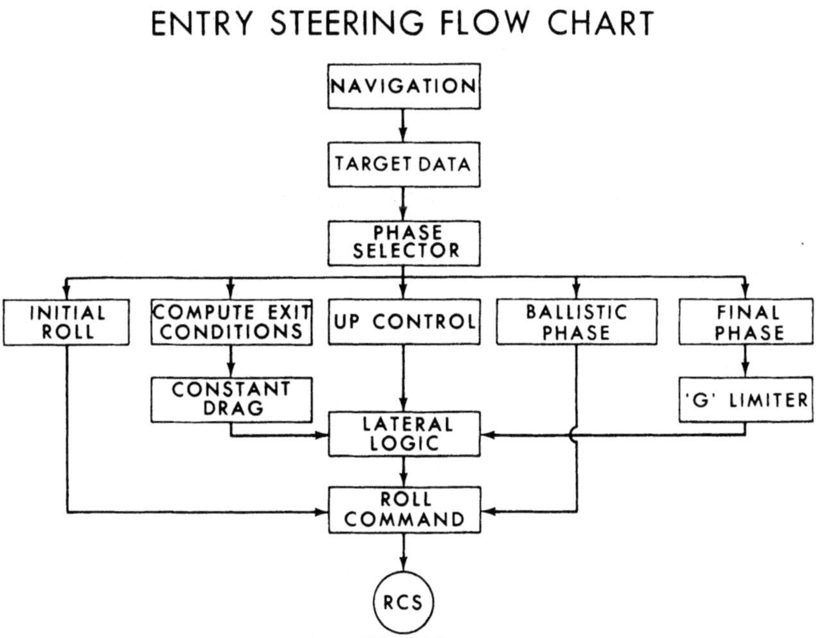


**Fig. 14-8.**

celeration is combined with a calculated gravitational acceleration in the equations of motion.

The TARGET DATA section calculates the distance between the position of the spacecraft obtained by the NAVIGATION section and the inertial position of the target at the estimated time of arrival. The inertial velocity of the spacecraft is used until the velocity becomes less than approximately one-half the circular velocity of a spacecraft in earth orbit at an altitude of 300,000 feet.

The PHASE SELECTOR directs the logic to the section which is concerned with the phase of the entry that the spacecraft is in at that time. Initially, the PHASE SELECTOR will direct the logic to the INITIAL ROLL section or to the BALLISTIC PHASE section. The correct path will depend on whether the spacecraft is returning at parabolic velocity or at circular velocity.

During the initial phase of entry, the bank angle is held constant until the drag exceeds 2 *g*'s. At that time the vehicle is rolled to a zero-bank angle which is held constant until the second phase is entered. The second phase is entered when the altitude rate of the spacecraft approaching the earth is greater than 700 feet per second.

The second phase is the heart of the Apollo entry guidance system. This section includes the logic which will decide the trajectory which will be flown in order to reach the target. There are three paths which the logic may follow. The three paths are via the CONSTANT DRAG section, the UP CONTROL section, or the FINAL PHASE section. If the predicted exit velocity is greater than circular orbital velocity or if the predicted range capability is greater than the actual range plus 25 nautical miles, the CONSTANT DRAG section is used. In this event the second phase will be entered again during the next computation cycle which is two seconds later. The logic will be directed to the FINAL PHASE only when the predicted exit velocity is less than 18,000 feet per second. The UP CONTROL section is entered only when the difference between the predicted range to the target and the range of the calculated trajectory is less than 25 nautical miles.

The CONSTANT DRAG section attempts to guide the spacecraft along a constant acceleration path. The value of the acceleration is calculated the first time the second section of the guidance logic is entered.

The UP CONTROL section attempts to guide the vehicle to the calculated exit conditions. The guidance system continues to use this section until the drag becomes less than some arbitrary value or the altitude rate becomes negative. If the drag falls below this arbitrary value, the guidance logic enters the BALLISTIC PHASE section. If the altitude rate becomes negative first, the FINAL PHASE section is entered.

During the ballistic lob, the BALLISTIC PHASE checks for the start of the final phase of the trajectory. The attitude of the spacecraft is controlled during the ballistic lob so that the sideslip is zero and the spacecraft is trimmed about its nominal angle of attack.

The FINAL PHASE section uses a stored table of values to attempt to steer the vehicle along a reference trajectory to the target. The steering command of the FINAL PHASE will be modified by the "$G$" LIMITER whenever the drag is greater than 5 $g$'s. The bank angle will be decreased in an attempt to keep the drag level below 10 $g$'s.

The output command of the CONSTANT DRAG, UP CONTROL, and FINAL PHASE sections is a value of $L/D$. The commanded roll angle is determined by the ratio of the commanded value of the $L/D$ to the maximum value of the $L/D$.

The LATERAL LOGIC section is used to decide to which side of the vertical plane the vehicle should be rolled. Basically, the vehicle is rolled toward the target. To avoid a large number of roll reversals, there is a deadband built into the logic; that is, lift may be directed away from the target if the predicted landing point is within limits. This limit was arbitrarily set at one-half the lateral range capability.

The final section of the guidance logic is the ROLL COMMAND. This section selects the direction of roll which will result in the shortest angle to be travelled. The command from this section is then transmitted to the reaction control system.

The entry of a spacecraft at near parabolic velocities presents a number of sources of danger to the crew. The chief dangers are excessive accelerations and exiting along a trajectory which would exceed the lifetime of any of a number of onboard systems such as life support, power, and attitude control fuel. It is desirable to have an independent and reliable system for monitoring the primary guidance system during the entry phase.

The entry monitoring system is sufficiently accurate to detect impending unacceptable trajectory characteristics such as excessive accelerations or an uncontrolled atmospheric skip in sufficient time that preventive actions can be taken. The entry monitor system does not unnecessarily restrict the performance of the primary guidance system, and it is an order of magnitude more reliable than the primary guidance system. The entry monitor system does not have the capability for precise range control, although it is capable of gross range control.

The entry monitor system contains four basic subsystems: (1) an entry threshold indicator, (2) a corridor indicator, (3) a bank indicator, and (4) a flight monitor. The entry threshold indicator is an on-off signal that is excited when the sensed acceleration is greater than some nominal value. The corridor indicator consists of two signals which are used to indicate whether the entry is at the top or the bottom of the corridor. The signals result from comparing the sensed acceleration to a nominal mid-corridor

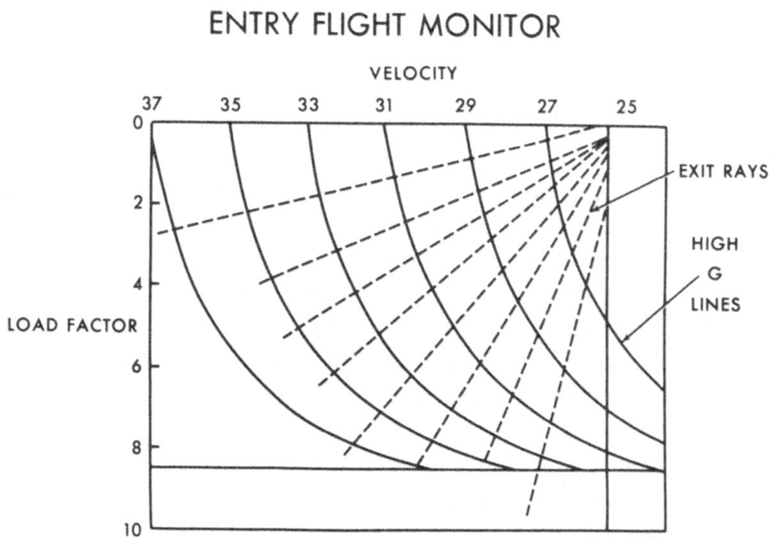


Fig. 14-9.

value at a discrete time interval after the threshold indicator is turned on. The bank attitude indicator is a meter which indicates angular rotation about the approximate stability axes.

The flight monitor which is planned for the Apollo vehicle is a rectilinear plotter which presents a trace of the variation of total acceleration with the velocity of the vehicle. The astronaut compares this trace with information which is presented on the face of the plotter. Figure 14-9 shows a simple version of the information which is presented. Two families of curves are presented on the plotter. One family serves to indicate if the acceleration limit is being exceeded and the second family is used to warn of an uncontrolled skip from the atmosphere. The high $g$ lines are of interest only if the vehicle is not at zero bank angle. Whenever there is a danger of excessive accelerations the vehicle should be rolled to a zero bank angle.

The second family of curves are the most important since these are used to prevent an uncontrollable skip. For every combination of acceleration and velocity there exists a limiting rate at which safe atmospheric exits can be made. If the flight trace is compared to a set of rays which emanate from approximately zero $g$ and an exit velocity which is less than the local circular orbital velocity, a safe limiting exit rate can be defined by tangency of the actual flight trace and the ray. These rays are shown as straight lines in Figure 14-9. In actual Apollo flight hardware the rays are curved lines.

## [14-7] RE-ENTRY PROGRAMS BEYOND APOLLO

The most significant difference between the reentry guidance problems of Gemini and Apollo is the difference between the entry velocities of the two spacecraft. The most likely major NASA manned space flight program to follow Apollo, the Apollo Applications Program (AAP), involves long duration missions in the vicinity of the Earth and the Moon for the purposes of carrying out scientific and biological experiments. Here the guidance, navigation and control problems are similar to Apollo except for the possibility of very long duration missions (up to one year in length). This, of course, places a stringent requirement on the reliability of the guidance and control system, which, in turn, is reflected in a more comprehensive ground-test program to establish the operational reliability for the mission.

Figure 14-10 shows the mission duration, tracking and communications range, earth reentry speed, and navigation techniques for Gemini, Apollo and the AAP Program. Additionally, shown here is the probable future manned space flight program beyond AAP, inter-planetary exploration. When compared to AAP, inter-planetary exploration requires a further in-

INTERPLANETARY MISSIONS COMPARED WITH APOLLO AND GEMINI MISSIONS

| | GEMINI | APOLLO | AAP | INTERPLANETARY |
|---|---|---|---|---|
| MISSION DURATION | 2 WEEKS | 2 WEEKS | 1 YEAR | 2 YEARS |
| TRACKING AND COMMUNICATIONS RANGE | $10^3$ MILES | $250\times10^3$ MILES | $250\times10^3$ MILES | $100\text{-}200\times10^6$ MILES |
| EARTH REENTY SPEED | $26\times10^3$ FT/SEC | $36\times10^3$ FT/SEC | $36\times10^3$ FT/SEC | $45\text{-}60\times10^3$ FT/SEC |
| NAVIGATION | EARTH BASED TRACKING (EBT) | EBT PLUS SELF - CONTAINED NEAR MOON | EBT PLUS SELF - CONTAINED NEAR MOON | EBT PLUS SELF - CONTAINED |

**Fig. 14-10.**

crease in mission duration, considerable increase in tracking and communication ranges, 50% or more increase in earth reentry speed, and a greater emphasis on self-contained onboard navigation. The basic guidance and control techniques currently being used in Apollo will be applicable in both AAP and inter-planetary exploration.

## [14-8] THE APOLLO MISSION

From the very inception of the Apollo program there were many discussions in NASA concerning the best trajectory to fly to the Moon. There were basically three schools of thought: (1) Lunar orbital rendezvous, (2) Earth orbital rendezvous, and (3) Direct flight from the Earth to the Moon and return without rendezvous. After extended studies, the lunar orbital concept was chosen.

Figure 14-11 shows the Apollo landing area on the Moon. This area is $\pm5$ degrees in latitude from the lunar equator and $\pm45$ degrees east and west longitude as measured from the zenith point connecting the Earth's center with the Moon's center. Ranger, Lunar Orbiter, and Surveyor photographs have indicated that there are many choice landing sites in this particular area. The area was chosen as a compromise among many mission objectives such as launch window, fuel requirements, terrain characteristics, sun angles, and surface visibility characteristics.

**Fig. 14-11.**

## [14-9] APOLLO SPACECRAFT

Figure 14-12 shows the Apollo spacecraft. The spacecraft is a five stage vehicle which rides atop the three stage Saturn V booster. A total of eight stages are involved in the lunar mission. Figure 14-12 shows only the top five stages involved. The total Saturn-Apollo vehicle stands 360 feet high at launch. The first stage is the Saturn IC booster, the second stage Saturn II, and the third stage Saturn IV-B. These stages comprise what is commonly called the Saturn V booster system. S-IC is manufactured by the Boeing Company and uses five Rocketdyne F-1 engines. Each F-1 engine, burning RP-1 and liquid oxygen produces 1.5 million pounds of thrust for an overall first stage boost of 7.5 million pounds of thrust. The S-II stage is manufactured by the Space and Information Systems Division of North American Aviation. It employs five Rocketdyne J-2 engines. Each J-2 engine, burning liquid hydrogen and liquid oxygen, produces 200,000 pounds of thrust for an overall second stage boost of 1 million pounds of thrust. The third stage is an S-IVB stage which produces 200,000 pounds of thrust. This employs a single Rocketdyne J-2 engine which is fueled with liquid hydrogen and liquid oxygen.

The total spacecraft weight which is injected toward the Moon via this

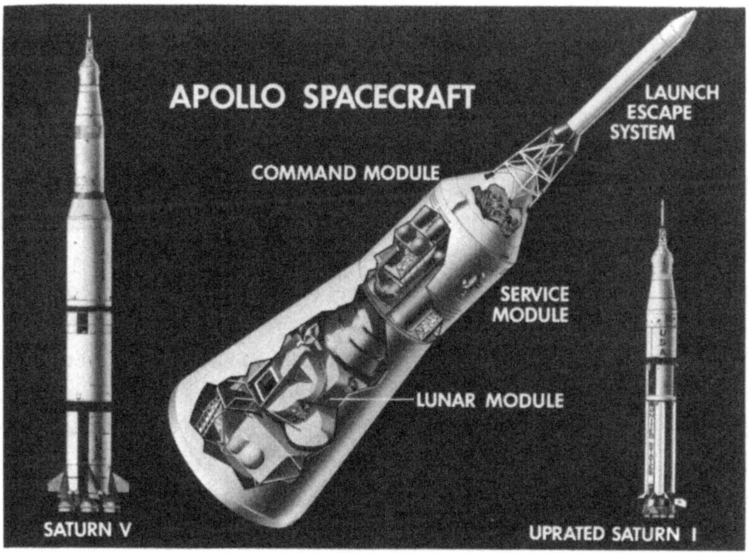


**Fig. 14-12.**

powerful rocket system is approximately 100,000 pounds. The five stages of the spacecraft are: (1) Launch Escape System, (2) Command Module, (3) Service Module, (4) Lunar Module Ascent Stage, and (5) Lunar Module Descent Stage.

The launch escape system is used solely to propel the Command Module a safe distance from the launch pad or away from the path of the launch vehicle in the event of a pad abort or an abort shortly after launch. Regardless of the cause of the abort, the Command Module will be propelled by the launch escape system to a safe altitude or lateral distance away from the danger area for effective operation of the landing system. During a successful launch, the launch escape system is jettisoned after reaching a prescribed altitude. The launch escape system is manufactured by Lockheed Aircraft Corporation and uses a solid star grain propellant with 150,000 pounds of thrust.

The Command Module is a spacecraft control center which houses the flight crew and all equipment necessary to control and monitor spacecraft systems and provide habitability, comfort, and safety for the crew. It is an irregular shaped cone with the primary structure encompassed by three heat shields. The forward and aft heat shield structures are coated with ablative material and joined to the primary structure by fasteners and insulation. The crew compartment heat shield consists of externally applied segments which are coated with ablative material and mechanically fastened to the primary structure. The relative position of the heat shield with respect to the primary structure divides the Command Module into three sections: the forward compartment, the crew compartment, and the aft compartment.

The Service Module is a cylinder formed by six panels of one-inch aluminum honeycomb. This cylinder is divided into six sectors, two of which house auxiliary equipment, two of which contain an oxidizer tank, and two of which contain fuel. The service module propulsion system uses a hypergolic propellant. The engine assembly is gimbal mounted to align the engine thrust with the spacecraft center of gravity to prevent spacecraft tumbling.

The Lunar Module will carry two members in the Apollo crew from the Command Module in orbit around the Moon to the lunar surface. The descent from lunar orbit is initiated by the Lunar Module descent engine. This is a gimballed, throttleable engine which is used for braking and hovering to a landing on the lunar surface. The Lunar Module will provide a base of operations for lunar exploration. Food, water, electrical power, environmental control, and communications will sustain the crew members for a period of one to two days. During the lunar exploration period one crew member will remain aboard the Lunar Module at all times. Upon completion of the lunar stay period the Lunar Module will ascend to rendezvous and dock with the orbiting Command Module. The descent stage will remain on the lunar surface as a launching platform. Upon rendezvousing with the Command Module, the crew will transfer from the Lunar Module ascent stage to the Command Module prior to the return journey to Earth.

Figure 14-13 is a pictorial representation of the Command Module, Service Module, and the Lunar Module in orbit. The Command Module and the Lunar Module are connected by a docking tunnel through which the astronaut can crawl from one vehicle to the other. You will note that the Lunar Module in this picture is attached to the apex of the conical Command Module while in the previous picture showing the vehicle at launch the Lunar Module was behind the cylindrical Service Module. The vehicles are placed in the configuration shown in Figure 14-13 shortly after the Apollo spacecraft is injected from earth orbit toward the Moon. This injection maneuver is called translunar injection. After the S-IVB (upper stage of the Saturn) has propelled the spacecraft toward the Moon, the transposition and docking maneuver is performed. In this maneuver, the Command and Service Modules are propelled away from the Lunar Module and S-IVB under astronaut manual control, turned around, and the Command Module docks with the Lunar Module as shown in this picture. The Lunar Module is then pulled away from the S-IVB. After the transposition and docking maneuver, the S-IVB is no longer of value to the mission.

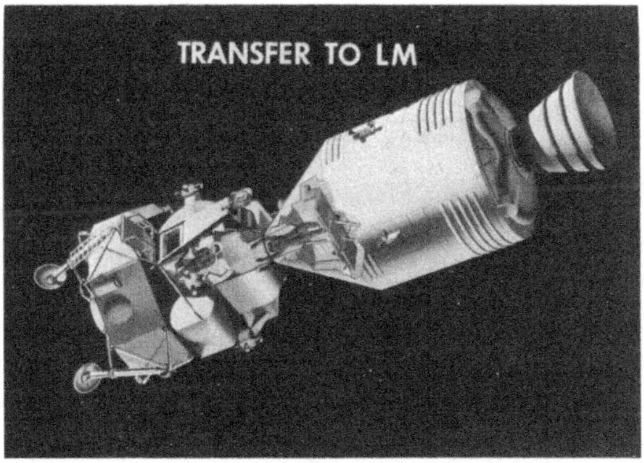


**Fig. 14-13.**

## [14-10] APOLLO GUIDANCE AND CONTROL

The purpose of the guidance system is to control the position and velocity of the vehicle. The navigation process involves the *determination* and *indication* of position and velocity, and the guidance process involves *controlling* these quantities in a closed-loop fashion. Figure 14-14 shows a generalized functional diagram of the guidance and control system. In order to minimize guidance errors the system must reduce the effect of interfering quantities, and it must respond quickly to command signals. An inertial guidance system is fundamentally mechanized as a specific force measuring system using single axis accelerometers which operate in coordinates that are determined by gyros.

The guidance system operates as a force-vector control system, i.e., the system must change the direction and magnitude of controllable forces (lift, drag, and thrust) in such a way that the vehicle reaches its desired point in space and time. It is usual in the theory of dynamics of rigid bodies in three dimensions to separate the motion of the center of mass from the motion of the body around the center of mass. Guidance is the process of moving the center of mass of the vehicle along some desired path. Stability and control are associated with motions about the center of mass.

The guidance and control system for all manned spacecraft have involved a mix of spacecraft systems and ground systems. The guidelines in designing the guidance and control systems for Apollo were:

1) It is mandatory that there be a ground navigation capability which can function in earth orbit, cislunar space, lunar orbit, during the lunar landing phases, and during the lunar rendezvous phases.

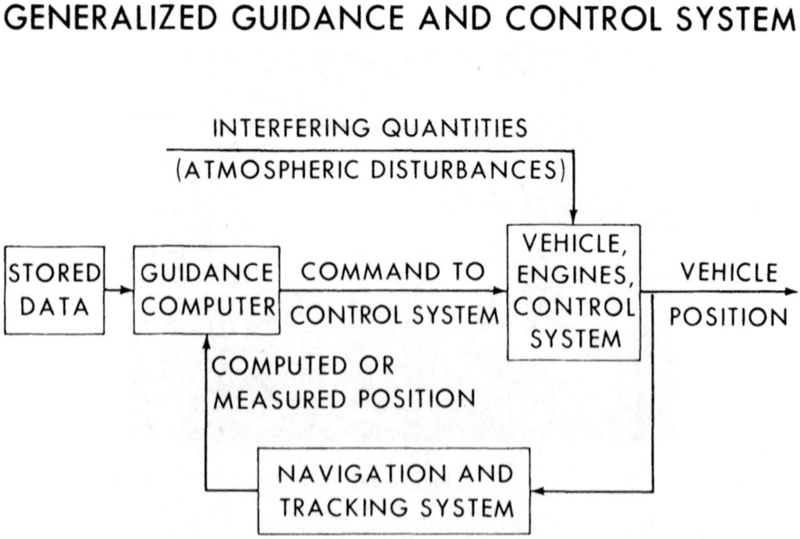


**Fig. 14-14.**

2) It is mandatory that the spacecraft contain onboard a completely self-contained navigation, guidance, and control capability to be used in the event that the data link with the ground is lost.

3) The onboard system must be designed in such a way to take maximum advantage of the ground system and to include all necessary interfaces.

Figure 14-15 shows the navigation, guidance and control system which evolved for the Command Module. The Lunar Module system is very similar and is discussed later in this chapter. The primary *navigation* system is the ground system. This consists of the manned space flight network (MSFN), which involves a number of tracking stations around the world operating in conjunction with the Houston Mission Control Center (MCC). This system is connected to the onboard system by way of the updata link and voice communications. The updata link provides the navigation state vector to the Apollo guidance computer (AGC). The primary guidance and control system consists of the AGC, the inertial measurement unit (IMU), the scanning telescope (SCT), sextant (SXT), and the display and keyboard assembly (DSKY).

The primary guidance and control system operates the reaction control system (RCS) which is used primarily for attitude control in space and during reentry. The AGC also activates the gimbal servos to drive the service propulsion (SPS) engines. In the event the primary control system has a failure, the backup system (labeled in Figure 14-15 the Stabilization System) can also drive the reaction control system and the SPS gimbals. The SCS (stabilization and control system) provides an attitude reference. It also has

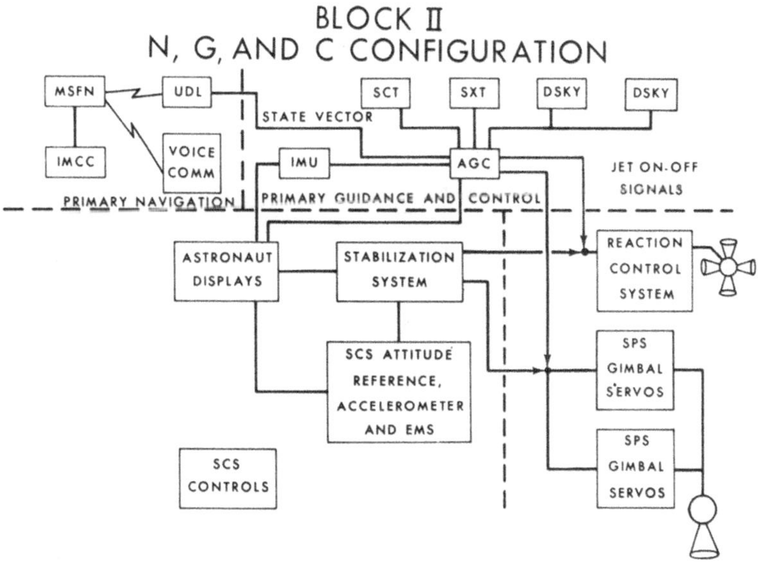


**Fig. 14-15.**

accelerometers. The entry monitor system (EMS) is a backup guidance system to be used during the entry phase in the event of failure of the primary system as discussed in Section 14-6. An integral part of both the primary system and the backup stabilization system is the astronaut. He obtains information from the computer by the DSKY and from the display panel. He communicates with the computer through the DSKY and is able to control the system through the use of throttle and hand controller.

The stabilization system is shown in block diagram form in Figure 14-16. The basic functions of this system are as follows: (1) Drive the jet drivers which turn on and off the small reaction thrusters, and (2) direct the gimbals of the Service Module engine to orientate properly the thrust vector of the main engine. Attitude information comes either from the G&N system (guidance and navigation system) or the AGAP (attitude gyro accelerometer package). Rate information comes from the rate gyro package (RGP) and is displayed on the display panel. Rate and attitude information is used in conjunction with the manual controller to drive through the two engine paths, one path via a deadband limiter, pseudo-rate logic, and jet select logic to the jet drivers. The other path is via thrust vector control electronics to the gimbal servo driver. The term pseudo-rate means that the output of the switching amplifier (an on-off device) commands vehicle accelerations which do not account for reaction jet time delays and dynamics. The short-period output of this signal through a lag filter is indicative of vehicle rates. The

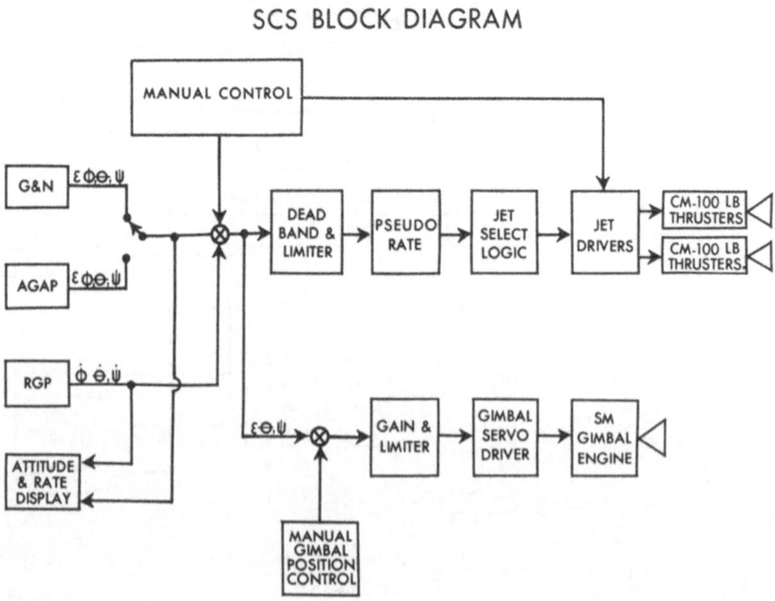


Fig. 14-16.

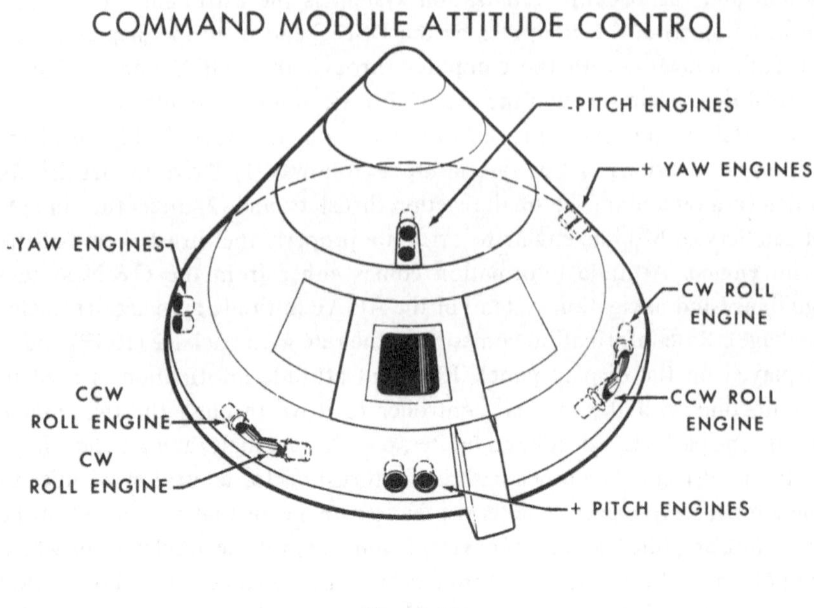


Fig. 14-17.

time constant of the lag network determines the interval over which the output is a valid indication of vehicle rates. The gains in time constant have been selected for the Apollo SCS to provide the desired signal for an average vehicle inertia during the lunar mission. The configuration of the pseudo-rate feedback for the Apollo SCS provides for limit cycle operation. During maneuvers the feedback signals pulse the jets prior to the commanded maneuver rate being achieved, thus resulting in an overdamped response. To avoid this the pseudo-rate feedback is switched out during manual maneuvers.

Figure 14-17 shows the Command Module attitude control system. Shown are the pitch engines, yaw engines, and the clockwise and counter-clockwise roll engines. Each of these engines develops 100 pounds of thrust.

The guidance and navigation system is located in the lower equipment bay of the spacecraft, Figure 14-18.

GUIDANCE & NAVIGATION
SPACECRAFT LOCATION

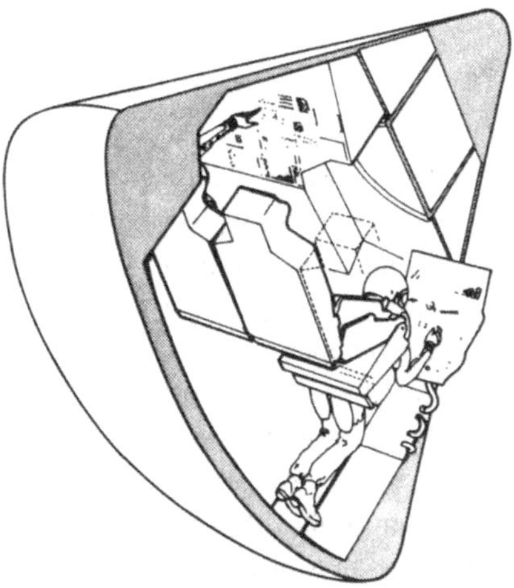

**Fig. 14-18.**

The guidance and navigation equipment is shown in a handling fixture in Figure 14-19. The primary components of this system are the display and keyboard assembly, the gimbal position indicators, sextant, scanning telescope, displays and controls, power and servo assembly, and computer. The inertial measurement unit is behind the panel and is aligned to the optical system by a navigation base. A precise angular relation must be maintained between the optical system and the inertial measurement unit through the navigation base.

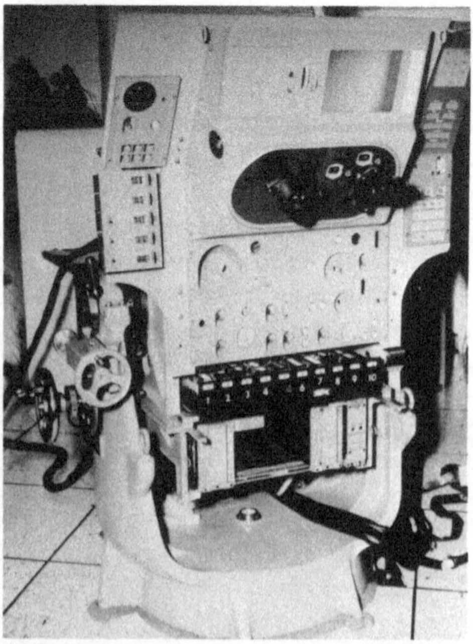


**Fig. 14-19.**

Figure 14-20 is a schematic diagram of the gimbals of the IMU. The stable member houses three single-degree-of-freedom 25 IRIG gyros and three 16 PIPA accelerometers. The gyros maintain an inertial reference system, and the accelerometers measure specific force in the three coordinate directions of this system. The acceleration measurements are integrated in the computer to give velocity and integrated again to give position. The platform is isolated from the spacecraft by the three-gimbal system shown in Figure 14-20.

Figure 14-21 shows the Apollo inertial measurement unit (IMU) with the resolvers on one of the outer gimbals removed. This unit is about the size of a basketball and is very similar to a Polaris platform. The corrugations on the outer portion are coolant lines through which the coolant fluid flows to maintain precision temperature control of the IMU.

Figure 14-22 shows the IMU with the top removed. Three gimbals, platform electronics, and the accelerometer and gyro package can be seen in this picture.

Figure 14-23 is a photograph of the inertial measurement unit and the optical system (scanning telescope and sextant) mounted on the precision navigation base which maintains accurate angular orientation between the two subsystems. The optical system is used to align the inertial system and for navigation in earth orbit, lunar orbit, and in cislunar space. The inertial

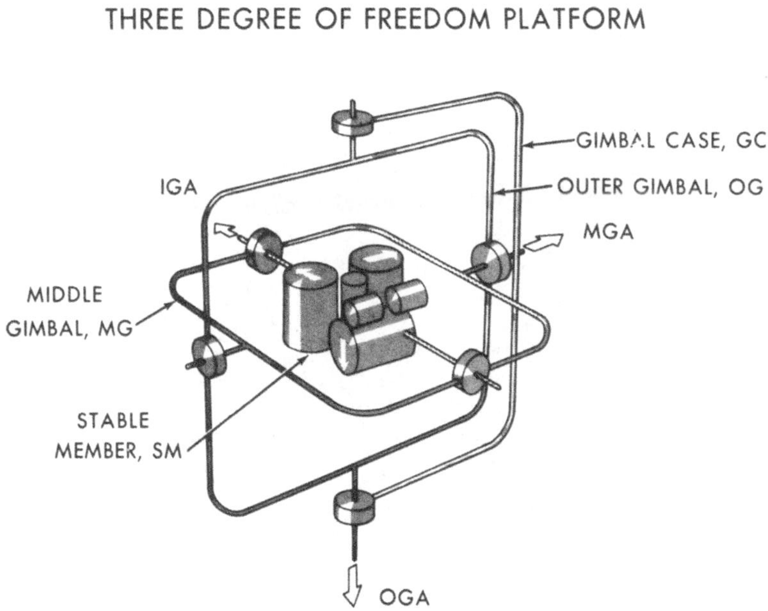


**Fig. 14-20.**

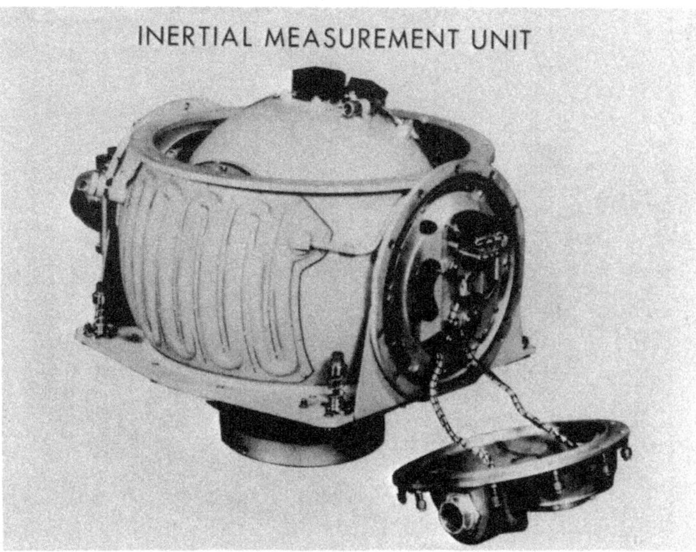


**Fig. 14-21.**

APOLLO
INNER, MIDDLE & OUTER GIMBAL ASSEMBLIES
IMU-5 FOR APOLLO G&N EQUIPMENT

**Fig. 14-22.**

THE IMU, NAVIGATION BASE
AND OPTICAL SUBSYSTEM

**Fig. 14-23.**

measurement unit is used as a primary attitude reference and is used for guidance purposes during all maneuvers and during reentry.

Figure 14-24 shows the display and keyboard (DSKY) consoles. In examining the operation of the guidance and control system it is obvious that the computer has a central and very demanding function. The computer in both the Command Module and Lunar Module are identical. The Apollo computer is a very powerful and lightweight computer with the largest memory of any airborne computer in history. It has a memory of 36,000 words (each of 16 bits). It is approximately equal to an IBM 704 in computational capability. Careful choice of components and packaging assures high reliability.

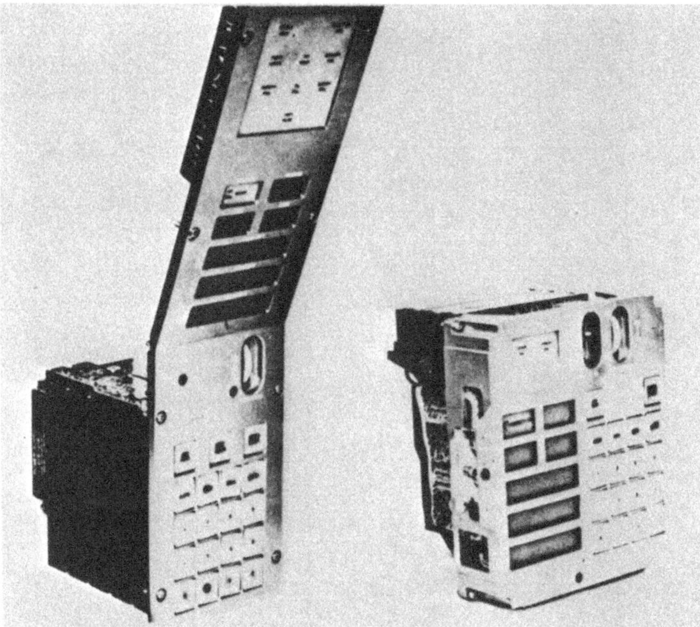

**Fig. 14-24.**

Figure 14-25 shows the face plate of the DSKY. This subsystem provides the communication link between the astronaut and the computer. Through the DSKY the astronaut can monitor system activity, alter parameters, and dictate system modes. In addition, the DSKY has indicator lights which display system and computer status and alarm. The computer display on the DSKY consists of three two-digit displays labeled "program," "verb," and "noun," and three five-digit general word readouts. The two-digit displays are coded for various modes. The program display indicates the major operating mode of the computer such as "lunar landing maneuver." The "verb" and "noun" displays are used together and coded to give numerous possibilities of meaningful phrases or instructions. Examples of typical "verb" and "noun" displays are:

| *Verb* | *Noun* |
|---|---|
| Display value | Velocity |
| Compute | Abort velocity |
| Read in | Landmark angle |

When the computer wishes to communicate a request for data or signal an alarm to the astronaut, the "verb" and "noun" numbers flash until the astronaut takes action. He enters data to the computer through the keyboard which is on the right hand side of the display seen in Figure 14-25.

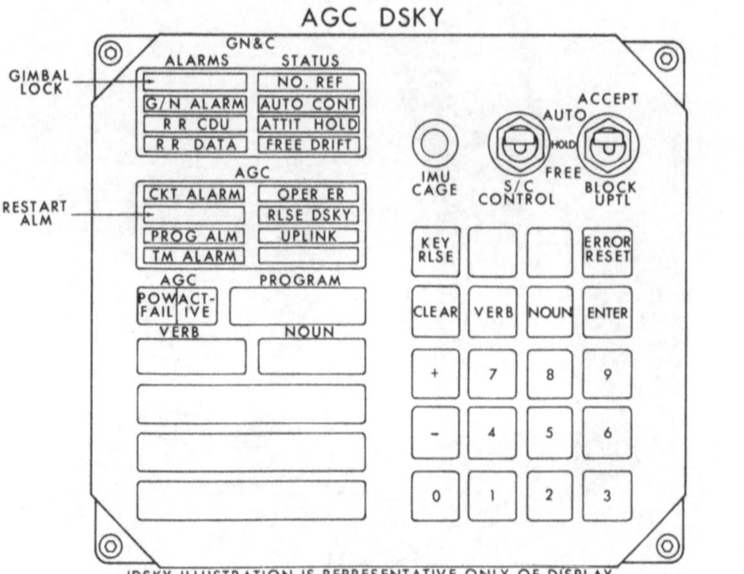


(DSKY ILLUSTRATION IS REPRESENTATIVE ONLY OF DISPLAY DATA AND CONTROL FUNCTIONS, NOT PHYSICAL CONFIGURATION AND ARRANGEMENT)

**Fig. 14-25.**

## [14-11] GROUND TRACKING SYSTEM

A schematic representation of the operation of the manned space flight network tracking system is shown in Figure 14-26. The vehicle is illuminated by an 85 ft antenna which provides range, angles, and velocity. This information is transmitted to the Mission Control Center in Houston from which navigation information is determined. The vehicle can also be tracked by 30 ft antennae which use three-way doppler information to provide position and velocity data.

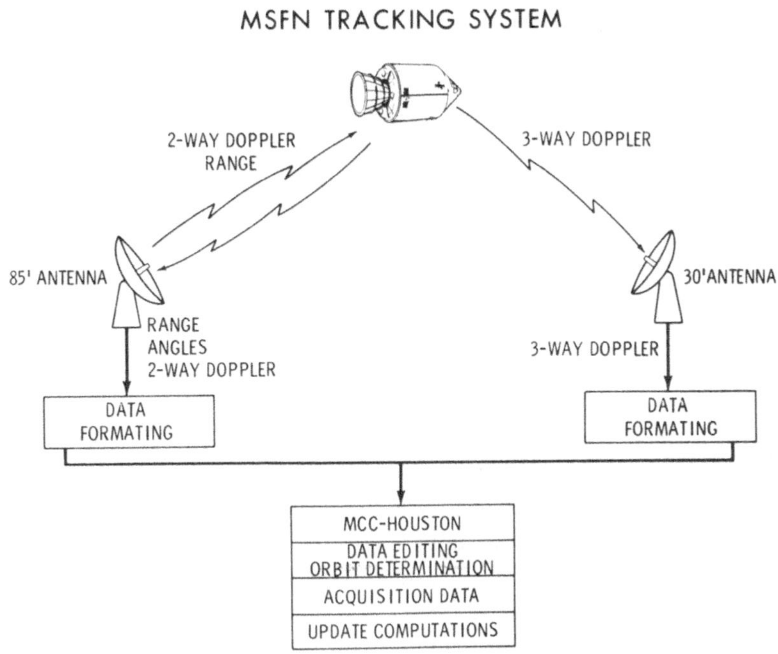


**Fig. 14-26.**

Distance is determined by modulating the carrier with random digits (0 and 1). The signal is retransmitted by the Command Module and the Lunar Module. The measurement of transit time of the signal is a measure of the distance of the spacecraft. Velocity is determined by measuring the doppler shift in the signal returned by the spacecraft.

Figure 14-27 shows the location of MSFN unified S-band station sites. There is overlap of coverage among the three 85 ft stations. These stations are located in the western United States (Goldstone, California); Madrid, Spain; and Canberra, Australia. Additionally, there are ten 30-ft. antennae spaced strategically around the world as shown in Figure 14-27 (Bermuda, Carnarvon, Guaymas, Hawaii, Cape Kennedy, Corpus Christi, Houston, Guam, Ascension, and Antigua).

## MSFN UNIFIED S-BAND STATION SITES

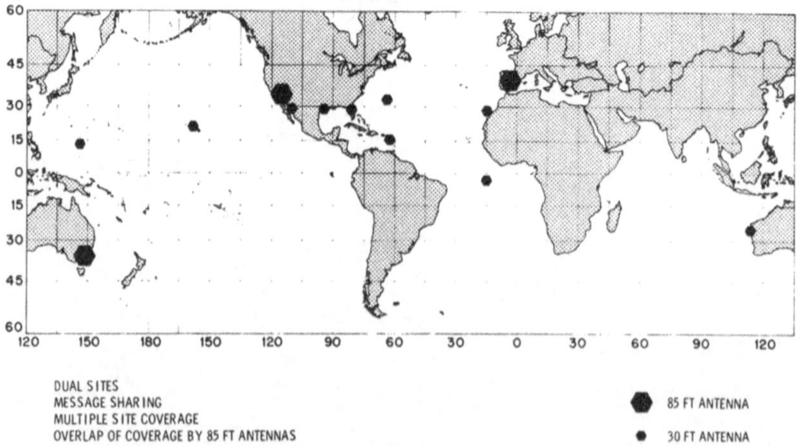


**Fig. 14-27.**

The one-sigma tracking accuracies of the MSFN tracking stations are as follows:

Range: Noise 60 ft., Bias 120 ft.

Angles: Noise 0.8 milliradians, Bias 1.6 milliradians

Two-way doppler: Noise 0.1 ft., Bias .07 ft.

Three-way doppler: Noise 0.1 ft., Bias 0.2 ft.

The frequence reference is a rubidium frequency standard with a stability of $5 \times 10^{-11}$ parts per part. The MSFN stations, with either the 30 or 85 ft. antenna, can track spacecraft at lunar distances using either the high-gain antenna or omnidirectional antenna.

During the earth orbital phase, the MSFN system uses C-band and measures range and two angles. During the cislunar phase, the system uses the unified S-band system which measures three-way doppler, range, and two angles. In lunar orbit, the system uses the unified S-band measuring three-way doppler.

Navigation data derived onboard the spacecraft is telemetered to the MSFN. The radar tracking data and the telemetered data are sent to the Mission Control Center at Houston and the guidance and navigation parameters are computed in the real-time computation center. After the guidance and navigation data is determined, it is telemetered to the Apollo guidance computer onboard the spacecraft.

In the navigation system performance data presented later in this chapter, it is assumed that there is an uncertainty (one sigma) in latitude and longitude of the MSFN stations of 1 to 6 arc seconds. The altitude error is assumed to be 100 to 200 ft. The uncertainty of the gravitational constant (one sigma) for the earth is assumed to be $106 \times 10^9$ ft$^3$/sec$^2$. For the Moon, this is assumed to be $6 \times 10^9$ft$^3$/sec$^2$. In the orbit determination program, the parameters estimated are three components of velocity, three components of position, and two three-way doppler biases. Using these assumptions, the accuracy in performance of the system is presented later in this chapter.

## [14-12] SPACECRAFT NAVIGATION TECHNIQUES

Let us look now at the techniques for navigating onboard the spacecraft. One technique for navigating is shown in Figure 14-28. Here a star is picked up with the scanning telescope and centered in the field of view, at which time the astronaut shifts to the sextant (a 28-power narrow field-of-vision instrument). The astronaut uses the sextant to position accurately the star over a landmark. When the star and the landmark are superimposed, the astronaut presses a button and the angles between the two, as well as time, are entered into the computer automatically. This information is entered by the astronaut via the DSKY.

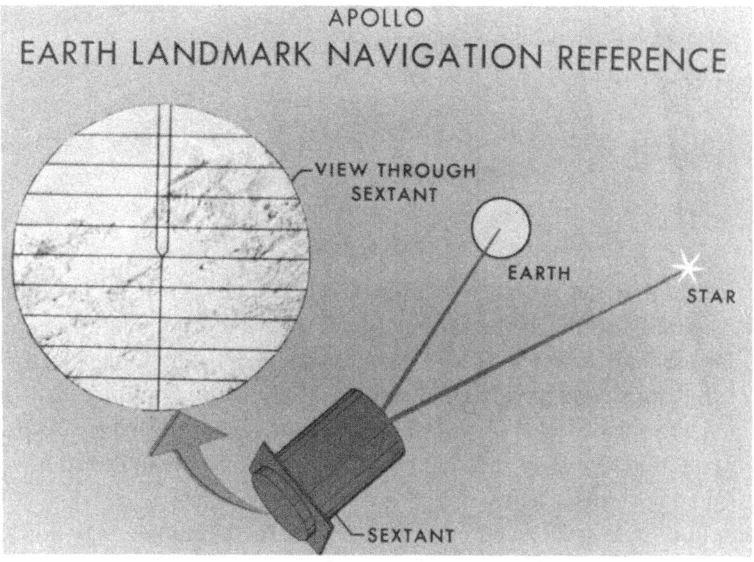


**Fig. 14-28.**

Figure 14-29 shows the angles involved between the star and the landmark. Here the astronaut has superimposed a star over the landmark and the computer uses the trunnion angle indicated in this diagram. In earth orbit, of course, the astronaut can measure his position by tracking known landmarks and by using Keplerian mechanics. It is also possible by knowing the error propagation equations to track unknown landmarks and to compute his position accurately in earth or lunar orbit while at the same time computing the geographic position of the landmark.

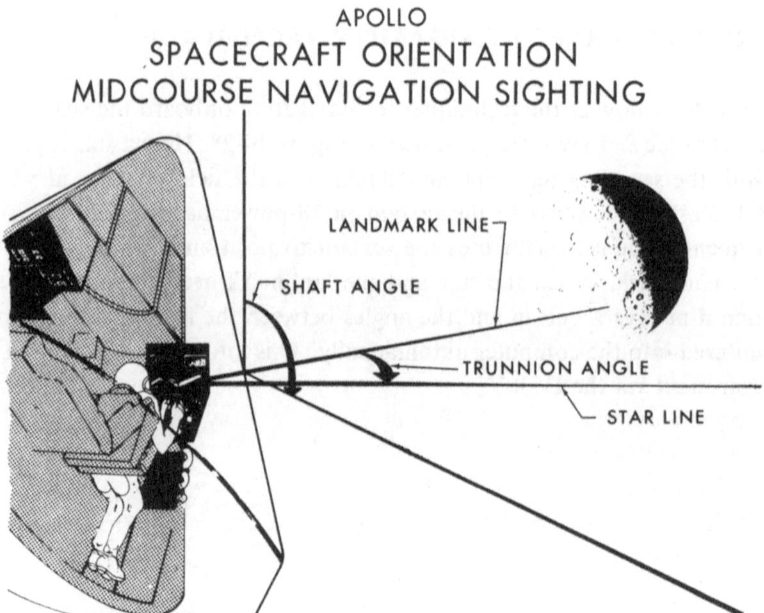


**Fig. 14-29.**

Let us look at the geometry of measuring a navigational fix in cislunar space. Figure 14-30 shows the various angles involved using the stars Fomalhaut, Deneb, and Antares. In this geometrical sketch, the lunar horizon is used with Antares and navigational landmarks are used with Fomalhaut and Deneb. The angles measured with the three stars form three cones in space. The intersection of two of these cones forms a line and the intersection with the third cone forms a point.

Computation is performed automatically in the computer. Apollo does not use the conventional method of computation performed by mariners for many centuries, i.e., computation based upon two or more star sightings and

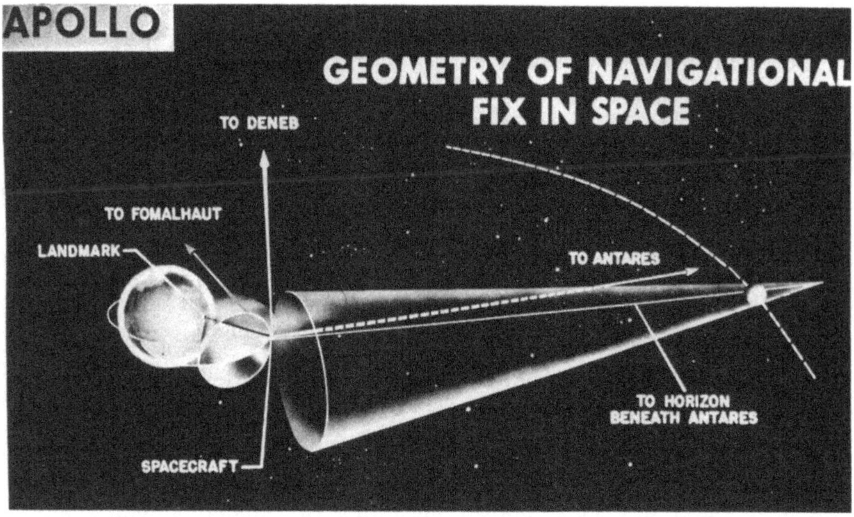


**Fig. 14-30.**

running the earlier sightings forward to the last sighting and computing a discrete fix. Instead, recursive navigation techniques are used involving Kalman filter theory. Under this concept the accuracy of position and velocity determination is improved as more and more sightings are taken and the uncertainties are reduced with each sighting. Fixes as they are known in maritime navigation are not performed.

During the earth orbital phase, scanning telescope uncertainties are assumed to be 4 milliradians and landmark position uncertainties are assumed to be 1000 ft. in the performance data presented later in this chapter. During the translunar and transearth phases, the sextant uncertainty is assumed to be 10 arc seconds and the uncertainties in the horizon are assumed to be 1 nautical mile for Earth and 0.5 nautical miles for Moon. Additionally, we assume that forty landmark sightings are made enroute to the Moon and enroute back to the Earth. The velocity corrections made enroute to the Moon and returning to the Earth are assumed to be accurate in magnitude to 1%. The pointing of the thrust vector is accurate to 10 milliradians. In the lunar orbital phase the scanning telescope uncertainty is assumed to be 4 milliradians and the landmark position uncertainties on the Moon are 1000 to 5500 feet in a horizontal direction and 2300 to 3000 feet in a vertical direction. Five landmarks on the earthside are chosen along the lunar orbit track and three sightings per landmarks are made on each of the first two orbits. Periodic updates are made thereafter.

## [14-13] GUIDANCE AND NAVIGATION PERFORMANCE

Let us look now at the performance of the system using the assumptions of the manned space flight network (MSFN) and the onboard system discussed previously. Figure 14-31 summarizes the results of a digital computer simulation program using the assumptions cited in Sections 14-11 and 14-12. The RMS position measured by the onboard system and by the MSFN is given in nautical miles and velocities are measured in feet per second. The accuracies are shown at injection as measured either by land stations or by ship. Also shown are accuracy at the first midcourse, second midcourse, and third midcourse corrections enroute to the Moon and at perilune. In lunar orbit, the accuracies are shown at Lunar Module separation and Lunar Module rendezvous with the Command Module. The return accuracies are shown for injection at the transearth phase and at first, second, and third midcourse corrections and at entry. It should be noted that at Earth reentry, the MSFN accuracy is 0.5 nautical miles while the onboard system accuracy is about 9 nautical miles. With the MSFN, velocity is accurate to 2.2 ft. per second while the onboard system is inaccurate to about 45 ft. per second.

COMPARISON OF MSFN/ON-BOARD NAVIGATIONAL UNCERTAINTIES

| PHASE | EVENT | RMS POSITION, NM | | RMS VELOCITY, FPS | |
|---|---|---|---|---|---|
| | | MSFN | ON-BOARD | MSFN | ON-BOARD |
| TRANSLUNAR | INJECTION (LAND) | 0.1 | 2.2 | 1.5 | 13.5 |
| | (SHIP) | 2.7 | | 37.1 | |
| | FIRST MIDCOURSE | 1.2 | 5.4 | 0.7 | 5.2 |
| | SECOND MIDCOURSE | 10.3 | 8.4 | 0.2 | 0.4 |
| | THIRD MIDCOURSE | 3.1 | 2.5 | 0.8 | 1.1 |
| | PERILUNE | 1.2 | 1.2 | 10.2 | 5.0 |
| LUNAR ORBIT | LM SEPARATION | 0.4 | 0.9 | 3.2 | 2.8 |
| | LM RENDEZVOUS | 0.2 | 0.3 | 0.6 | 1.2 |
| TRANSEARTH | INJECTION | 0.4 | 0.4 | 1.0 | 1.8 |
| | FIRST MIDCOURSE | 6.0 | 4.4 | 0.3 | 0.7 |
| | SECOND MIDCOURSE | 4.8 | 7.1 | 0.2 | 0.6 |
| | THIRD MIDCOURSE | 1.4 | 8.7 | 1.1 | 6.5 |
| | ENTRY | 0.5 | 8.7 | 2.2 | 45.3 |

**Fig. 14-31.**

Figure 14-32 is a comparison of the data as measured onboard and by the MSFN during the translunar phase. The top plot is position and the bottom plot is velocity as a function of time. The MSFN system is more accurate until about 35 hours after the spacecraft is injected toward the Moon. At that time the onboard system has somewhat greater accuracy. The midcourse corrections are indicated here by the diamonds labeled MCC. These are made at two hours after injection, 49 hours after injection, and 61 hours after injection.

Figure 14-33 gives results concerning navigational uncertainties during the lunar orbit phase. The earth-based manned space flight network is more accurate than the onboard system throughout the trajectory. Time is measured from the time of spacecraft insertion into lunar orbit. The solid lines show points of MSFN tracking and the dashed lines show periods during which telescope sightings of lunar landmarks are made. Of course, errors grow during the period between tracking or sightings and they immediately drop to more accurate values as greater data is accumulated.

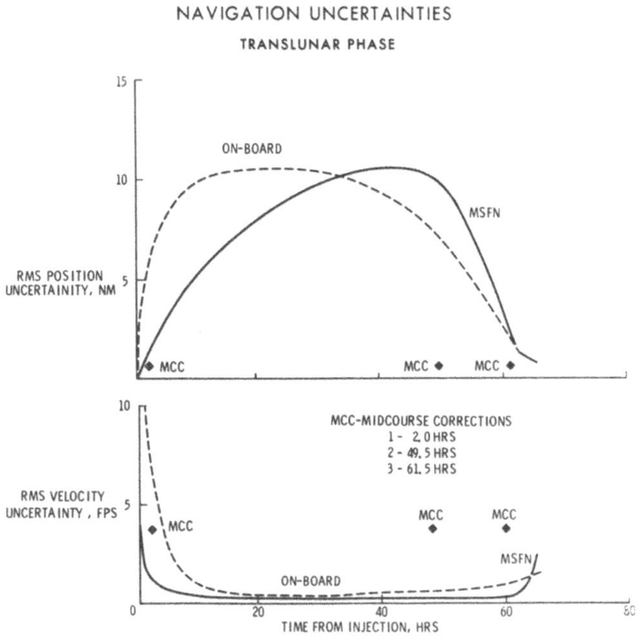


**Fig. 14-32.**

Figure 14-34 shows the midcourse velocity performance during the translunar phase. The bottom plot shows the ft/second uncertainty in velocity corrections which would be made as a function of time from injection to the first velocity correction as measured by the MSFN and the onboard system. The upper of the two figures shows the sum of the first two velocity corrections. There is a significant velocity savings by using MSFN data. These figures show that it is worthwhile to delay, from a fuel standpoint, before making velocity corrections. These figures are indicative of the fuel econo-

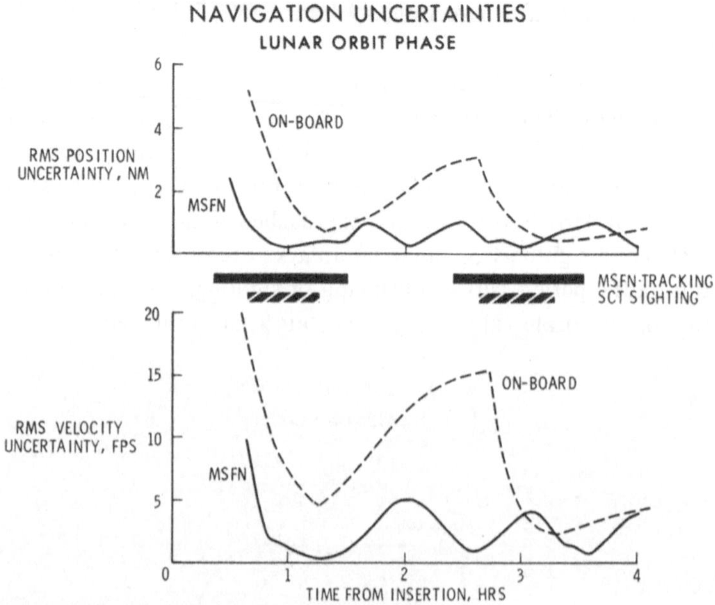


**Fig. 14-33.**

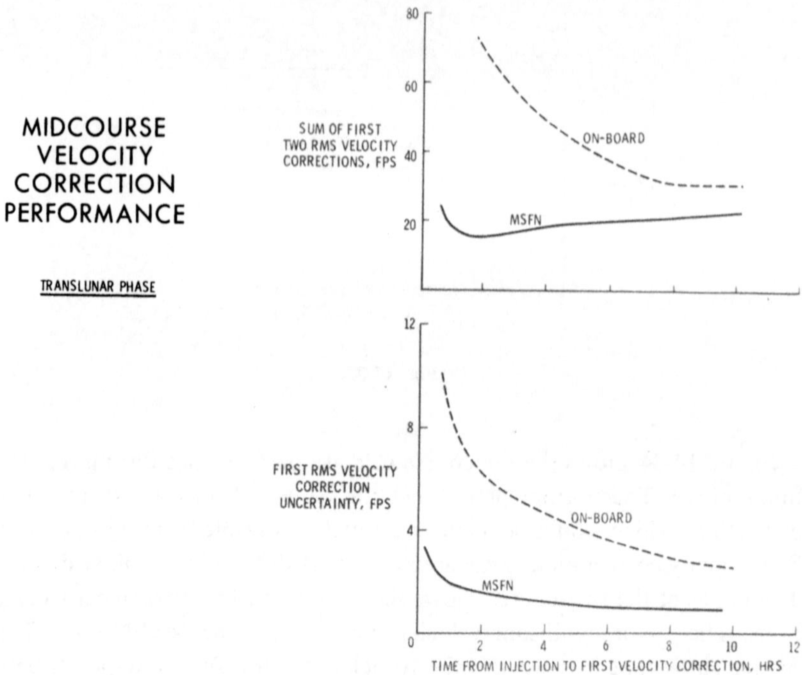


**Fig. 14-34.**

mies which could be made as the uncertainties in velocity corrections are reduced.

Figure 14-35 describes in simple form the interface between the onboard system and the manned space flight network. The computations for the MSFN are made by the Houston Mission Control Center. The interface with the computer is made via two routes, the up and down data link directly from the MSFN to the Apollo guidance computer and voice communications to the crew who enter the data through the display and keyboard assembly to the computer. The computer updata rate is 10 up-link words per second. The total navigation updata information required is a state vector consisting of six components and time. The updata frequency required is once prior to each guidance maneuver. Total transmission time is 60 seconds maximum (for a 99% probability of no errors). The navigational updata information is not time critical.

DESCRIPTION OF ON-BOARD/MSFN NAVIGATION INTERFACE

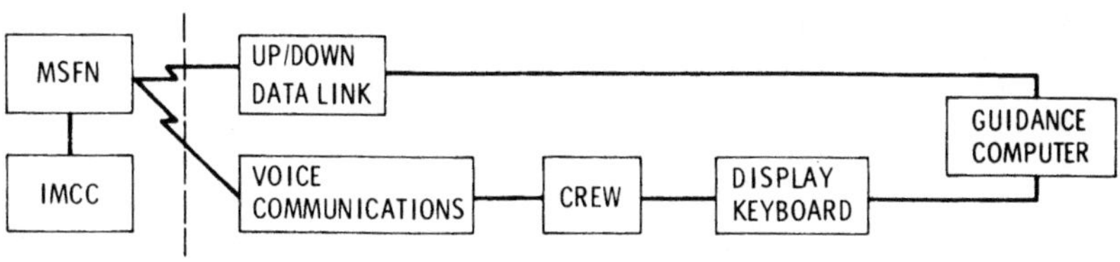


- COMPUTER UPDATE RATE - 10 UPLINK WORDS/SEC
- NAVIGATION UPDATE INFORMATION REQUIRED
  - STATE VECTOR - 6 COMPONENTS
  - TIME
  - AIM POINTS
  - COMPUTER TIME ALINE
- UPDATE FREQUENCY REQUIRED - ONCE PRIOR TO EACH GUIDANCE MANEUVER
- TRANSMISSION TIME
  - 60 SECONDS MAXIMUM FOR 99% PROBABILITY OF NO ERRORS
  - NAVIGATION UPDATE NOT TIME CRITICAL

**Fig. 14-35.**

## [14-14] APOLLO REENTRY

The end result of the lunar mission is a safe reentry into the earth's atmosphere. Figure 14-36 pictorially represents this problem. The guidance system must guide the spacecraft into a corridor approximately 26 miles deep. This accuracy is obtained in three midcourse corrections during the return flight from the Moon. Prior to entering the earth's atmosphere, the Service Module is jettisoned. If the spacecraft comes in above this entry corridor, too little energy will be transferred from the vehicle to the atmosphere and the spacecraft will enter an elliptical earth orbit. The eccentricity of this orbit is a function of how far above the upper entry boundary the spacecraft trajectory intersects the atmosphere. Since there is no significant propulsion onboard the Command Module except the 100 pound attitude jet system, there would be no way for the crew to return to earth from this elliptical orbit. It is therefore essential that this entry corridor be entered. If the spacecraft enters below the corridor, the heat capabilities of the heat shield or the allowable acceleration loads may be exceeded.

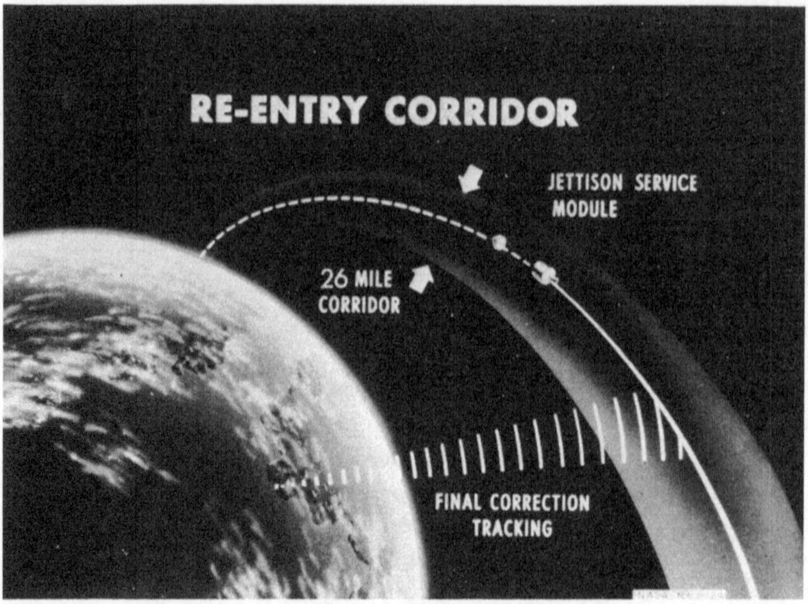


**Fig. 14-36.**

One of the most critical portions of the mission from the guidance and control standpoint is the reentry phase. In Apollo the earth's atmosphere is entered by making one skip on return from the lunar mission, as pointed out in Section 14-6. The total range of the vehicle from the point of entry to touchdown is nominally 2500 miles. The propulsion system is not used to remove energy from the vehicle; all of the energy is removed by the atmosphere.

## [14-15] LUNAR LANDING

The lunar landing is a very interesting planetary entry phase. The techniques that are used here are likely to be techniques used for landing on planets with little or no atmosphere. Because of the uncertainties raised in the knowledge of the density of the Martian atmosphere by Mariner, it is possible that a modified form of the Lunar Module entry trajectory may be used on Mars.

Figure 14-37 shows the Lunar Module guidance and control system. This is similar to the Command Module system, although the nomenclature is different. There are four major elements shown: (1) Guidance and navigation system, (2) Stabilization and control system, (3) Reaction Control System (attitude jets), and (4) Ascent and descent engines. The Lunar Module guidance and navigation system has a landing radar and a rendezvous radar. The computer is the central data processor as shown in Figure 14-37. It receives data from the radar, DSKY, alignment optical telescope (AOT), and the inertial measurement unit. Other inputs to the Lunar Module guidance computer are from the attitude controller (which is the control stick used by the astronaut) and the throttle command. The computer drives either the ascent or the descent engines and the reaction control jets.

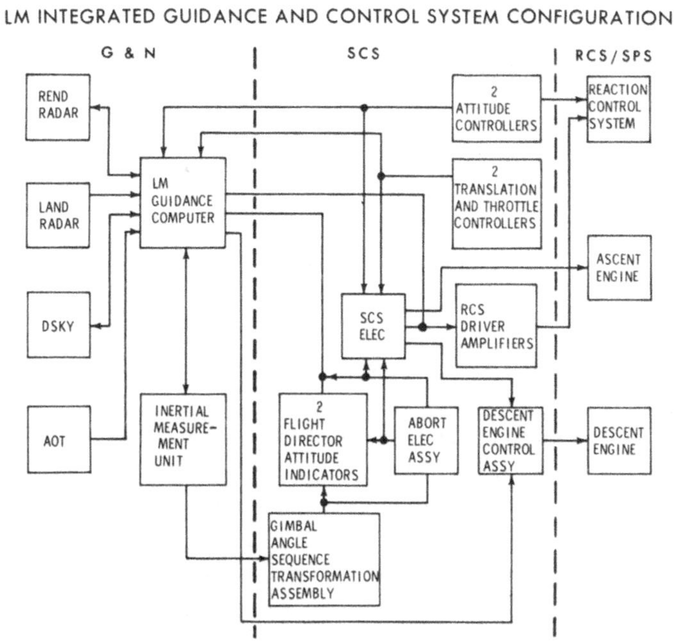


**Fig. 14-37.**

If the computer of the guidance and navigation system fails, the attitude controller and the throttle operate via the stabilization and control system to drive the ascent and descent engines. Either the stabilization and control system or the guidance and navigation system is used to provide attitude information to the FDAI (flight director attitude indicator) which is a gyro horizon. In the event the primary guidance and navigation system fails, the Lunar Module is equipped with an abort guidance system which is used to assure safe return to a rendezvous trajectory.

Figure 14-38 shows a more simplified block diagram of the Lunar Module stabilization and control system showing the flow of attitude data, range data, time engine commands, and other information required for control, guidance and navigation.

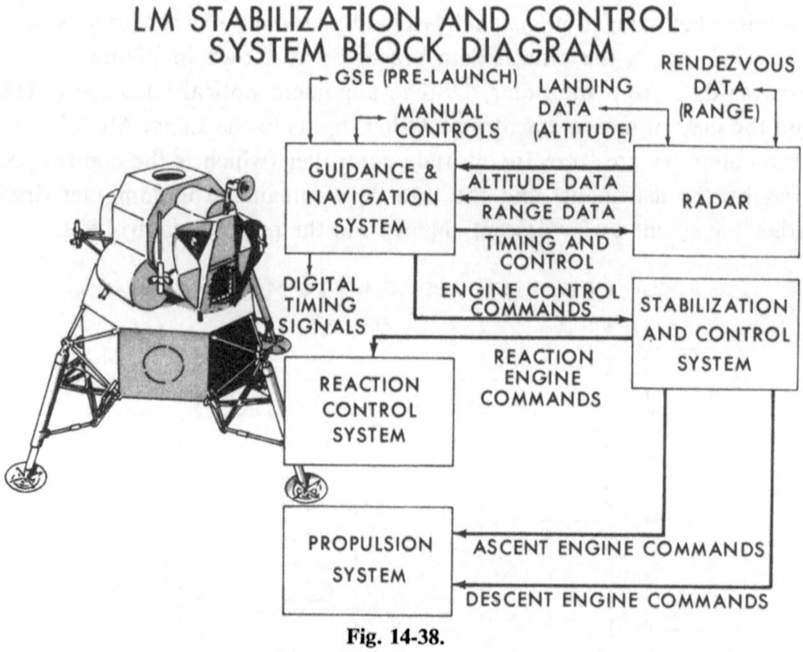


**Fig. 14-38.**

Let us look at some of the features of the trajectory from lunar orbit to the lunar surface. Figure 14-39 shows this trajectory. The Command Module, Service Module, and Lunar Module are initially in an 80-mile circular orbit around the Moon. The velocity of this orbit is approximately 6000 ft./sec. and the period is about 2 hours. This means that the spacecrafts are hidden from the Earth about 1 hour out of every 2 hours. The operation of the on-board guidance and control system is important during these periods. When the vehicle reaches the proper point for lunar entry, an impulse of approximately 100 ft./sec. is applied by the descent engine. The descent trajectory is a Hohmann trajectory. After the vehicle has traversed 180° around the Moon, it reaches a perilune of about 50,000 feet. The trajectory is monitored by the two astronauts in the Lunar Module who track the Command Module. It is also monitored by the astronaut remaining in the Command Module who tracks the Lunar Module with the Command Module optical system. Additionally, the earth-based ground tracking system tracks both vehicles.

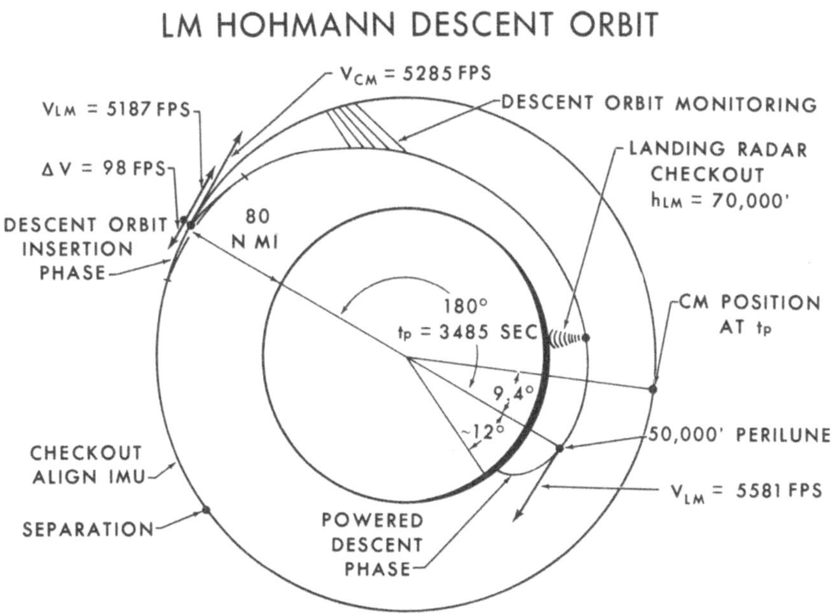


**Fig. 14-39.**

The next phase of the landing maneuver is shown in Figure 14-40. This phase of the trajectory covers a distance of about 240 miles across the lunar surface. During this phase the vehicle proceeds from an altitude of 50,000 down to an altitude of 8,600 ft. Here the Lunar Module descent engine is on for a period of about 450 seconds. After completion of this phase the vehicle enters a point known as "high gate."

Figure 14-41 shows the "constant attitude" phase of the mission following "high gate." This is the phase of the mission from about 8,600 feet down to 500 to 1000 feet. During this phase of the mission the descent engine is throttled back to about 50% and the vehicle is pitched up into an atititude of about 35° to 45° with respect to the horizontal. During this phase of the mission the astronaut is surveying his landing site out the window. He examines the landing area and chooses a particular point at which he wishes

LUNAR LANDING MANEUVER PHASES

EXAGGERATED SCALE

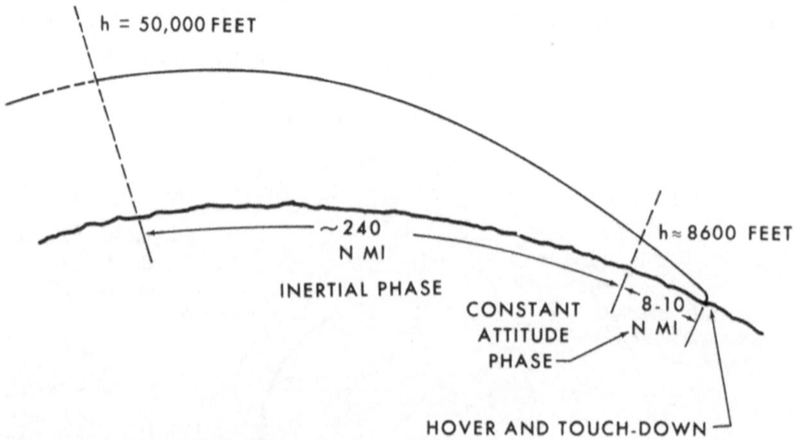


**Fig. 14-40.**

LUNAR LANDING MANEUVER
CONSTANT ATTITUDE PHASE

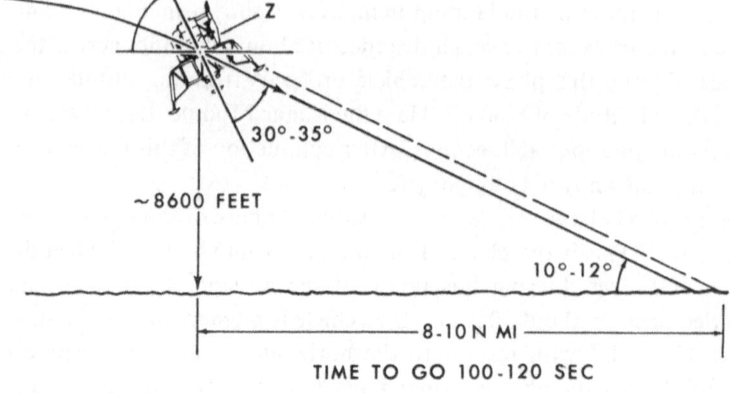


**Fig. 14-41.**

to land. He then rotates the vehicle so as to place a reticle on the window over the landing point and presses a button on the DSKY. The computer automatically guides the vehicle to make a landing at this point. If the command pilot should, on closer examination, desire to go to a different landing site, he can shift to this site by simply placing the reticle over the next landing point. The astronaut has a maneuver capability of approximately two miles before he runs into fuel difficulties.

One of the interesting problems associated with the lunar landing is that of visibility. If the sun angle is parallel to the trajectory path, i.e., if the sun is behind the astronaut, it tends to wash out the landing area such that he cannot discern characteristics of craters, etc. It is much like the reflection of the sun shining on the ocean. Therefore, the sun is placed so as to have an angular offset from the angle of the landing trajectory or else the crew will be required to make a dog-leg maneuver to a landing at a site which he can survey carefully off to the side.

After the spacecraft is maneuvered down to the "low-gate" point at 500 to 1000 feet, the command pilot takes over control of the vehicle manually. He flies to an altitude of about 100 feet, at which time he nulls out all accumulated drifts in the platform with the aid of his landing radar. He is then in a position to make an instrument landing or a visual landing depending

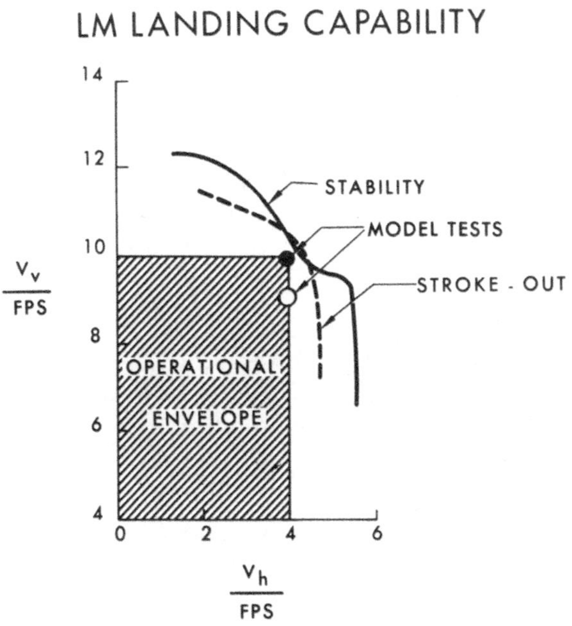


**Fig. 14-42.**

upon the degree of obscuration of his vision by a cloud of dust which might be created by his engine plume.

Figure 14-42 shows the envelope of the landing capability of the Lunar Module. The vehicle must be landed with something less than 10 ft/sec vertical velocity and something less than 4 ft/sec horizontal velocity. These values were chosen after many tradeoffs among lunar terrain characteristics, control characteristics of the vehicle, landing gear weight and size, and fuel requirements.

If the spacecraft should land with too great a horizontal velocity or if it should begin to tip over into a crater, the command pilot can press a button and abort the mission as long as he acts before the vehicle exceeds and angle of 45° measured with respect to the vertical. Similarly, if the bearing strength of the lunar surface is of such a nature as not to support the vehicle and he notices that the vehicle is sinking, he can press a button and abort the mission. Considerable redundancy and safety has been built inte the vehicle characteristics and into the landing trajectory to provide for crew safety.

**Fig. 14-43.**

## [14-16] RELIABILITY

In Apollo, two reliability criteria are applied in systems design and in mission planning. One is associated with crew safety and the other is associated with mission success. Figure 14-44 shows the reliability diagram of the navigation and guidance system and the control system from a crew safety standpoint. In the navigation and guidance system, the onboard system and the ground system are in parallel paths. The ground system consists of the Mission Control Center, communications system, and the tracking network. The onboard system consists of the Apollo guidance computer, the optics telescope, and the inertial measurement unit.

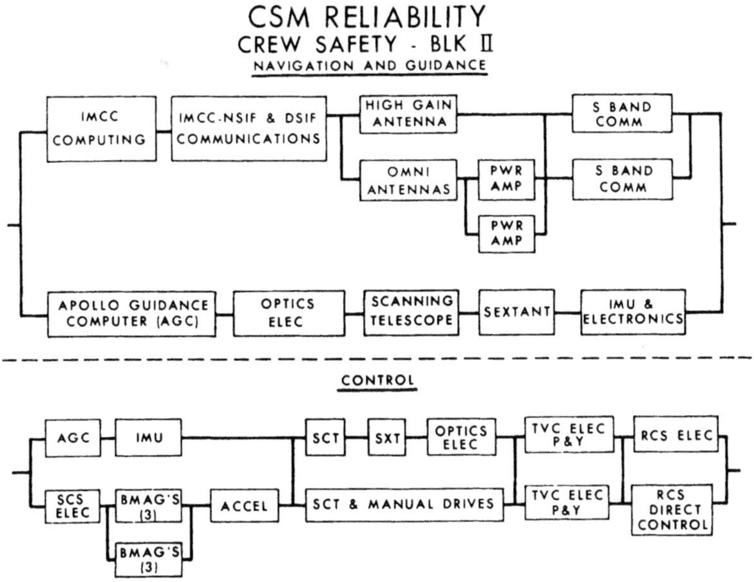


**Fig. 14-44.**

In the control system there is even greater redundancy from the crew safety standpoint. The computer and inertial measurement unit are connected in parallel with the body-mounted attitude gyros and accelerometers. Also in parallel are the sextant and scanning telescope operating automatically or manually and dual thrust-vector control electronics and dual reaction control jet electronics.

The mission criteria for crew safety is such that when equipment malfunctions have occurred to the point where one more malfunction of any kind would endanger the crew, then the mission should be aborted at that point.

Figure 14-45 shows the mission success block diagram. In this block diagram, there are series connections of the guidance and control system rather than the parallel paths shown in the crew safety diagram. The probability of mission success, of course, is somewhat less than that for crew safety.

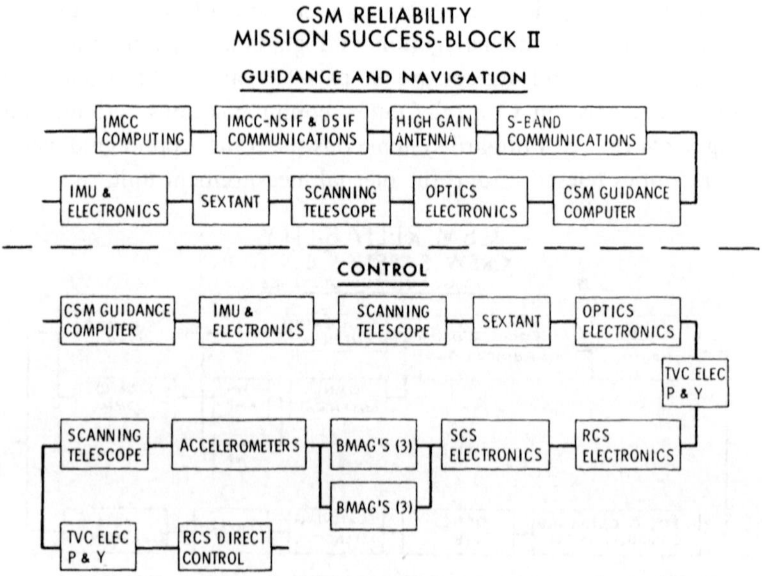


**Fig. 14-45.**

## [14-17] SUMMARY

The accuracy requirements of the Apollo guidance and control systems are largely based upon the reentry phase of the mission which is the most stringent from the standpoint of guidance system performance. The guidance system requirements for reentry are a function of the following:

1) Initial condition errors at the reentry interface
2) Initial stable member alignment accuracy
3) Inertial component accuracy
4) Guidance equation mechanization
5) Reentry range requirements
6) Allowable reentry dispersion.

Initial condition uncertainties at the reentry interface are dependent on the accuracy of midcourse navigation. The Apollo spacecraft utilizes the manned space flight network as the primary source of navigation data.

There is, however, an accurate onboard self-contained navigation system as well.

The alignment of the stable platform and the drift of the platform prior to reentry have a large effect on the reentry accuracy. Because of operational considerations, the platform must be aligned at least one hour prior to reaching the point of .05 $g$'s. The drift of the platform that occurs during this period of time must be considered in the reentry error budget. During atmospheric reentry the inertial component accuracies influence landing dispersion. Each inertial component error source contributes to this inaccuracy, of course.

Guidance equation mechanization is such that both nominal and off-nominal conditions contribute little to touchdown dispersion. The reentry range also places a requirement on the accuracy of the guidance system. The reentry range for the Apollo mission is 1500 nautical miles to 2500 nautical miles. The reentry requirements for Apollo have been relaxed considerably from that planned early in the program. This primarily results from the following changes:

1) The oceans (vs. land) are the prime landing areas. Water landings do not require the high precision of land landings.

2) Apollo entry range has been reduced from a maximum of 5000 nautical miles to 2500 nautical miles. This reduction also has been made possible by resorting to water landings.

The concepts for reentry developed in Apollo are applicable to reentry at high velocities into a dense atmosphere (as in the case of the Command Module reentering the earth's atmosphere from the Moon) or vacuum entry to a landing (as in the case of the lunar landing). Either of these concepts or a combination thereof are applicable to entry into any planetary atmosphere of the solar system. The hardware concepts and the mission concepts discussed in this chapter have established the technological base for manned exploration of the solar system during the next few decades.

## REFERENCES

Duncan, R. C.: *Dynamics of Atmospheric Entry* McGraw-Hill Book Company New York, 1962.

15

# Dynamics of Planetary Approach

WILLIAM T. THOMSON, Professor and Chairman
*Mechanical Engineering Department*
*University of California at Santa Barbara*
*Consultant TRW Systems*

## [15-1] INTRODUCTION

For a precise interplanetary orbit, the n-body problem including the sun and the nearby planets must be solved. However, for a preliminary analysis the method of patched-conics based on three distinct two-body problems is generally satisfactory. The orbit will then consist of a planetocentric segment near the departure planet, a heliocentric segment, and a final planetocentric segment near the destination planet. In this discussion we will be concerned mainly with the third phase dealing with the dynamics of planetary approach. In particular we will consider the Earth-Mars journey, in which case the approach trajectory is computed on the two-body Spacecraft-Mars assumption.

Certain simplifying assumptions can be made which can be justified. The Sun and the nearby planets can be neglected in that their influence on both the spacecraft and Mars is nearly equal, thus allowing a two-body assumption. The approach-trajectory is then hyperbolic with respect to the planet Mars. Further simplification results by placing the center of the coordinate system at the center of Mars and examining the impact point on a plane normal to the asymptote of the incoming hyperbola.

## [15-2] TRAJECTORY PARAMETERS OF IMPORTANCE

During a planetary approach or landing, one of the more important considerations is the relative geometry between the spacecraft, the capsule, the target planet, the Earth and the Sun. The communication to Earth is most efficient when the capsule is near the sub-Earth point (a point on the planet's surface pierced by the line between the centers of the Earth and the planet).

If a landing is to be attempted, it is desirable to land on the sunlit side near the morning terminator (line dividing the dark and light sides of the planet). Pertinent to the problem are:

a) Latitude of the sub-solar point.
b) Latitude of the sub-Earth point.
c) Sun-Mars-Earth angle.
d) Latitude of the impact point.
e) The angle between the incoming asymptote and the Mars-Sun vector (this angle is designated as ZAP by the Jet Propulsion Lab).
f) The hyperbolic excess velocity vector $v_\infty$ (defined as the vector difference between the target planet's heliocentric velocity at encounter, and the heliocentric velocity of the spacecraft as it enters the gravitational influence of the planet).
g) The miss parameter B (distance from the origin of the planet-centered coordinate system to the impact point of the asymptote, measured on the plane perpendicular to the incoming asymptote.

The first three quantities are independent of the trajectory adopted, and are functions only of the calendar date of arrival. These are shown in Figs. 15-1, 2, and 3 for Mars during the 1971–1972 opportunity.

The eccentricity of the Earth orbit is 0.017 whereas that of Mars is 0.093. Thus the Sun-Mars-Earth angle may have a maximum between 37° and 45° depending on whether the encounter date coincides with Mars pericenter or apocenter. See Figure 15-4.

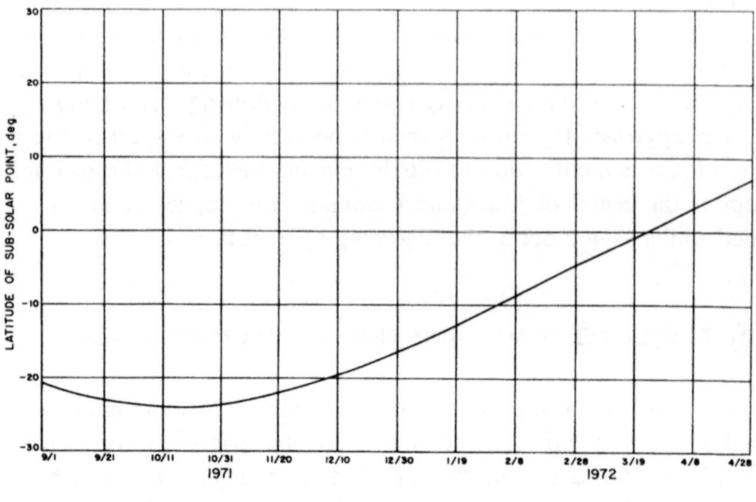


**Fig. 15-1.** Latitude of sub-solar point versus arrival date.

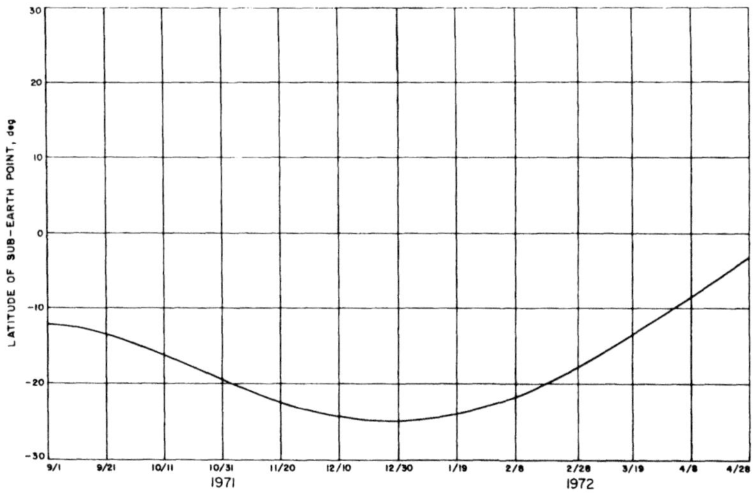


**Fig. 15-2.** Latitude of sub-earth point versus arrival date.

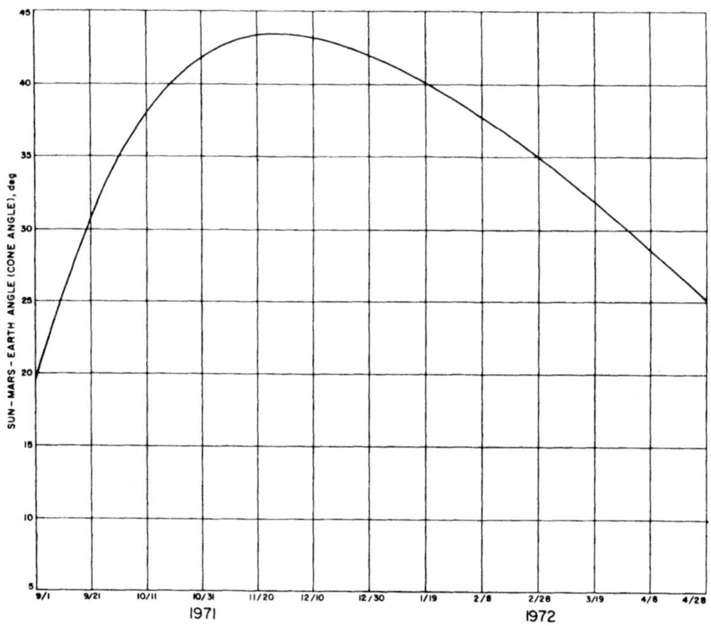


**Fig. 15-3.** Sun-Mars-Earth angle versus arrival date.

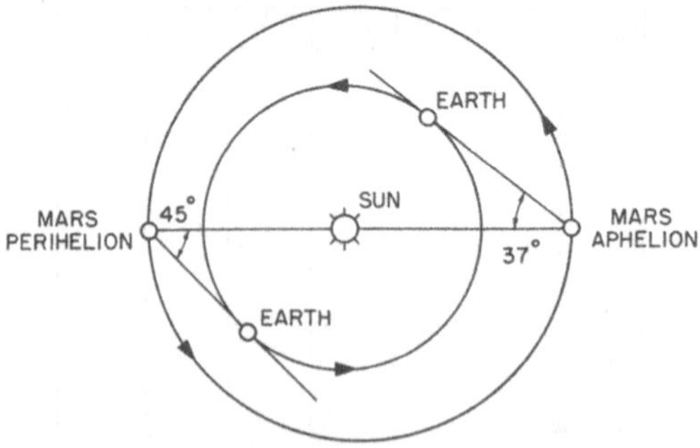


**Fig. 15-4.**

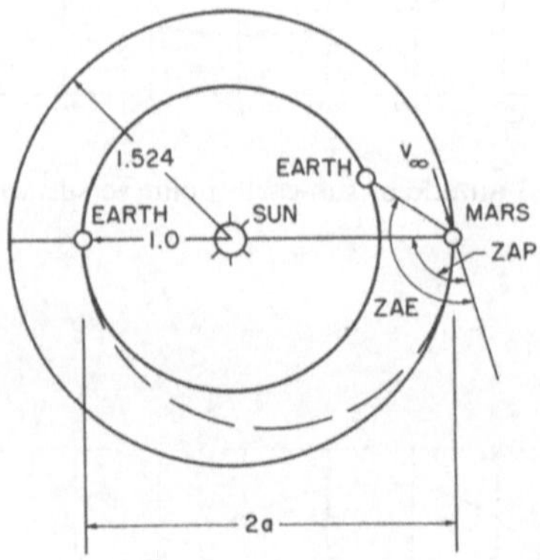


**Fig. 15-5.**

## [15-3] THE TRANSFER ORBIT

It is well at this point to review briefly the requirements for the interplanetary mission; i.e. Earth to Mars. The attraction of the Sun predominates over most of the trajectory which is a heliocentric ellipse. The energy equation (See Section 15-12, Appendix)

$$E = \frac{v^2}{2} - \frac{K}{r} = \mp \frac{K}{2a} \quad \begin{matrix} - \text{ for ellipse} \\ + \text{ for hyperbola} \end{matrix} \tag{15-1}$$

arranged in the form

$$v^2 = K\left(\frac{2}{r} - \frac{1}{a}\right) \tag{15-2}$$

can be considered in terms of astronautical units which greatly simplifies the arithmetic. For example we can assume the orbit of Earth to be a circle of

1 astronautical unit a.u., and the velocity of Earth to be unity. Then

$$1^2 = K\left(\frac{2}{1} - \frac{1}{1}\right), \qquad K = 1.0$$

On this basis the velocity of Mars is

$$v^2 = 1\left(\frac{1}{1.524}\right) = 0.656, \qquad v = 0.809$$

Since the mean velocity of Earth is 29.8 km/sec=97,800 ft/sec, the mean velocity of Mars is

$$0.809 \times 29.8 = 24.1 \text{ km/sec} = 79{,}000 \text{ ft/sec}.$$

For the Hohmann transfer orbit between Earth and Mars, we first compute the semi-major axis, see Figure 15-5.

$$a = \tfrac{1}{2}(1 + 1.524) = 1.262.$$

so that the velocity equation becomes

$$v^2 = \left(\frac{2}{r} - \frac{1}{1.262}\right).$$

The required velocities at departure and arrival then becomes

$$v_{\text{dep}} = \sqrt{\frac{2}{1} - \frac{1}{1.262}} = 1.10 \;= 32.8 \text{ km/sec} = 104{,}300 \text{ ft/sec}.$$

$$v_{\text{arr}} = \sqrt{\frac{2}{1.524} - \frac{1}{1.262}} = 0.712 = 21.2 \text{ km/sec} = \;69{,}600 \text{ ft/sec}.$$

and the velocity of the spacecraft relative to Mars is

$$v_\infty = 24.1 - 21.2 = 2.9 \text{ km/sec} = 9{,}520 \text{ ft/sec}.$$

In general the aphelion of the transfer orbit is chosen somewhat outside of the Mars' orbit, which increases the value of the semi-major axis "a" and results in a larger arrival velocity which is not colinear with the velocity of Mars.

## [15-4] REFERENCE PLANE AND APPROACH DIRECTION

The problem of the approach phase is greatly simplified by using a reference plane centered in the target planet which is oriented perpendicular to the incoming asymptote of the spacecraft hyperbolic orbit, as shown in Figure 15-6. *S* is in the direction of the asymptote, and the *TR* plane is the aim plane normal to *S*. The *T* axis is generally chosen in the ecliptic plane of the target planet.

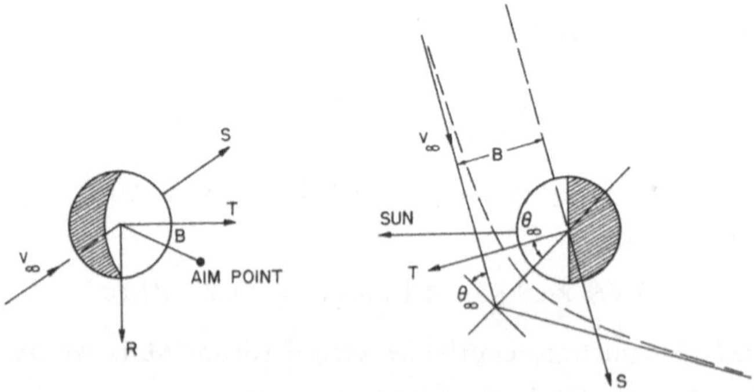


**Fig. 15-6.**

As shown in the previous section, the heliocentric velocity of the planet is greater than the velocity of the spacecraft in the vicinity of the target planet, and the hyperbolic excess velocity, defined as the vector difference between the planet heliocentric velocity and the spacecraft heliocentric velocity, is the velocity of the spacecraft relative to the destination planet.

*abs. vel. of planet + vel. of s.c. rel. to planet = vel. of s.c.*

Thus the spacecraft will approach the planet from its leading side. Figure 15-7 shows three possible cases.

In the Type I trajectory, the spacecraft encounters the planet before its aphelion, and the approach is from the sunny side of the leading direction. In the Type II trajectory the encounter takes place after aphelion of the

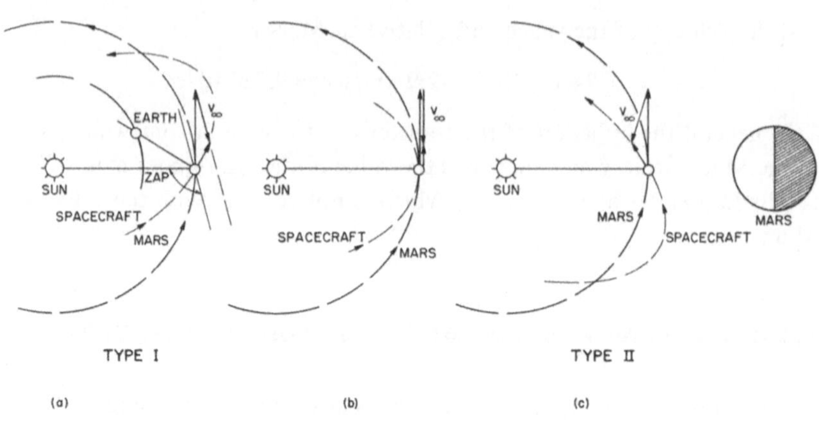


**Fig. 15-7.**

spacecraft heliocentric orbit, and the approach is from the leading dark side. If the encounter coincides with the aphelion, the approach direction of $\mathbf{v}_\infty$ is 90° from the Sun line. Generally the aphelion is chosen outside the planet radius, which results in either the Type I or Type II encounter.

The angle between $\mathbf{v}_\infty$ and the Mars-Sun line is designated as ZAP and the angle between $\mathbf{v}_\infty$ and the Mars-Earth vector is ZAE. See Figure 15-5. If $\mathbf{v}_\infty$ lies in the ecliptic plane and if ZAP is less than 90°, the spacecraft landing will be in the dark. If $\mathbf{v}_\infty$ lies in the ecliptic plane and if the angle ZAE is less than 90°, the Earth will not be visible to the spacecraft on landing.

## [15-5] THEORY OF THE CORRIDOR

Due to the attraction of the planet, any vehicle approaching the planet within a certain effective cross section will strike the planet, see Figure 15-8. On the basis of the two-body problem (between the planet and the vehicle) the effective cross section or corridor can be established from the principles of conservation of moment of momentum and total energy.

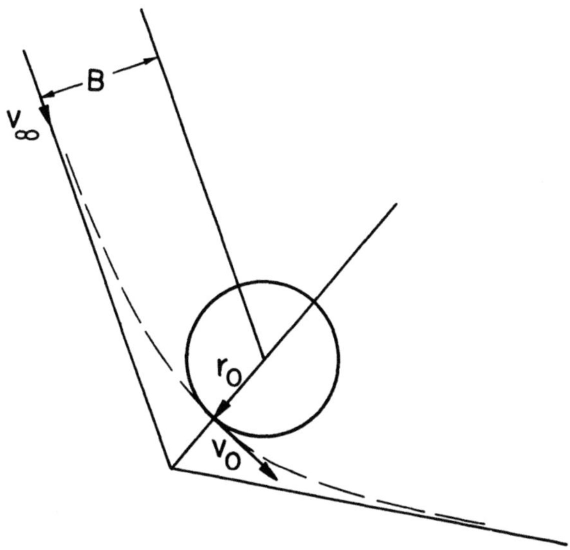


**Fig. 15-8.**

Equating the moment of momentum of the vehicle at a great distance from the planet (generally 2 to 3 days before encounter with Mars) to that at grazing encounter with the surface of the planet, we have

$$Bv_\infty = r_0 v_0 \tag{15–3}$$

where $v_\infty$ is the excess hyperbolic (or asymptotic) speed of the spacecraft.

From the energy equation

$$E_p = \frac{v^2}{2} - \frac{K_p}{r} \tag{15–4}$$

applied at $r=\infty$ and $r=r_0$, we have

$$\frac{v_\infty^2}{2} = \frac{v_0^2}{2} - \frac{K_p}{r_0} \tag{15–5}$$

or

$$\left(\frac{v_0}{v_\infty}\right)^2 = 1 + 2\left(\frac{K_p}{r_0 v_\infty^2}\right) \tag{15–6}$$

Eliminating $v_0$ from the moment of momentum equation, the radius of the effective cross section is found to be

$$\frac{B}{r_0} = \sqrt{1 + 2\left(\frac{K_p}{r_0 v_\infty^2}\right)} \tag{15–7}$$

It is evident from this equation that when $v_\infty = \infty$, the value of $B$ is $r_0$, the radius of the planet. As $v_\infty$ approaches zero, $B$ becomes infinite so that all bodies with zero relative velocity with respect to the planet will be eventually

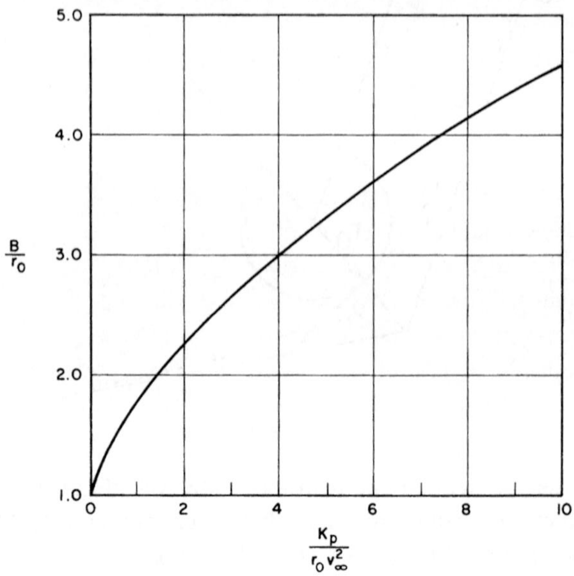


**Fig. 15-9.**

attracted to it. Thus in all actual cases of $v_\infty$ finite, the effective cross section is greater than the radius of the planet. The relationship of Eq. (15–7) is shown graphically in Figure 15-9.

## [15-6] ORBIT BENDING DUE TO THE GRAVITATIONAL ATTRACTION OF THE PLANET

Figure 15-10 shows the target planet and the coordinate plane perpendicular to the incoming asymptote. The miss parameter $B$ is the distance between the center of the planet and the impact point of the asymptote on the aim plane. The parameters $B$ and $v_\infty$ are sufficient to establish two quantities of interest, namely the periapsis height $(r_p - r_0)$ and the bending angle $\theta_\infty$ through which the hyperbolic trajectory turns. The relationship between these are available again from the principles of conservation of moment of momentum and conservation of energy.

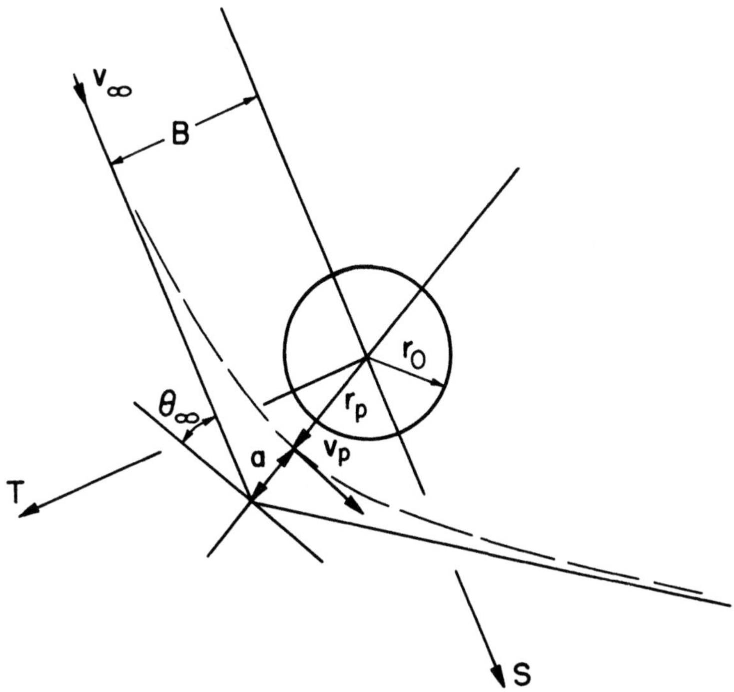


**Fig. 15-10.**

Equating the moment of momentum at periapsis $r_p$ and at a great distance from the planet, we have

$$Bv_\infty = r_p v_p \tag{15–8}$$

Substituting into the energy equation

$$\frac{v_p^2}{2} - \frac{K_p}{r_p} = \frac{v_\infty^2}{2} \tag{15–9}$$

we obtain

$$r_p = -\frac{K_p}{v_\infty^2} + \sqrt{\left(\frac{K_p}{v_\infty^2}\right)^2 + B^2} \tag{15–10}$$

or in nondimensional form

$$\frac{r_p v_\infty^2}{K_p} = -1 + \sqrt{1 + \left(\frac{Bv_\infty^2}{K_p}\right)^2} \tag{15–11}$$

Figure 15-11 shows a plot of $(r_p - r_0)$, the periapsis altitude vs $B$ with $v_\infty$ as parameter, as computed from Eq. (15-10) for the planet Mars. The curves

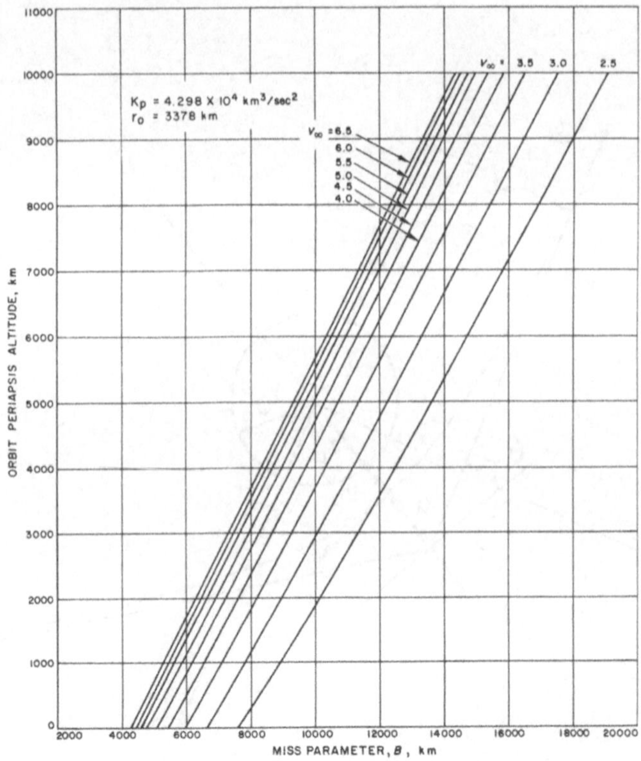


**Fig. 15-11.** Orbiter periapsis altitude versus miss parameter for Mars.

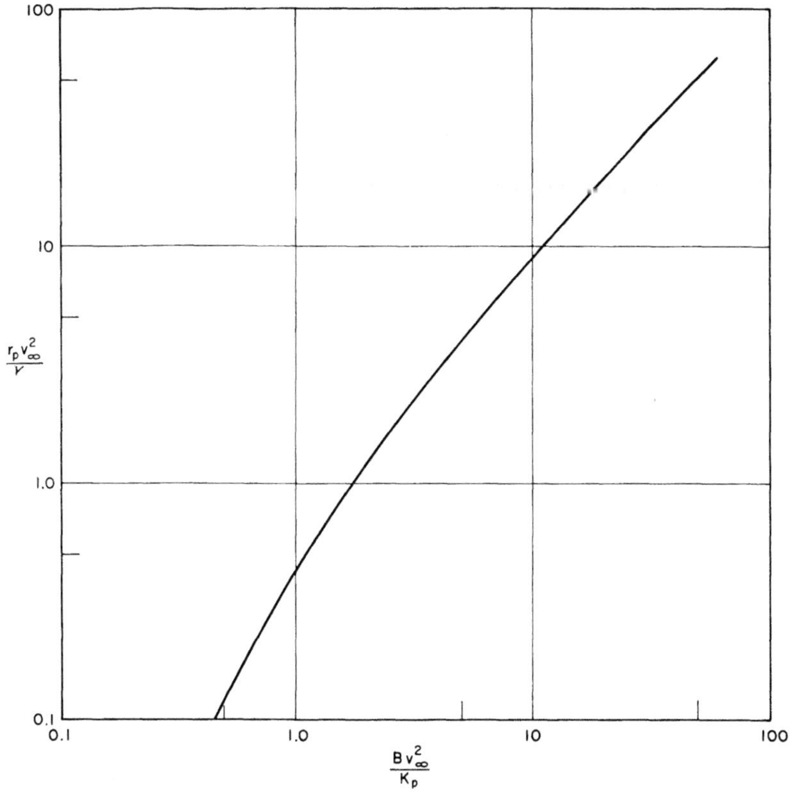


**Fig. 15-12.** Nondimensional periapsis distance vs miss parameter (Eq. 11).

of Figure 15-11 can be collapsed into a single curve applicable to any planet by Eq. (15–11) which is plotted in Figure 15-12.

The bending angle $\theta_\infty$ is simply found from the geometry of Figure 15-10 as

$$\cos\theta_\infty = \frac{B}{r_p + a} \tag{15–12}$$

But from the energy equation (for hyperbolic orbit)

$$v^2 = K_p\left(\frac{2}{r} + \frac{1}{a}\right) \tag{15–13}$$

and by letting $r \to \infty$ and $v \to v_\infty$, we obtain

$$a = \frac{K_p}{v_\infty^2} \tag{15–14}$$

Thus

$$\cos\theta_\infty = \frac{B}{r_p + \dfrac{K_p}{v_\infty^2}} = \frac{B}{\sqrt{\left(\dfrac{K_p}{v_\infty^2}\right)^2 + B^2}} = \frac{1}{\sqrt{1 + \left(\dfrac{K_p}{Bv_\infty^2}\right)^2}} \tag{15–15}$$

For large $v_\infty$ the above equation reduces to $\cos\theta_\infty \cong 1.0$. Figure 15-13 shows a plot of $\theta_\infty$ vs $(Bv_\infty^2/K_p)$ which applies to any planet.

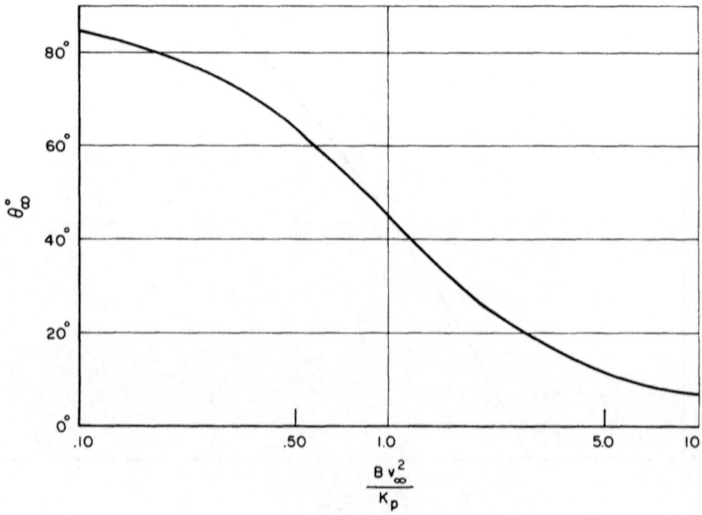


**Fig. 15-13.**

## [15-7] ENTRY ANGLE

The entry angle $\theta_E$ is the angle between the velocity vector and the tangent to the circle at atmospheric encounter. Its equation is developed with the aid of Figure 15-14 as follows:

$$rv\cos\theta_E = Bv_\infty \qquad \text{(moment of momentum)} \tag{15–16}$$

$$\frac{v^2}{2} - \frac{K_p}{r} = \frac{v_\infty^2}{2} \qquad \text{(energy)} \tag{15–17}$$

$$\frac{1}{2}\left(\frac{Bv_\infty}{r\cos\theta_E}\right)^2 - \frac{K_p}{r} = \frac{v_\infty^2}{2}$$

$$\cos\theta_E = \frac{Bv_\infty}{r\sqrt{2\left(\dfrac{K_p}{r} + \dfrac{v_\infty^2}{2}\right)}} = \frac{\left(\dfrac{B}{r}\right)}{\sqrt{1 + 2\left(\dfrac{K_p}{rv_\infty^2}\right)}} \tag{15–18}$$

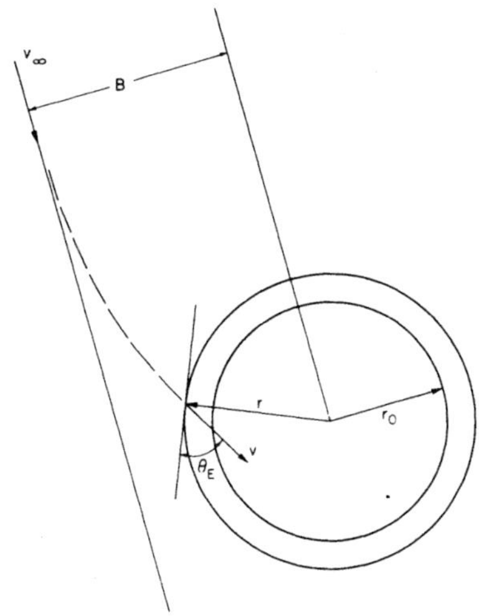


**Fig. 15-14.**

In Figure 15-15 $r/B \cos \theta_E$ is plotted against $(rv_\infty^2/K_p)$. For example, if in a Mars approach $v_\infty = 4$ km/sec, $K_p = 4.298 \times 10^4$ km$^3$/sec$^2$ and $r = 3621$ km, we have $rv_\infty^2/K_p = 1.35$ and from Figure 15-15 $r/B \cos \theta_E = .632$. Thus if $B =$ 5000 km, $\cos \theta_E = .632 \times \frac{5000}{3621} = .872$ and $\theta_E = 29.3°$.

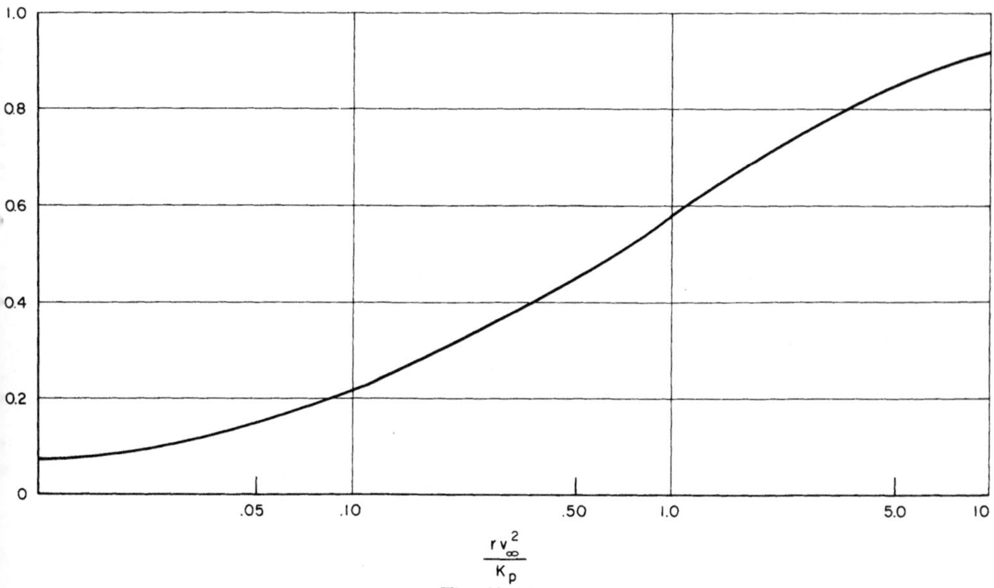


**Fig. 15-15.**

It is again seen that for a given $r/B$, $\theta_E$ depends only on $rv_\infty^2/K_p$ which applies to any planet.

## [15-8] RANGE ANGLE

If $B$ and $v_\infty$ are known, $\theta_\infty$ is also known (i.e. Eq. (15–15)) and the following relationship exists from the geometry of Figure 15-16.

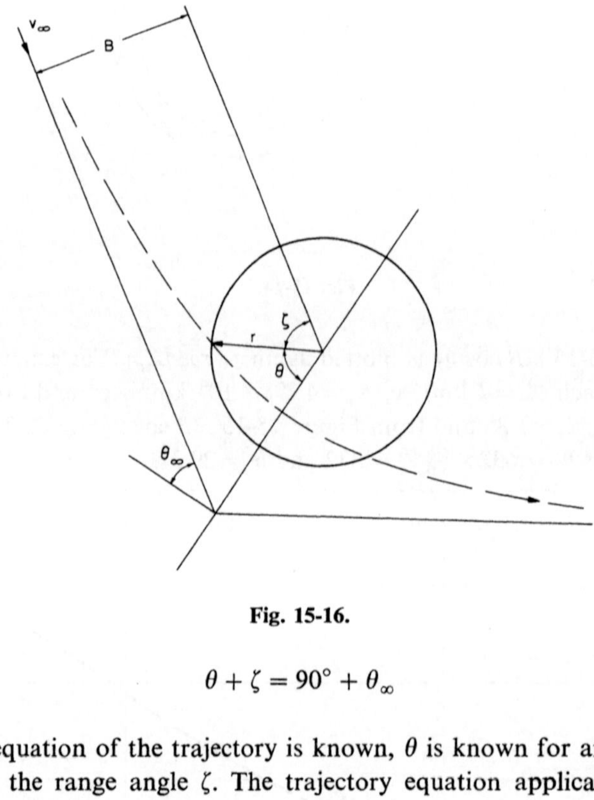


**Fig. 15-16.**

$$\theta + \zeta = 90° + \theta_\infty \tag{15–19}$$

Since the equation of the trajectory is known, $\theta$ is known for any $r$, which establishes the range angle $\zeta$. The trajectory equation applicable for the hyperbolic orbit is

$$\frac{r}{r_p} = \frac{(1+e)}{1 + e\cos\theta} \tag{15–20}$$

When $\theta = 90 + \theta_\infty$, $r = \infty$, so that $e = 1/\sin\theta_\infty$. Thus the above equation may be written as

$$\cos\theta = \frac{r_p}{r}(\sin\theta_\infty + 1) - \sin\theta_\infty$$

Substituting Eq. (15–10) for $r_p$ and the relation

$$\sin\theta_\infty = \frac{\left(\dfrac{K_p}{Bv_\infty^2}\right)}{\sqrt{1+\left(\dfrac{K_p}{Bv_\infty^2}\right)^2}} = \frac{\left(\dfrac{K_p}{rv_\infty^2}\right)}{\sqrt{\left(\dfrac{B}{r}\right)^2+\left(\dfrac{K_p}{rv_\infty^2}\right)^2}} \tag{15–21}$$

which may be obtained from Eq. (15–15), the equation for cos $\theta$ becomes

$$\cos\theta = \frac{\left(\dfrac{B}{r}\right)^2-\left(\dfrac{K_p}{rv_\infty^2}\right)}{\sqrt{\left(\dfrac{B}{r}\right)^2+\left(\dfrac{K_p}{rv_\infty^2}\right)^2}} \tag{15–22}$$

Thus knowing $rv_\infty^2/K_p$ and $B/r$ the angle $\theta$ can be determined. Letting $r=$ radius of the planet, the range angle $\zeta$ then becomes known. In Figure 15-17, $\zeta$ is plotted against $B/r$ with $rv_\infty^2/K_p$ as parameter.

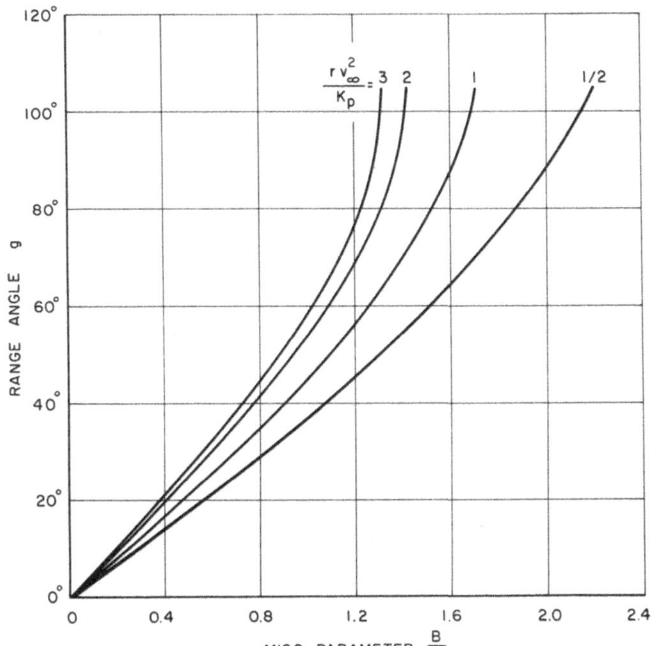


**Fig. 15-17.**

As an example consider Mars with $B=5000$ km, and $v_\infty=4$ km/sec. The radius of Mars is $r = 3378$ km and $K_p = 4.298 \times 10^4$. Thus

$$\frac{B}{r} = 1.48\,, \qquad \frac{rv_\infty^2}{K_p} = 1.26$$

and Eq. (15–22) gives

$$\cos\theta = 0.830\,, \qquad \theta = 33°54'$$

From Eq. (15–15) $\theta_\infty = 28°20'$ and $\zeta$ is found from Eq. (15–19) as

$$\zeta = 84°26'$$

## [15-9] BENDING ANGLE AT ATMOSPHERIC ENCOUNTER

The bending angle $\theta_B$ at atmospheric entry is found from the relationship (see Figure 15-18).

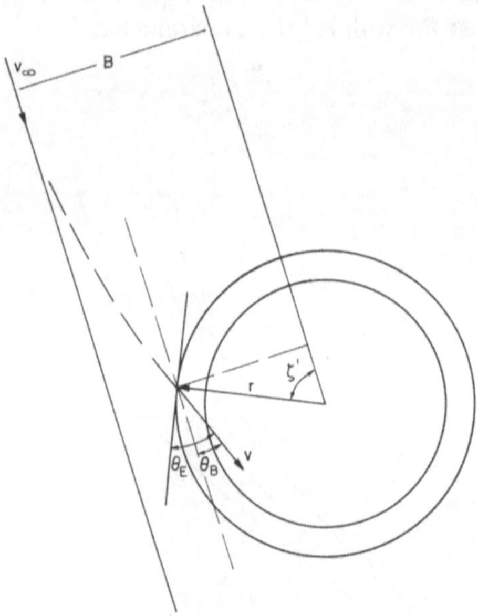


**Fig. 15-18.**

$$\theta_B = \theta_E + \zeta' - 90° \qquad (15\text{–}23)$$

where $\zeta'$ is here the range angle corresponding to $r$ measured to the limit of the atmosphere. It is evident then that $\theta_B$ can also be plotted against $B/r$ ($r$ = radius to limit of atmosphere) with $rv_\infty^2/K_p$ as parameter. Figure 15-19 is a plot of $\theta_B$ vs $B/r$ with $rv_\infty^2/K_p$ as parameter.

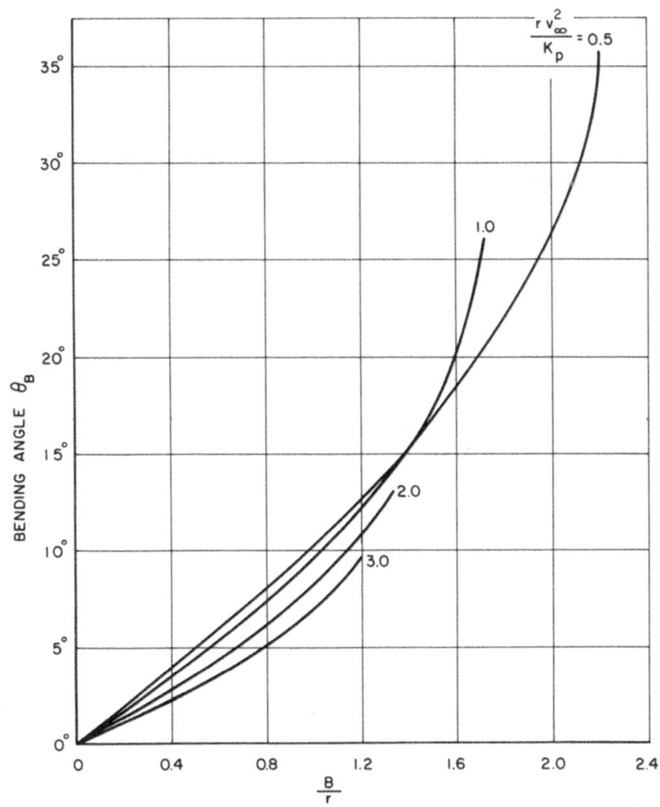


**Fig. 15-19.**

## [15-10] CAPSULE MANEUVER

A capsule to land on the planet may be separated from the spacecraft (bus) by a maneuver initiated with a velocity increment $\Delta \mathbf{v}$. Since the maneuver velocity makes only a small perturbation on the approach speed, the capsule deflection can be computed with sufficient accuracy on the basis of rectilinear motion (i.e. assuming the planet to be massless), as shown in Figure 15-20.

It is desirable for the capsule to land (to hit the aim plane) before the bus reaches the aim point, in order to maintain communication between the capsule and the bus. At capsule encounter, the distance between the bus and capsule is $X$ and the difference in time of arrival to the aim plane is $\Delta t$. We then have

$$X = \sqrt{D^2 + (v_\infty \Delta t)^2} \tag{15–24}$$

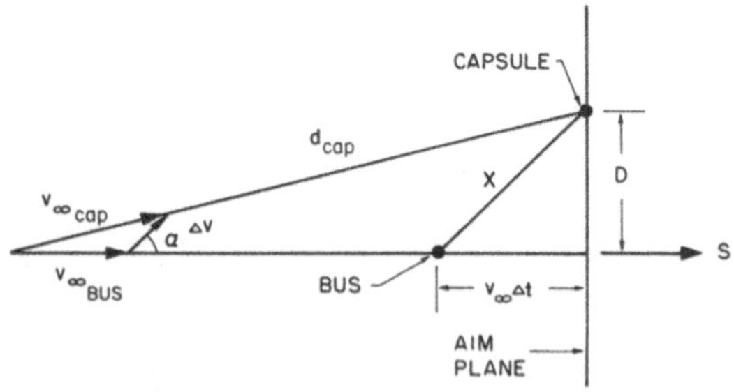


**Fig. 15-20.**

which indicates that this distance is independent of the time of separation.

From the similar triangles of Figure 15-20,

$$\frac{\Delta v}{v_{\infty\,\mathrm{capsule}}} = \frac{X}{d_{\mathrm{capsule}}} = \frac{X}{v_{\infty} t_e} \qquad (15\text{–}25)$$

where $d_{\mathrm{cap}}$ is the distance from the planet at which the maneuver takes place, and $t_e$ is its time before encounter. Thus the velocity increment required is

$$\Delta v = \frac{X}{t_e} \qquad (15\text{–}26)$$

which varies inversely with the maneuver execution time before encounter. The firing angle $\alpha$ of the velocity increment is

$$\alpha = \tan^{-1} \frac{D}{v_{\infty} \Delta t} \qquad (15\text{–}27)$$

## [15-11] GEOMETRY OF MARS LANDING AND COMMUNICATION

It was pointed out in section 15-4 that the general direction of the approaching vehicle is from the forward direction of the target planet and the aim plane was established perpendicular to the incoming asymptote. Figures 15-21 and 15-22 represent views into the aim plane with Mars as the planet to be approached. In Figure 15-21 the approach direction is nearly perpendicular to the Sun-Mars line (i.e. ZAP$\cong$90°). In Figure 15-22 ZAP is greater than 90° so that the approach is from the direction inside of the Mars orbit as in Figure 15-7a. Shown in these drawings are the sub-polar point, the sub-earth point, the orbiter aim point, the capsule aim point, the morning terminator and the 50° cone angle about the sub-earth point.

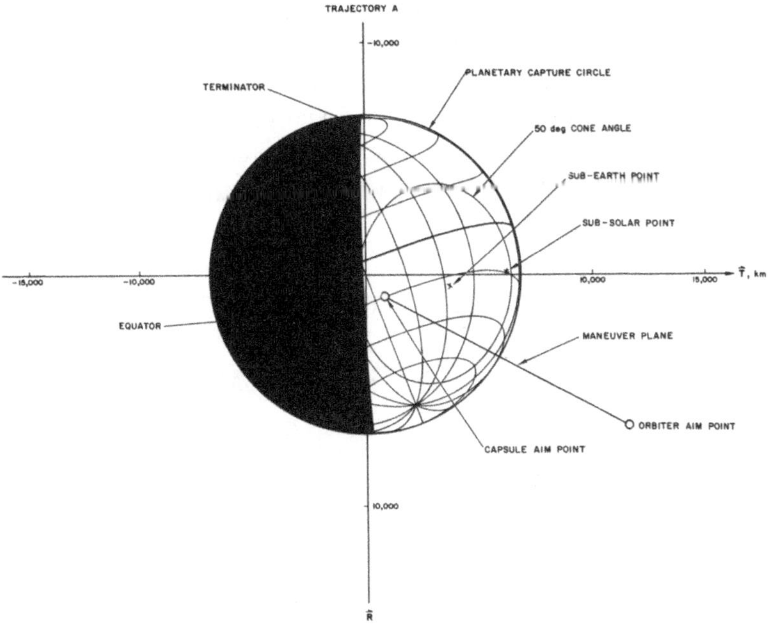


Fig. 15-21. Aim diagram: trajectory A.

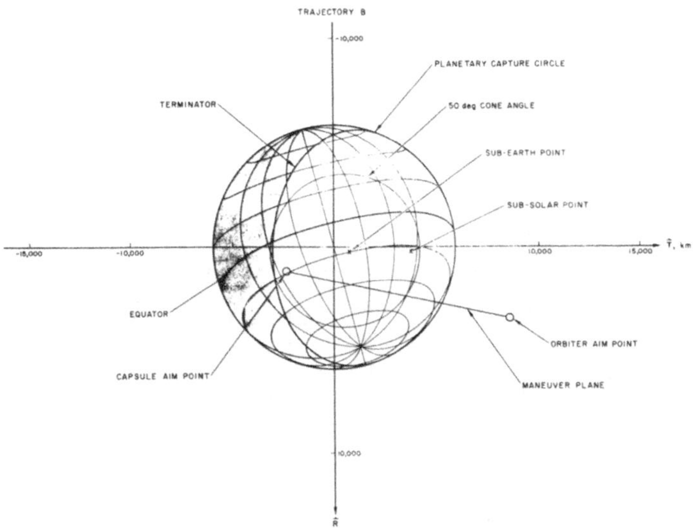


Fig. 15-22. Aim diagram: trajectory B.

The latitude of the landing point of the capsule establishes the view period for communication between the capsule and Earth. The greater the latitude deviation from the latitude of the sub-earth point, the shorter will be the communication view period with Earth. Figures 15-23, 15-24 and 15-25 show the relationship of the view period for a Mars landing.

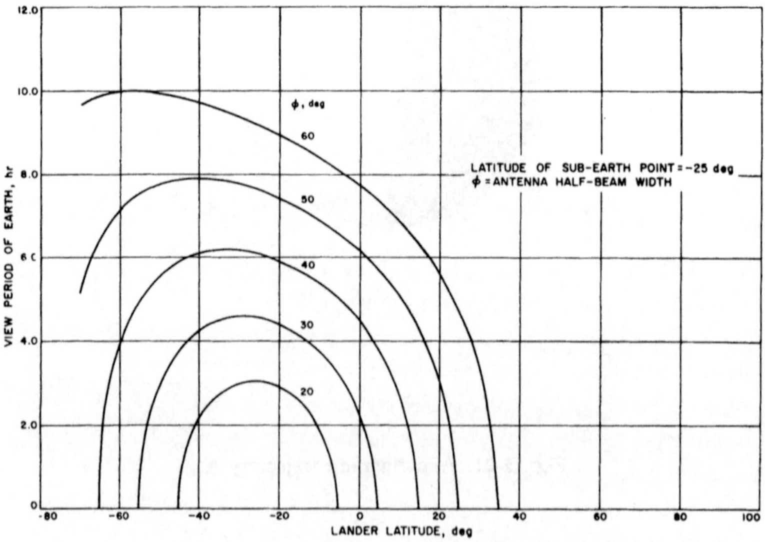


**Fig. 15-23.** Communication view period versus lander latitude, $\lambda_e = -25°$.

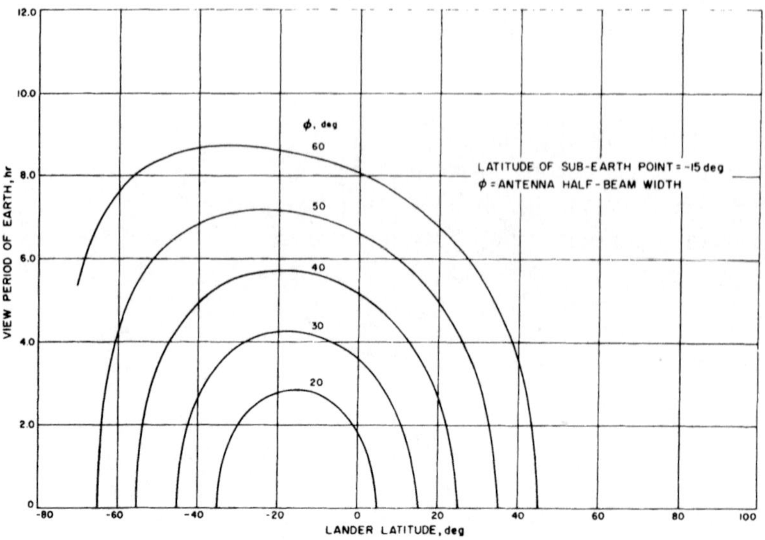


**Fig. 15-24.** Communication view period versus lander latitude, $\lambda_e = -15°$.

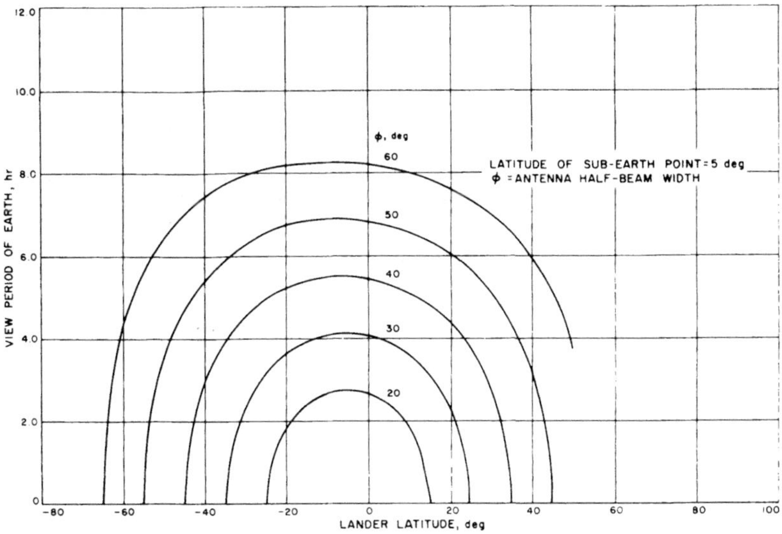


**Fig. 15-25.** Communication view period versus lander latitude, $\lambda_e = -5°$.

## [15-12] APPENDIX

*A. Elliptic Orbits*

1. Ratio $x/r$ for any point on the conic curve is constant and equal to $e$. Choosing points 1, 2, 3,

$$e = \frac{r_p}{c} = \frac{r_a}{2a + c} = \frac{a}{a + c}$$

Thus

$$\begin{aligned} c &= \frac{a}{e}(1 - e) \\ r_a &= a(1 + e) \\ r_p &= a(1 - e) \end{aligned} \tag{15–28}$$

2. Conservation of moment of momentum. Choose points 1 and 3

$$H = r_p v_p = r_a v_a \tag{15–29}$$

3. Conservation of Energy

$$E = \tfrac{1}{2}v_p^2 - \frac{K}{r_p} = \tfrac{1}{2}v_a^2 - \frac{K}{r_a} = \tfrac{1}{2}v^2 - \frac{K}{r} \tag{15–30}$$

4. Equations relating above quantities. Eliminate $v_a$ and $v_p$ from (15–29)

$$E = \frac{1}{2}\frac{H^2}{r_p^2} - \frac{K}{r_p} = \frac{1}{2}\frac{H^2}{r_a^2} - \frac{K}{r_a}$$

$$\frac{H^2}{2K}\left(\frac{1}{r_p^2} - \frac{1}{r_a^2}\right) = \left(\frac{1}{r_p} - \frac{1}{r_a}\right) \therefore \frac{H^2}{2K}\left(\frac{1}{r_p} + \frac{1}{r_a}\right) = 1 \qquad (15\text{–}31)$$

Substitute $r_a$ and $r_p$ from (15–28) into (15–31)

$$H = \sqrt{Ka(1 - e^2)} \qquad (15\text{–}32)$$

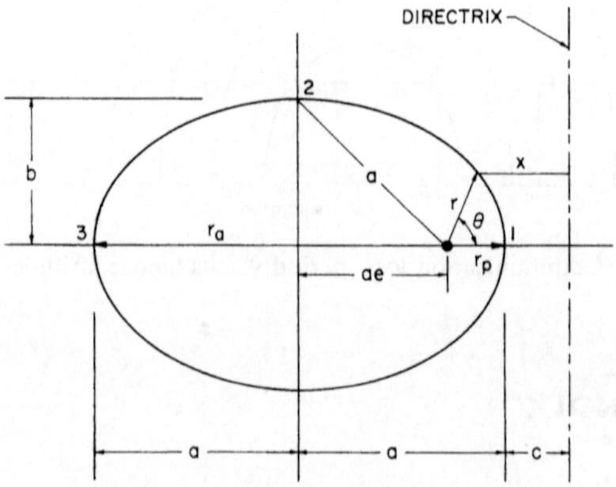


**Fig. 15-26.** Geometry of elliptic orbit.

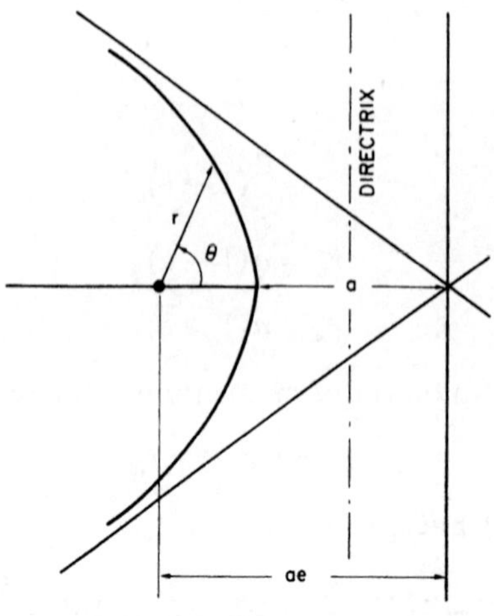


**Fig. 15-27.** Geometry of hyperbolic orbit.

Substitute (15–32) into $E$.

$$E = \frac{-K}{2a} \tag{15–33}$$

*B. Hyperbolic Orbits*

Development follows that of elliptic orbits. Results differ as follows:

$$r_p = -a(1 - e) \tag{15–28′}$$

$$H = \sqrt{Ka(e^2 - 1)} \tag{15–32′}$$

$$E = \frac{K}{2a} \tag{15–33′}$$

## REFERENCES

1. "Trajectory Selection Considerations for Voyager Mission to Mars During 1971–72". Jet Propulsion Lab., Pasadena, Calif. Document No. 281, Sept. 15, 1965.
2. Private Communications with E. Kane Casani, Jet Propulsion Lab.
3. Thomson, W. T.: Introduction to Space Dynamics, John Wiley & Sons, New York, 1961, Chapter 4.

# Part Five

# *Advanced Entry Programs*

16

# Entry Research and Development Program Management

JOSEPH F. SHEA, Manager
*Apollo Spacecraft Program Office*
*Manned Spacecraft Center*
*National Aeronautics and Space Administration*

## [16-1] INTRODUCTION

Long before entry into a planet's atmosphere or the subsequent return to Earth, the ponderous mechanism of "Research and Development"—"R and D"—as we today know it in this country, must be set in motion. Inevitably it will involve almost every element of our society—from Capitol Hill to the least supplier of raw materials or piece parts. The program will be so large that government funding is mandatory, and the complex process by which such vast enterprises get started must work itself out. The program will probably follow the awkward, often antagonistic pattern of management by a government technical agency and implementation by several tiers of industrial contractors.

Barring a minor miracle it will progress down the well-worn path of such programs—initial optimism bordering on euphoria, the hard time of development difficulties, cost escalation, schedule slippage (or adjustment, to use the more popular euphemism) and ultimate accomplishment in which the sins of the development program are washed away in the flush of success and the golden glow of national publicity.

The question of whether the seemingly inevitable mid-program problems can be avoided is of more than casual interest—although, by and large, they have become so commonplace in today's aerospace industry that many people, both in government and industry, feel that some schedule slippage and cost overrun is, in fact, inevitable.

Perhaps I'm naive, but I refuse to accept such a conclusion. A program, even a very large one, soundly planned and intelligently executed, should be "doable" within the commitments of the cost, schedule and performance projections which formed the basis for the decision to proceed with the program and which should be accepted by both government and industry participants as moral, as well as contractual obligations.

The incidence of research and development programs which have not met their original commitments is, however, so large that the above statement of faith needs considerable expansion before it can be considered credible. Unfortunately, the potential solution to the problem of poor program performance involves no startling new management technique, no revolutionary set of principles or actions which will, if only invoked with suitable magic incantations, instantly solve all problems.

Quite the contrary, the most promising approach is probably a return to old-fashioned common sense in both the planning and implementation of the program; above all, a recognition that, all fads notwithstanding, the best management system is competent people motivated to consider what can be referred to as the "total management" problem.

At the risk of sounding trite—for common sense translated to a set of comments intended to illustrate some of the flaws in present ways of doing business sounds uncomfortably like motherhood—this paper shall attempt to be somewhat more specific in five areas:

1) Initial program definition.
2) Government—industry roles and relationships.
3) Technical approach and implementation problems.
4) Organizational approach.
5) Management systems.

Although not all-encompassing, the discussion of these topics should provide a reasonably complete introduction to common sense management.

## [16-2] INITIAL DEFINITION

The initial program definition covers a broad and often frustrating period. In reality, there are two distinct phases to program definition. The first, and most nebulous, culminates with the decision to move ahead with the program. For a program as large as a planetary expedition, this obviously comes after years of advanced studies and involves not just the agency which will carry the responsibility, but the Executive Branch (particularly the Bureau of the Budget and the White House, and to varying degrees other interested technical agencies and whatever scientific advisory committees may be in vogue at the time) and, ultimately, the Congress.

The criteria for the enabling decision are, obviously, not purely technical.

A large program decision must consider the balance of the national budget, the continuation (or appropriate change) of the given agency's funding level, the economic health of the industry involved, the public posture toward the program, the national need it fulfills (prestige, technology, science, potential) and the political climate, to name the most obvious factors over and above technical feasibility. Indeed, after the past several decades of technological advance, the technical feasibility of almost any proposed program is today pretty well taken for granted by non-technical decision makers.

But despite all of the above factors, the decision to go ahead is, after all, based on technical data which purport to define the approach, the achievable schedule and the ultimate cost. This is really the first point at which commitments are made, and is the first point at which the program becomes vulnerable to the competence and judgment of the individuals who represent the project in these deliberations.

By and large, at this stage, these people are within the government. The data from which they work have evolved during the years of advanced study devoted to the program by both government and industry, and whatever background of experience they have gleaned from observation of, or participation in, previous programs. Considering the general quality of advanced studies, in which, by and large, all new developments come out as "no sweat," the experience from earlier programs is far and away the best guide to schedule and cost commitments. The pitfall which must be avoided, however, is a conviction that "to sell" the program the initial estimates must be "optimistic."

In effect, this argument says that, "After all, getting the program started is the really important thing, and, once it is going, changes in cost and schedule will be more easily accepted. Hasn't this been the way most other programs have gotten started? And, perhaps this one really will be different and have all the pieces fall into place easily and stay put."

As a matter of fact, programs of the type we are discussing are rarely sold on the basis of cost and schedule, except in the mind of the seller. Common sense would dictate being conservative in initial estimates, leaving, if possible, some maneuvering room to improve on the estimates and, hopefully, insure continuing support throughout the program by demonstrating an ability to manage within initial cost and schedule commitments.

A similar situation, though much more chronic, obtains in the second phase of initial program definition. The problem here is twosided—contractors bidding on the program are often convinced that early schedule and low cost are the prime things the government is after. Since the process of contractor selection for major programs which will run for several years is essentially irreversible after the momentum has built up in the first year or so, and since volume of business is as much a measure of an aerospace company success as the actual profit earned, shaving of cost and schedule estimates in the interest of improving the chance of winning the contract is extremely tempting. This tendency can be compounded by government contracting personnel who feel that once a contractor is formally committed to a cost and schedule with attendant fee agreement, any increases in cost or schedule slips will not increase, and perhaps even decrease, the fee, thus obtaining a better "buy" for the government than if fee had originally been calculated on a "realistic" program estimate.

Here, again, common sense argues that the least expensive program is the one properly defined from the outset. Fair fee should be paid for competent work done. Excessive attempts to restrict fee result in decreased contractor motivation and invite—indeed almost force—the contractor to make up for fee reduction by contractual haggling—in effect, interminable arguments about changes and whether they are within or outside scope. Not only does this practice divert management resources from more important elements of program control, it can also jeopardize what should be purely technical decision by forcing action intended to protect some critical contract position.

The problems in this area have been recognized by several agencies in recent years, and elaborate, sequential phases in the definition of a program have been defined. Here again, the critical problem is the judgment of the responsible individuals on the government and contractor teams as to what is realistic. The problem is compounded if new competition is held at the end of each definition phase. Fundamentally, it is impossible to transfer the results of one study or definition of approach to an organization other than the originator without a serious time loss and perturbation in the approach. Hence, the only useful sequential definition phases are those in which only the contractors in any particular phase are considered for the next contract, and the inherent competition is used to prevent the "inevitable commitment" from becoming a problem before the program is sufficiently defined to enter a long range contract.

## [16-3] GOVERNMENT-INDUSTRY RELATIONSHIPS

Up to the point of awarding the major development contracts, the role of government and industry in the program are fairly clear. The government agency defines the requirements in response to which contractor studies result in systems concepts and preliminary designs of critical hardware elements. These are reviewed and modified where deemed necessary by the government. The contractor then proposes a program cost and schedule to implement the desired plan, and, after suitable negotiations, the contract is definitized and the program is off and running. By and large, the interaction between government and industry up to contract award does not pose major problems.

However, the respective roles played in the implementation phase are a potential problem source unless the principles for the particular situation are clearly enunciated and followed. The appropriate distribution of function depends on the strengths and weaknesses of the organizations involved. Since it is extremely difficult for an organization to view its assets objectively, the proper balance is difficult to achieve.

The contractor usually has larger manpower resources and should be able to develop across the board competence. The government manpower is often more restricted, but hopefully includes a cadre of competent individuals who can provide the necessary strengths.

The government function should be divided into in-line, or "generating" functions and out-of-line, or check and balance functions. Typical of generating functions for a complex manned space flight program are:

1) Definition of approach
2) Development or approval of system specifications
3) Test program definition
4) Test program certification
5) Operation of missions
6) Integration of associate contractors' efforts.

These functions, in essence, reserve the overall mission planning and program level system engineering to the government agency.

The test functions represent a point of major check and balance in the program. In effect, a properly designed qualification test program for components and sub-system hardware is the ultimate means, short of flight, to assure that the design and development task has been properly executed.

In addition to the in-line functions, the government contractor interface is an area of major check and balance on the progress of the program. In the technical areas, this takes the form of design reviews and continual program surveillance throughout the contract. In principle, the government specifications should define requirements, not details of implementation. The contractor should be responsible for design of the necessary system elements, production and test. The government reviews should be aimed at establishing whether the design and production approaches being taken by the contractor represent competent practice and have a reasonably high chance of producing the desired product.

It is obvious that in most system areas several different design approaches *can* meet a given requirement, often with essentially equal chance of success, a priori, since at the beginning of a design the detailed, daily development problems which will crop up are hidden from the eye of the beholder. It is also true, unfortunately, that individual technical people frequently differ on details of approach. Such differences can be, too often, a question of taste rather than a fundamental flaw in either approach. In conducting the design reviews, the government people must remember that, in questions of technical taste, the contractor should be allowed to pursue the course that he has chosen. Otherwise, the contract intent, that the contractor produce hardware which fulfills the specification requirements, will not be met and the government will become responsible for the detailed design implementation.

When technical feeling runs high, this is a difficult path to trod, and requires mature judgments on both sides as to what issues are essential.

The inherent difficulty of the design review process is another reason why a rigorous test program is so essential to program success. By comparison with the difficult judgments involved in certifying a given design, the relative unambiguity of test results leaves little room for argument and usually spotlights the areas which must be fixed to meet the requirements.

In summary, a large program in space or defense areas will almost certainly involve technical organizations within both industry and government. Each has a role to play, and, in addition, the government organization must act as an explicit check and balance on the contractor's implementation of the program. Some of the antagonism inherent in this relationship can be eased by clear definition of "who does what" and adherence to the definition of responsibilities arrived at in the contract.

## [16-4] TECHNICAL APPROACH AND IMPLEMENTATION

The above sections have discussed the framework in which the program is carried out, independent of content. This section will attempt to enunciate a series of principles which are intended to make the program implementation as efficient as human judgment will permit. In the definition and design phase, these levels can be summed up within the phrase: "Do each part of the system as simply as possible, and plan to meet all the requirements in the first production article." In the manufacturing and test phase: "Avoid incremental steps, tool for production from the first prototype systems, and test at the highest practical assembly level." In the program planning: "Force the initial designs to the hardware test phase early, but allow adequate time between initial testing and flight tests to redesign at least once to solve problems found in test."

In the definition and design phase, the characteristic problem is a tendency to use the most advanced state of the art available in each discipline area. The tendency is understandable, because engineers, by and large, like to work on the forefront of knowledge and because they have probably been impressed by how difficult the performance and reliability requirements are for this new "challenge to man's ingenuity." However, as a matter of actual fact, the program stands the best chance of success if the state of the art is pushed only in those areas in which new devices or techniques are mandatory to enable completion of the mission within weight and performance constraints. Each area in which old art can be used is one less potential source of unexpected problems. One lesson, hard learned over years of development programs, is that each advance in techniquc (i.e., from tubes to transistors to microcircuitry in electronics) has been accompanied by infancy problems not envisioned as the devices were selected for their initial applications. It is not coincidental that the Volkswagen is probably the most reliable automobile and also seems to have the fewest design changes from year to year. Since research and development should not be a customer preference market, novelty without cause is wasteful of program resources.

A second prevalent attitude in design groups is the evolutionary approach to meeting the ultimate requirements. Although it may be tempting to proceed from breadboard layouts through several stages of prototypes to preproduction and ultimately production units, on the theory that each small step will enable the problems peculiar to it to be more readily solved, the shortest path from concept to usable hardware involves designing all but the first one or two breadboards as final production units, and starting the manufacturing process early enough to support the engineering and test requirements. The units so built and tested will find all the problems at the same time, and, painful as the process of introducing changes into the manufacturing cycle may be, the result will be the desired hardware at an earlier date than that possible with a stepped approach.

The planning of the program to force early design release is also a key to effective total program schedules. This can only be done if the program has been carefully specified so that each subsystem area knows its requirements and its interaction with all other related areas (interfaces). The discipline of individual design areas conforming to performance and interface specifications is essential to the implementation of a large program where individual specialty groups must work almost independently on system elements which must ultimately be integrated into the whole.

In the absence of early design release schedules, the designers have a tendency to incrementally improve their design concepts and can spend much effort solving paper problems which later turn out to be unimportant. By reducing the design to hardware and forcing it into test, both functional and environmental, the real problems in the approach will be uncovered. These can be solved by one of three paths:

1) Redesign.
2) Change of specification to relieve requirements which may be too severe change in design.
3) Change in design or specification of related systems which relieve overly severe requirements.

Here, an effective system engineering group which recognizes that the initial requirements may, in many instances be arbitrary, and which can effect redistribution of requirements among subsystems to more evenly "share the pain" is indispensable to the success of the program.

Finally, if the program planning has recognized that, despite the best of intentions of the best of organizations, requirements are rarely translated to correct hardware the first time through the design cycle, and has provided sufficient schedule flexibility to solve one round of problems, the program can approach the flight phase reasonably close to schedule. This presumes that sufficient management discipline is maintained so that the initial design phase remains on schedule and does not encroach on the "design fix" contingency time which seems to be an almost constant requirement independent of the amount of effort expended on the initial engineering releases.

## [16-5] ORGANIZATIONAL APPROACH

The primary elements of a research and development program are the technical content, cost, and schedule. Of the three, for a program such as a planetary mission, the technical content tends to dominate and really establishes the other two. As discussed above, the key to successful program management is to keep the technical program in balance—in effect, keep it from being too elaborate or "over sophisticated."

To do this, the engineering people must be convinced that, literally, "things can be too good" both in the initial design approach and in the continual incremental changes which good engineers can almost always come up with to improve their original designs. There is reported to be an inscription over the entrance to the Soviet Academy of Sciences which perhaps illustrates the problem succinctly: "The better is the enemy of the good." Many programs have floundered striving for a presumed "better" instead of settling for an adequate "good."

Management must seek to convince the engineering people that what is really desired is not, the "world's best" program, but rather, the "world's most adequate" program, in which "most adequate" means enough, but not too much. One way to do this is to recognize that in a program with high technical content, the conventional organizational structure which separates the engineering function from the business functions such as material (or subcontractor management) and program control (cost and scheduling) does not make sense. Whatever the organizational subterfuge (such as contracting officers within the government), the technical people really determine what has to be done.

The logical step is to focus the total management responsibility with the technical people and motivate them to recognize that their job is to deliver a working product at a given time for a given cost. The balances necessary to keep cost and schedule in line act as a strong counterforce to the inherent tendency to optimize the system beyond all reasonable needs.

This type of focus can be achieved in a variety of conventional organizational structures. The key is to be sure the business and schedule elements realize that they must support the appropriate engineering managers and the engineering manager, in turn, really does properly consider the three elements fully in his decisions.

## [16-6] MANAGEMENT SYSTEMS

It has become the vogue for major national programs to attempt to solve some of the fundamental problems which plague large numbers of people attempting to work toward a complex end by inventing management systems which are intended to compensate for individual competence by forcing stereotyped decision processes and standard, fool-proof data gathering methods. In several instances, the systems are more useful than the above comments might indicate—although it must be recognized that individual competence at key points in the program is still the ingredient for which there is no substitute.

In the technical area, the "in" documents these days are the Air Force 375 series documents and the NASA-500 series. In the main they call for early definition of program requirements, development of specifications and timely review of design. All are essential to program success.

There are, however, at least two problems. One is a presumption that every requirement can be generated from a logical base. Although in principle this should be possible, in practice the use of "reasonable" requirements based on judgment or proven "intuition" is often as good as the results of the most elaborate studies—and in some instances, the only way to reach definition in a reasonable period of time. As one frustrated engineer was heard to say, "I know a wheel is the best device to put under a car so it can move down the road, but how do I prove it?"

The second problem is that, as the specification tree progresses from overall program requirements to lower level hardware specifications, more and more numbers appear. Some of them represent key program requirements, but many have been selected somewhat arbitrarily and are not really significant when considered from a total system performance point of view. There is an unfortunate tendency, however, to treat specification numbers as sacred, particularly in lower tier contracts.

It has been my experience that something over half of the "failures" listed, for instance, in the qualification programs for subsystems on a major program represent not an outright failure (no go), but rather a failure to meet one of these arbitrary numerical requirements, which has insignificant effect on the intended mission. In such cases, it is essential that the responsible people understand the "real" requirement and be in a position to waive and/or change it. The organization must understand that situations frequently arise in which waivers are actually good for the program, and be encouraged to act accordingly.

Incidentally, the solution to this problem is not to go back and be sure that every numerical requirement is properly validated before it is set. Such a process would extend the definition phase unnecessarily. Fortunately, the number of problems which have to be considered in this area is usually small enough to be handled efficiently at the time of qualification if sufficient visibility and understanding exists in the program structure.

In the cost and schedule area, the program problem is having sufficient visibility to see problems developing in time to take corrective action. For schedules a PERT-like system is satisfactory. It must be supplemented by more detailed statusing for problems near at hand and, in general, is only as good as its original logic definition and updating input. Perhaps most important, it should be the only scheduling system in use for the bulk of the program, else a situation arises where people report against PERT, but work to their own, different system.

Cost reporting and projections also must be done in a system consistent from level to level if control is to be maintained. Formal attempts to link PERT and cost reporting in a single system format have not been too successful. To date, any necessary integration is best performed by program management with access to both sets of data.

In addition to program definition and information gathering systems, recent years have seen the introduction of special check and balance organizational elements to compensate for specific program deficiencies. As emphasis changes, the old organizations have a tendency to remain in place and perhaps even expand. Such areas of activity as value engineering, maintenance analysis, quality assurance and to a lesser extent reliability, can burden the program with people reviewing the work of the "in line" organization elements with marginal return to the program.

This problem is usually initiated by the government, but is procreated by industry once the initial momentum is built up. The solution is to reemphasize the need for total awareness of requirements at the generating point, and the use of the normal in-line supervisory function as the prime check and balance, rather than out-of-line organizational elements, which must compete with the in-line organization for competent people.

## [16-7] SUMMARY

The above sections have attempted to discuss in very general terms the problems in large research and development programs, and indicate directions in which the solution to these problems lies. Although the solutions suggested may sound naively simple, the challenge to program management is to interpret the specifics of their program in light of such a set of general principles—rather than assume that "this program is different" and warrants exceptions.

The planetary programs, which will still be several years in definition, represent a unique opportunity to do a program correctly from the start. In the long run, the philosophic approach applied to the program management may far transcend, in program impact, the detailed mechanization of the reentry system.

GPSR Compliance

*The European Union's (EU) General Product Safety Regulation (GPSR) is a set of rules that requires consumer products to be safe and our obligations to ensure this.*

*If you have any concerns about our products, you can contact us on ProductSafety@springernature.com*

In case Publisher is established outside the EU, the EU authorized representative is:

Springer Nature Customer Service Center GmbH
Europaplatz 3
69115 Heidelberg, Germany

**Batch number: 10620163**

Printed by Printforce, the Netherlands